ANTIMONIDE-BASED INFRARED DETECTORS

A New Perspective

ANTIMONIDE-BASED INFRARED DETECTORS
A New Perspective

Antoni Rogalski, Małgorzata Kopytko, and Piotr Martyniuk

SPIE PRESS
Bellingham, Washington USA

Library of Congress Cataloging-in-Publication Data

Names: Rogalski, Antoni, author. | Kopytko, Małgorzata, author. | Martyniuk, Piotr, author.
Title: Antimonide-based infrared detectors : a new perspective / Antoni Rogalski, Malgorzata Kopytko, and Piotr Martyniuk.
Description: Bellingham, Washington, USA : SPIE Press, [2018] | Includes bibliographical references and index.
Identifiers: LCCN 2017013783 | ISBN 9781510611399 (print ; alk. paper) | ISBN 1510611398 (print ; alk. paper) | ISBN 9781510611405 (PDF) | ISBN 1510611401 (PDF) | ISBN 9781510611412 (ePub) | ISBN 151061141X (ePub) | ISBN 9781510611429 (Kindle/Mobi) | ISBN 1510611428 (Kindle/Mobi)
Subjects: LCSH: Infrared detectors. | Infrared technology–Materials. | Semiconductors. | Antimonides. | Superlattices as materials.
Classification: LCC TA1573 .R64 2017 | DDC 681/.25–dc23 LC record available at https://lccn.loc.gov/2017013783

Published by

SPIE
P.O. Box 10
Bellingham, Washington 98227-0010 USA
Phone: +1 360.676.3290
Fax: +1 360.647.1445
Email: books@spie.org
Web: http://spie.org

The content of this book reflects the work and thought of the authors. Every effort has been made to publish reliable and accurate information herein, but the publisher is not responsible for the validity of the information or for any outcomes resulting from reliance thereon.

Printed in the United States of America.
First Printing.
For updates to this book, visit http://spie.org and type "PM280" in the search field.

Contents

Preface

Among the many materials investigated in the infrared (IR) field, narrow-gap semiconductors are the most important in the IR photon detector family. Although the first widely used narrow-gap materials were lead salts (during the 1950s, IR detectors were built using single-element-cooled PbS and PbSe photoconductive detectors, primarily for anti-missile seekers), this semiconductor family was not well distinguished. This situation seems to have resulted from two reasons: the preparation process of lead salt photoconductive polycrystalline detectors was not well understood and could only be reproduced with well-tried recipes; and the theory of narrow-gap semiconductor bandgap structure was not well known for correct interpretation of the measured transport and photoelectrical properties of these materials.

The discovery of the transistor stimulated a considerable improvement in the growth and material purification techniques. At the same time, rapid advances were being made in the newly discovered III-V compound semiconductor family. One such material was InSb from which the first practical photovoltaic detector was fabricated in 1955 [G. R. Mitchell, A. E. Goldberg, and S. W. Kurnick, *Phys. Rev.* **97**, 239 (1955)]. In 1957, P. W. Kane [*J. Phys. Chem. Solids* **1**, 249 (1957)] using a method of quantum perturbation theory (the so-called $\mathbf{k}\cdot\mathbf{p}$ method), correctly described the band structure of InSb. Since that time, the Kane band model has been of considerable importance for narrow-gap semiconductor materials.

The end of the 1950s saw the introduction of narrow-gap semiconductor alloys in III-V (InAsSb), IV-VI (PbSnTe, PbSnSe), and II-VI (HgCdTe) material systems. These alloys allowed the bandgap of the semiconductor and hence the spectral response of the detector to be custom tailored for specific applications. Discovery of variable bandgap $Hg_{1-x}Cd_xTe$ (HgCdTe) ternary alloy in 1959 by Lawson and co-workers [*J. Phys. Chem. Solids* **9**, 325 (1959)] triggered an unprecedented degree of freedom in infrared detector design. Over the next five decades, this material system successfully fought off major challenges from different material systems, but despite that, it has more competitors today than ever before. It is interesting, however, that none of these competitors can compete in terms of fundamental properties. They may promise to be more manufacturable, but never to provide higher performance or, with the exception of thermal detectors, to operate at higher temperature.

Recently, there has been considerable progress towards III-V antimonide-based, low-dimensional solids development, and device design innovations. Their development results from two primary motivations: the perceived challenges of reproducibly fabricating high-operability HgCdTe focal plane arrays (FPAs) at reasonable cost and theoretical predictions of lower Auger recombination for type-II superlattice (T2SL) detectors compared to HgCdTe. Lower Auger recombination translates into a fundamental advantage for T2SL over HgCdTe in terms of lower dark current and/or higher operating temperature provided other parameters such as Shockley–Read–Hall lifetimes are equal.

In fact, investigations of antimonide-based materials began at about the same time as HgCdTe—in the 1950s, and the apparent rapid success of their technology, especially low-dimensional solids, depends on the previous five decades of III-V materials and device research. However, the sophisticated physics associated with the antimonide-based bandgap engineering concept started at the beginning of the 1990s gave a new impact and interest in development of infrared detector structures within academic and national laboratories. In addition, implementation of barriers in photoconductor structures, in so-called barrier detectors, prevents current flow in the majority carrier band of a detector's absorber but allows unimpeded flow in the minority carrier band. As a result, this concept resurrects the performance of antimonide-based focal plane arrays and gives a new perspective in their applications. A new emerging strategy includes antimonide-based T2SLs, barrier structures such as the nBn detector with lower generation-recombination leakage mechanisms, photon trapping detectors, and multi-stage/cascade infrared devices.

This book describes the present status of new concepts of antimonide-based infrared detectors. The intent is to focus on designs having the largest impact on the mainstream of infrared detector technologies today. A secondary aim is to outline the evolution of detector technologies showing why certain device designs and architectures have emerged recently as alternative technologies to the HgCdTe ternary alloy. The third goal is to emphasize the applicability of detectors in the design of FPAs. This is especially addressed to the InAsSb ternary alloys system and T2SL materials. It seems to be clear that some of these solutions have emerged as real competitors of HgCdTe photodetectors. Special efforts are directed on the physical limits of detector performance and the performance comparison of antimonide-based detectors with the current stage of HgCdTe photodiodes. The reader should gain a good understanding of the similarities and contrasts and the strengths and weaknesses of a multitude of approaches that have been developed over two last decades as an effort to improve our ability to sense infrared radiation.

The level of this book is suitable for graduate students in physics and engineering who have received preparation in modern solid-state physics and electronic circuits. This book will be of interest to individuals working with aerospace sensors and systems, remote sensing, thermal imaging, military imaging, optical telecommunications, infrared spectroscopy, and lidar, as well. To satisfy all these needs, each chapter first discusses the principles needed to understand the chapter topic as well as some historical background before presenting the reader with the most recent information available. For those currently in the field, this book can be used as a collection of useful data, as a guide to literature in the field, and as an overview of topics in the field. The book also could be used as a reference for participants of educational short courses, such as those organized by SPIE.

The book is divided into ten chapters. The introduction (Chapter 1) gives a short historical overview of the development of IR detectors with antimonide-based materials and describes the detector classification and figures of merit of infrared detectors. The main topics in crystal growth technology, both bulk materials and epitaxial layers, as well their physical properties are given in Chapter 2. Special emphasis is paid to the modern epitaxy technologies such as molecular beam epitaxy and metalorganic chemical vapor deposition. Chapter 3 provides similar information about type-II superlattices. The next two chapters concern technology and performance of both bulk as well as superlattice antimonide-based infrared detectors. New classes of infrared detectors called barrier detectors, trapping detectors, and cascade detectors are covered in three succeeding chapters: Chapters 6, 7, and 8. An overview of antimonide-based FPA architectures is given in Chapter 9. Finally, remarks are included in the last chapter.

Antoni Rogalski
Małgorzata Kopytko
Piotr Martyniuk
March 2018

Acknowledgments

In the course of writing this book, many people have assisted us and offered their support. We would like, first, to express appreciation to the management of the Institute of Applied Physics, Military University of Technology, for providing the environment in which we worked on the book. The writing of the book has been partially done under financial support of the Polish Ministry of Sciences and Higher Education, The National Science Centre (Poland) - (Grant nos. 2015/17/B/ST5/01753 and 2015/19/B/ST7/02200) and The National Centre for Research and Development (Poland) - (Grant no. POIR.04.01.04-00-0027/16-00).

The authors have benefited from the kind cooperation of many scientists who are actively working in infrared detector technologies. The preparation of this book was aided by many informative and stimulating discussions that the authors had with their colleagues at the Institute of Applied Physics, Military University of Technology in Warsaw. The authors thank the following individuals for providing preprints, unpublished information, and in some cases original figures, which are used in preparing the book: Dr. M.A. Kinch (DRS Infrared Technologies, Dallas), Drs. S.D. Gunapala and D.Z.-Y. Ting (California Institute of Technology, Pasadena), Dr. M. Kimata (Ritsumeikan University, Shiga), Dr. M. Razeghi (Northwestern University, Evanston), Drs. M.Z. Tidrow and P. Norton (U.S. Army RDECOM CERDEC NVESD, Ft. Belvoir), Dr. S. Krishna (University of New Mexico, Albuquerque), and Prof. J. Piotrowski (Vigo System Ltd., Ożarów Mazowiecki). Thanks also to SPIE Press, especially Ms. Nicole Harris, for her cooperation and care in publishing this edition.

Chapter 1
Infrared Detector Characterization

Over the past several hundreds of years, optical systems (telescopes, microscopes, eyeglasses, cameras, etc.) have formed their optical image on the human retina, photographic plate, or film. The birth of photodetectors dates back to 1873 when Smith discovered photoconductivity in selenium. Progress was slow until 1905, when Einstein explained the newly observed photoelectric effect in metals, and Planck solved the blackbody emission puzzle by introducing the quantum hypothesis. Applications and new devices soon flourished, pushed by the dawning technology of vacuum tube sensors developed in the 1920s and 1930s, culminating in the advent of television. Zworykin and Morton, the celebrated fathers of videonics, on the last page of their legendary book *Television* (1939) concluded that: *"when rockets will fly to the moon and to other celestial bodies, the first images we will see of them will be those taken by camera tubes, which will open to mankind new horizons."* Their foresight became a reality with the Apollo and Explorer missions. Photolithography enabled the fabrication of silicon monolithic imaging focal planes for the visible spectrum beginning in the early 1960s. Some of these early developments were intended for a videophone, and other efforts were for television cameras, satellite surveillance, and digital imaging. Infrared imaging has been vigorously pursued in parallel with visible imaging because of its utility in military applications. More recently (1997), the charged-coupled device (CCD) camera aboard the Hubble Space Telescope delivered a deep-space picture, a result of 10 day's integration, featuring galaxies of the 30th magnitude—an unimaginable figure, even for astronomers of our generation. Thus, photodetectors continue to open to humanity the most amazing new horizons.

1.1 Introduction

Many materials have been investigated in the infrared (IR) field. Figure 1.1 gives approximate dates of significant developmental efforts for infrared

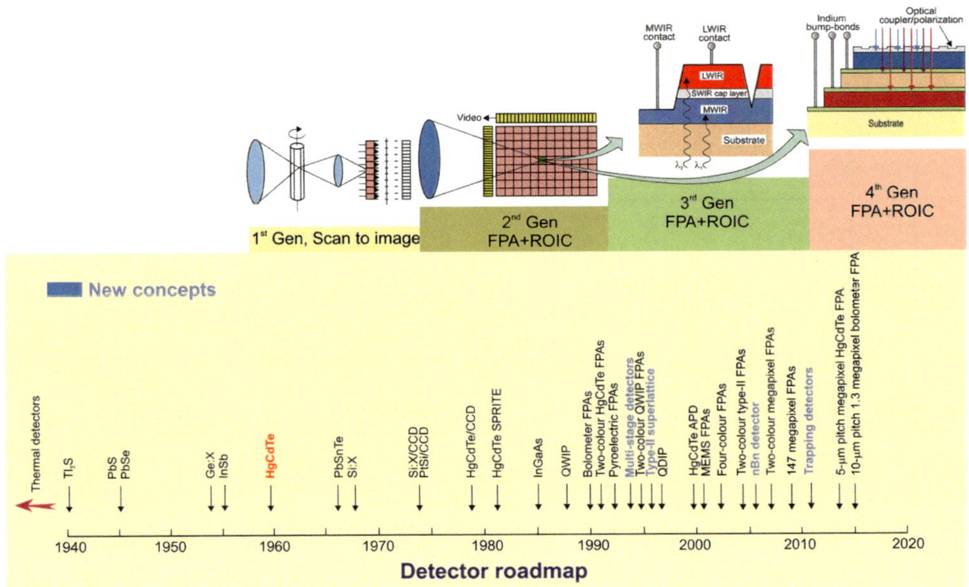

Figure 1.1 History of the development of infrared detectors and systems. New concepts of detectors developed in last two decades are marked in blue. Four generations of systems can be considered for principal military and civilian applications: first generation (scanning systems), second generation (staring systems with electronic scanning), third generation (staring systems with a large number of pixels and two-color functionality), and fourth generation (staring systems with a very large number of pixels, multi-color functionality, and other on-chip functions; e.g., better radiation/pixel coupling, avalanche multiplication in pixels, and polarization/phase sensitivity) (adapted from Ref. 3).

materials. During the 1950s, IR detectors were built using single-element-cooled lead salt photodetectors, primarily for anti-air-missile seekers. Usually lead salt detectors were polycrystalline and were produced by vacuum evaporation and chemical deposition from a solution, followed by a post-growth sensitization process.[1] The first extrinsic photoconductive detectors were reported in the early 1950s[2] after the discovery of the transistor, which stimulated a considerable improvement and growth of material purification techniques. Since the techniques for controlled impurity introduction became available for germanium at an earlier date, the first high-performance extrinsic detectors were based on Ge:Hg with activation energy for the Hg acceptor of 0.089 eV. Extrinsic photoconductive response from copper, zinc, and gold impurity levels in germanium gave rise to devices using the 8- to 14-µm long wavelength IR (LWIR) spectral window and beyond to the 14- to 30-µm very long wavelength IR (VLWIR) region.

In 1967 the first comprehensive extrinsic Si detector-oriented paper was published by Soref.[4] However, the state of extrinsic Si was not changed significantly. Although Si has several advantages over Ge (namely, a lower dielectric constant giving shorter dielectric relaxation time and lower capacitance, higher dopant solubility and larger photoionization cross section for higher quantum efficiency, and lower refractive index for lower reflectance), these were not sufficient to warrant the necessary development efforts needed to bring it to the level of the, by then, highly developed Ge detectors. After being dormant for about ten years, extrinsic Si was reconsidered after the invention of CCDs by Boyle and Smith.[5] In 1973, Shepherd and Yang[6] proposed the metal-silicide/ silicon Schottky barrier detectors. For the first time it became possible to have much more sophisticated readout schemes—both detection and readout could be implemented in one common silicon chip.

At the same time, rapid advances were being made in narrow bandgap semiconductors that would later prove useful in extending wavelength capabilities and improving sensitivity. The fundamental properties of narrow-gap semiconductors (high optical absorption coefficient, high electron mobility and low thermal generation rate), together with the capability for bandgap engineering, make these alloy systems almost ideal for a wide range of IR detectors. The first such material was InSb,[7] a member of the newly discovered III-V compound semiconductor family, but its operation is limited to the mid-wavelength IR (MWIR) spectral range. The perceived requirement for detection in LWIR band led to development of narrow-gap ternary alloy systems such as InAsSb, PbSnTe, and HgCdTe.[8–10]

For 10 years during the late 1960s to the mid-1970s, HgCdTe alloy detectors were in serious competition with IV-VI alloy devices (mainly PbSnTe) for developing photodiodes because of the latter's production and storage problems.[9] However, development of PbSnTe photodiodes was discontinued because the chalcogenides suffered from two significant drawbacks: very high thermal coefficient of expansion (a factor of 7 higher than Si) and short Shockley–Read–Hall (SRH) carrier lifetime. A large thermal coefficient of expansion lead to failure of the indium bonds in hybrid structure (between silicon readout and the detector array) after repeated thermal cycling from room temperature to the cryogenic temperature of operation. In addition, the high dielectric constant of PbSnTe ($\sim$500) resulted in *RC*-response times that were too slow for LWIR scanning systems under development at that time. However, for two-dimensional (2D) staring imaging systems, which are currently under development, this would not be such a significant issue.

HgCdTe has inspired the development of the four "generations" of detector devices (see Fig. 1.1). In the late 1960s and early 1970s, first-generation linear photoconductor arrays were developed. The first generation scanning system does not include multiplexing functions in the infrared focal plane (IRFP). In the mid-1970s attention turned to the photodiodes for passive IR imaging applications. In contrast to photoconductors, photodiodes with their very low power dissipation, inherently high impedance, negligible $1/f$ noise, and easy multiplexing on a focal plane silicon chip, can be assembled in 2D arrays containing more than megapixel elements, limited only by existing technologies. After the invention of CCDs by Boyle and Smith,[5] the idea of an all-solid-state electronically scanned 2D IR detector array caused attention to be turned to

HgCdTe photodiodes. In the end of the 1970s the emphasis was directed toward large photovoltaic HgCdTe arrays in the MWIR and LWIR spectral bands for thermal imaging. Recent efforts have been extended to short wavelengths, e.g., for starlight imaging in the short wavelength IR (SWIR) range, as well as to VLWIR spaceborne remote sensing beyond 15 μm.

The third-generation HgCdTe and type-II superlattice (T2SL) systems continue to be developed, and concept development towards the so-called fourth generation systems was also recently initiated. The definition of fourth-generation systems is not well established. These systems provide enhanced capabilities in terms of greater number of pixels, higher frame rates, and better thermal resolution, as well as multicolor functionality and other on-chip functions. Multicolor capabilities are highly desirable for advanced IR systems. Collection of data in distinct IR spectral bands can discriminate for both the absolute temperature and the unique signature of objects within the scene. By providing this new dimension of contrast, multiband detection also offers advanced color processing algorithms to further improve sensitivity compared to that of single-color devices. It is expected that the functionalities of fourth-generation systems could manifest themselves as spectral, polarization, phase, or dynamic range signatures that could extract more information from a given scene.[11]

At the beginning of the 1990s, several national agencies (e.g., in U.S., Germany, and France) switched their research emphasis to III-V low-dimensional solid materials (quantum wells and superlattices), as an alternative technology option to HgCdTe, to attain their stated goal of inexpensive large-area IR focal plane arrays (FPAs) amenable to fabrication by the horizontal integration of material foundries and processing centers of excellence. There has been considerable progress towards the materials development and device design innovations. Several new concepts for improvement of the performance of photodetectors have been proposed (see bottom part of Fig. 1.1), where approximate data of significant development efforts are marked in blue. In particular, significant advances have been made in the bandgap engineering of various compound III-V semiconductors that has led to new detector architectures. New emerging strategies include T2SLs, barrier structures such as nBn detectors with lower generation–recombination leakage mechanisms, photon trapping detectors, and multi-stage/cascade infrared devices. The barrier-structure detector concept has recently been applied to resurrect the performance of III-V FPAs, allowing them to operate at considerably higher temperatures than their photodiode counterparts simply by the elimination of depletion regions in the absorber volume. At present, the trade-offs between both competing III-V and II-VI IR materials technologies is observed. It is expected that these two significant schools of thought with regard to the ultimate in photon detection, namely, operation at room temperature, might play a crucial role in the future developments.

1.2 Classification of Infrared Detectors

Optical radiation is considered to be radiation ranging from vacuum ultraviolet to submillimeter wavelengths (25 nm to 3000 μm). The terahertz (THz) region of electromagnetic spectrum (see Fig. 1.2) is often described as the final unexplored area of the spectrum and still presents a challenge for both electronic and photonic technologies. It is frequently treated as the spectral region within the frequency range of $\nu \approx 0.1$–10 THz ($\lambda \approx 3$ mm – 30 μm) and is partly overlapping with the loosely treated submillimeter (sub-mm) wavelength band $\nu \approx 0.1$–3 THz ($\lambda \approx 3$ mm – 100 μm).

The majority of optical detectors can be classified in two broad categories: photon detectors (also called quantum detectors) and thermal detectors.

1.2.1 Photon detectors

In photon detectors the radiation is absorbed within the material by interaction with electrons either bound to lattice atoms or to impurity atoms or with free electrons. The observed electrical output signal results from the changed electronic energy distribution. The fundamental optical excitation processes in semiconductors are illustrated in Fig. 1.3. In quantum wells [Fig. 1.3(b)] the intersubband absorption takes place between the energy levels of a quantum well associated with the conduction band (n-doped) or valence band (p-doped). In the case of type-II InAs/GaSb superlattice [Fig. 1.3(c)] the superlattice bandgap is determined by the energy difference between the electron miniband E1 and the first heavy-hole state HH1 at the Brillouin zone center. A consequence of the type-II band alignment is spatial separation of electrons and holes.

Relative response of infrared detectors is plotted as a function of wavelength with either a vertical scale of W^{-1} or photon^{-1} (see Fig. 1.4). The

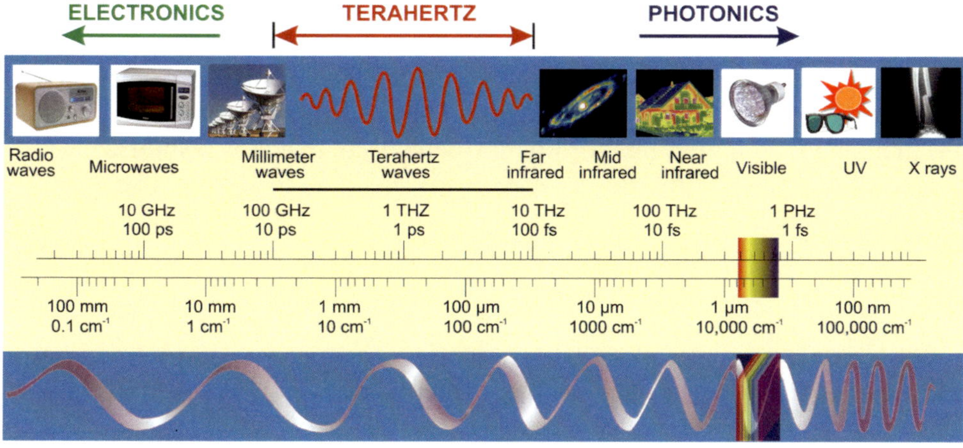

Figure 1.2   The electromagnetic spectrum (adapted from Ref. 12).

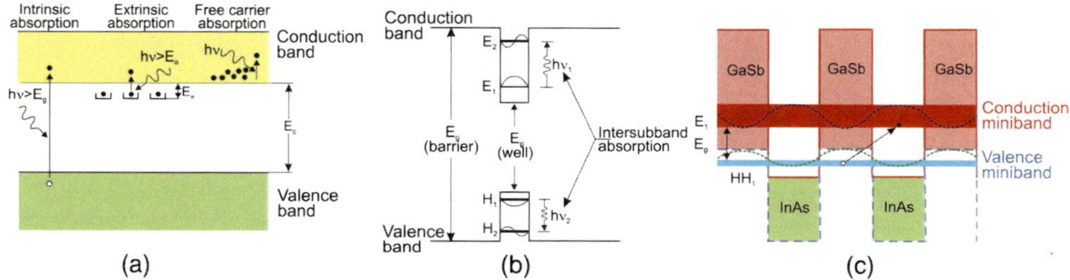

Figure 1.3 Optical excitation processes in: (a) bulk semiconductors, (b) quantum wells, and (c) type-II InAs/GaSb superlattices.

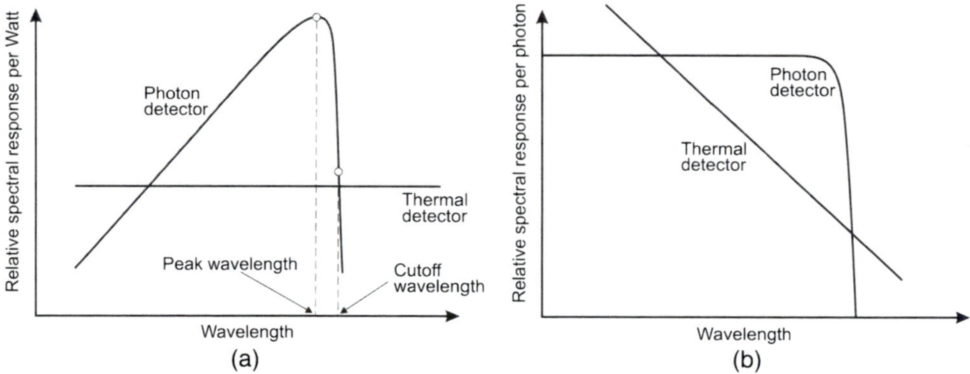

Figure 1.4 Relative spectral response for a photon and thermal detector for (a) constant incident radiant power and (b) photon flux, respectively.

photon detectors show a selective wavelength dependence of response per unit incident radiation power. Their response is proportional to the rate of arrival photons as the energy per photon is inversely proportional to wavelength. In consequence, the spectral response increases linearly with increasing wavelength [see Fig. 1.4(a)], until the cutoff wavelength is reached, which is determined by the detector material. The cutoff wavelength is usually specified as the long wavelength point at which the detector responsivity falls to 50% of the peak responsivity.

Thermal detectors tend to be spectrally flat in the first case (their response is proportional to the energy absorbed), thus they exhibit a flat spectral response [see Fig 1.4(a)], while photon detectors are generally flat in the second case [see Fig. 1.4(b)].

Photon detectors exhibit both good signal-to-noise performance and a very fast response. But to achieve this, the photon IR detectors may require cryogenic cooling. This is necessary to prevent the thermal generation of charge carriers. The thermal transitions compete with the optical ones, making non-cooled devices very noisy.

Depending on the nature of the interaction, the class of photon detectors is further sub-divided into different types. The most important are: intrinsic detectors, extrinsic detectors, and photoemissive detectors (Schottky barriers).[3] Different types of detectors are briefly characterized in Table 1.1.

A key difference between intrinsic and extrinsic detectors is that extrinsic detectors require much cooling to achieve high sensitivity at a given spectral response cutoff in comparison with intrinsic detectors. Low-temperature operation is associated with longer-wavelength sensitivity in order to suppress noise due to thermally induced transitions between close-lying energy levels.

There is a fundamental relationship between the temperature of the background viewed by the detector and the lower temperature at which the detector must operate to achieve background-limited performance (BLIP). HgCdTe photodetectors with a cutoff wavelength of 12.4 μm operate at 77 K. One can scale the results of this example to other temperatures and cutoff wavelengths by noting that for a given level of detector performance, $T\lambda_c \approx$ constant;[13] i.e., the longer λ_c, the lower is T while their product remains roughly constant. This relation holds because quantities that determine detector performance vary mainly as an exponential of $E_{exc}/kT = hc/kT\lambda_c$, where E_{exc} is the excitation energy, k is Boltzmann's constant, h is Planck's constant, and c is the velocity of light.

The long wavelength cutoff can be approximated as

$$T_{\max} = \frac{300\text{K}}{\lambda_c[\mu\text{m}]}. \qquad (1.1)$$

The general trend is illustrated in Fig. 1.5 for six high-performance detector materials suitable for low-background applications: Si, InGaAs, InSb, HgCdTe photodiodes, and Si:As blocked impurity band (BIB) detectors; and extrinsic Ge:Ga unstressed and stressed detectors. Terahertz photoconductors are operated in extrinsic mode.

The most widely used photovoltaic detector is the p–n junction, where a strong internal electric field exists across the junction even in the absence of radiation. Photons incident on the junction produce free hole–electron pairs that are separated by the internal electric field across the junction, causing a change in voltage across the open-circuit cell or a current to flow in the short-circuited case. Due to the absence of recombination noise, the limiting p–n junction's noise level can ideally be $\sqrt{2}$ times lower than that of the photoconductor.

Photoconductors that utilize excitation of an electron from the valence to conduction band are called intrinsic detectors. Instead those that operate by exciting electrons into the conduction band or holes into the valence band from impurity states within the band (impurity-bound states in energy gap, quantum wells, or quantum dots), are called extrinsic detectors. A key

Table 1.1 Photon detectors.

Mode of operation	Schematic of detector	Operation and properties
Photoconductor		This is essentially a radiation-sensitive resistor, generally a semiconductor, either in thin film or bulk form. A photon may release an electron–hole pair or an impurity-bound charge carrier, thereby increasing the electrical conductivity. In almost all cases the change in conductivity is measured by means of electrodes attached to the sample. For low-resistance material, the photoconductor is usually operated in a constant current circuit. For high-resistance photoconductors, a constant voltage circuit is preferred, and the signal is detected as a change in current in the bias circuit.
Blocked impurity band (BIB) detector		The active region of a BIB detector structure, usually based on epitaxially grown n-type material, is sandwiched between a higher-doped degenerate substrate electrode and an undoped blocking layer. Doping of the active layer is high enough for the onset of an impurity band in order to display a high quantum efficiency for impurity ionization (in the case of Si:As BIB, the active layer is doped to $\approx 5 \times 10^{17}$ cm^{-3}). The device exhibits a diode-like characteristic, except that photoexcitation of electrons takes place between the donor impurity and the conduction band. The heavily doped n-type IR-active layer has a small concentration of negatively charged compensating acceptor impurities. In the absence of an applied bias, charge neutrality requires an equal concentration of ionized donors. Whereas the negative charges are fixed at acceptor sites, the positive charges associated with ionized donor sites (D^+ charges) are mobile and can propagate through the IR-active layer via the mechanism of hopping between occupied (D^0) and vacant (D^+) neighboring sites. A positive bias to the transparent contact creates a field that drives the pre-existing D^+ charges towards the substrate, while the undoped blocking layer prevents the injection of new D^+ charges. A region depleted of D^+ charges is therefore created, with a width depending on the applied bias and on the compensating acceptor concentration.
p–n junction photodiode		This is the most widely used photovoltaic detector but is rather rarely used as a THz detector. Photons with energy greater than the energy gap create electron–hole pairs in the material on both sides of the junction. By diffusion, the electrons and holes generated within a diffusion length from the junction reach the space-charge region where they are separated by the strong electric field; minority carriers become majority carriers on the other side. This way a photocurrent is generated causing a change in voltage across the open-circuit cell or a current to flow in the short-circuited case. The limiting noise level of photodiodes can ideally be $\sqrt{2}$ times lower than that of the photoconductor, due to the absence of recombination noise. Response times are generally limited by device capacitance and detector-circuit resistance.

nBn detector

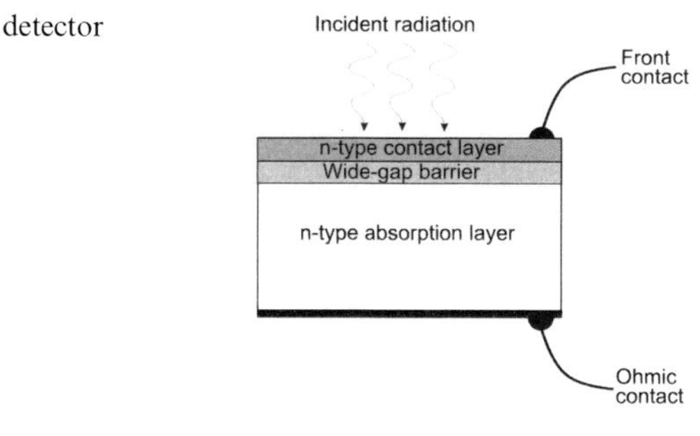

The nBn detector consists of a narrow-gap n-type absorber layer (AL), a thin wide-gap barrier layer (BL), and a narrow-gap n-type contact layer (CL). The thin wide-gap BL presents a large barrier in the conduction band that eliminates electron flow. Current through the nBn detector relies on transport of mobile holes through drift and diffusion in the BL between the two n-type narrow-gap regions. Effectively, the nBn detector is designed to reduce the dark current (generation–recombination current originating within the depletion layer) and noise without impeding the photocurrent (signal). In particular, the barrier serves to reduce the surface leakage current. The nBn detector operates as a unipolar unity-gain detector, and this design can be stated as a hybrid between photoconductor and photodiode.

Metal-insulator-semiconductor (MIS) photodiode

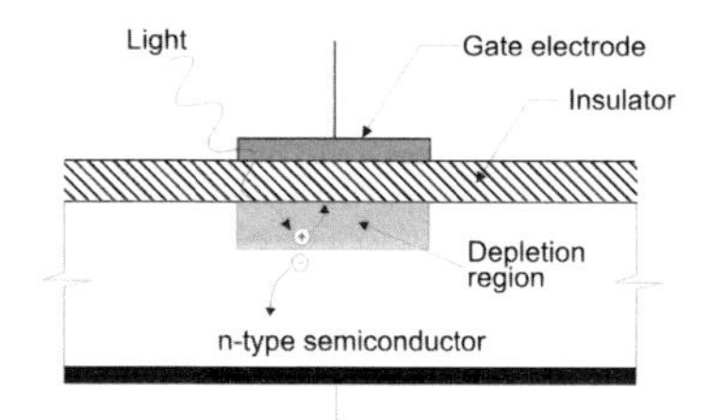

The MIS device consists of a metal gate separated from a semiconductor surface by an insulator (I). By applying a negative voltage to the metal electrode, electrons are repelled from the I–S interface, creating a depletion region. When incident photons create hole–electron pairs, the minority carriers drift away to the depletion region and the volume of the depletion region shrinks. The total amount of charge that a photogate can collect is defined as its well capacity. The total well capacity is decided by the gate bias, the insulator thickness, the area of the electrodes, and the background doping of the semiconductor. Numerous such photogates with proper clocking sequence form a CCD imaging array.

Schottky barrier photodiode

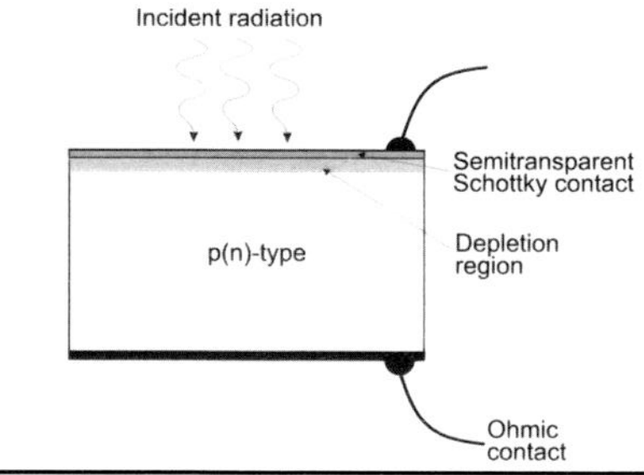

Schottky barrier photodiodes reveal some advantages over p–n junction photodiodes: fabrication simplicity (deposition of metal barrier on n(p)-semiconductor), absence of high-temperature diffusion processes, and high speed of response. Since it is a majority carrier device, minority carrier storage and removal problems do not exist, and therefore higher bandwidths can be expected.

The thermionic emission process in a Schottky barrier is much more efficient than the diffusion process, and therefore for a given built-in voltage, the saturation current in a Schottky diode is several orders of magnitude higher than in the p–n junction.

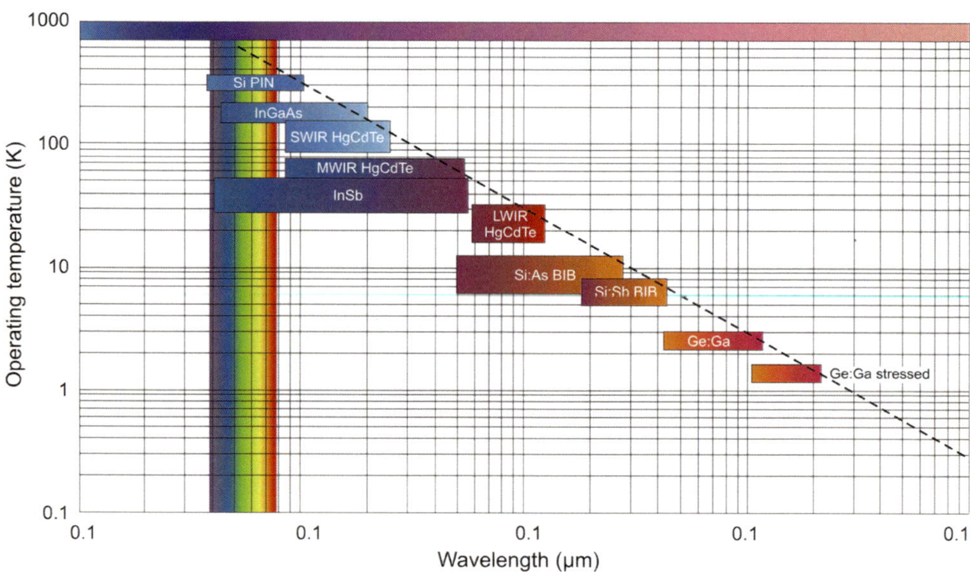

Figure 1.5 Operating temperatures for low-background material systems with their spectral band of greatest sensitivity. The dashed line indicates the trend toward lower operating temperature for longer-wavelength detection (adapted from Ref. 3).

difference between intrinsic and extrinsic detectors is that extrinsic detectors require much cooling to achieve high sensitivity at a given spectral response cutoff in comparison with intrinsic detectors. Low-temperature operation is associated with longer-wavelength sensitivity in order to suppress noise due to thermally induced transitions between close-lying energy levels. Intrinsic detectors are most common at the short wavelengths, below 20 μm. In the longer-wavelength region the photoconductors are operated in extrinsic mode. One advantage of photoconductors is their current gain, which is equal to the recombination time divided by the majority-carrier transit time. This current gain leads to higher responsivity than is possible with nonavalanching photovoltaic detectors. However, serious problem of photoconductors operated at low temperature is nonuniformity of detector element due to recombination mechanisms at the electrical contacts and its dependence on electrical bias.

Recently, interfacial workfunction internal photoemission detectors, quantum well and quantum dot detectors, which can be included to extrinsic photoconductors, have been proposed especially for IR and THz spectral bands.[3] The very fast time response of quantum well and quantum dot semiconductor detectors make them attractive for heterodyne detection.

1.2.2 Thermal detectors

The second class of detectors is composed of thermal detectors. In a thermal detector shown schematically in Fig. 1.6, the incident radiation is absorbed to

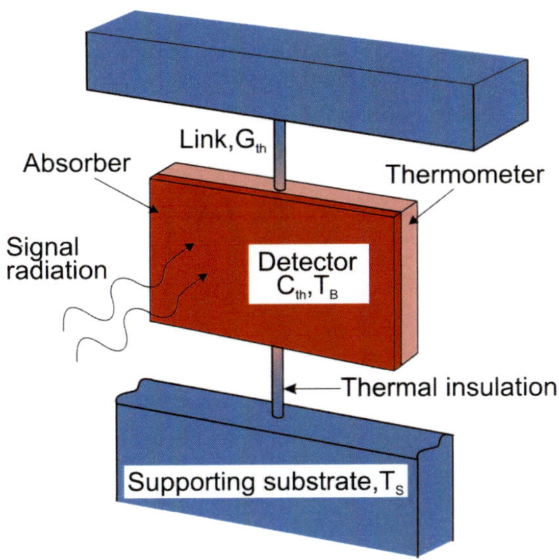

Figure 1.6 Schematic diagram of thermal detector (adapted from Ref. 3).

change the material temperature, and the resultant change in some physical property is used to generate an electrical output. The detector is suspended on lags, which are connected to the heat sink. The signal does not depend upon the photonic nature of the incident radiation. Thus, thermal effects are generally wavelength independent [see Fig. 1.4(a)]; the signal depends upon the radiant power (or its rate of change) but not upon its spectral content. Since the radiation can be absorbed in a black surface coating, the spectral response can be very broad. Attention is directed toward three approaches that have found the greatest utility in infrared technology, namely, bolometers, pyroelectric, and thermoelectric effects. The thermopile is one of the oldest IR detectors, and is a collection of thermocouples connected in series in order to achieve better temperature sensitivity. In pyroelectric detectors a change in the internal electrical polarization is measured, whereas in the case of thermistor bolometers a change in the electrical resistance is measured. For a long time, thermopiles were slow, insensitive, bulky, and costly devices. But with developments in semiconductor technology, thermopiles can be optimized for specific applications. Recently, thanks to conventional complementary metal-oxide semiconductor (CMOS) processes, the thermopile's on-chip circuitry technology has opened the door to mass production.

Usually a bolometer is a thin, blackened flake or slab, whose impedance is highly temperature dependent. Bolometers may be divided into several types. The most commonly used are the metal, the thermistor, and the semiconductor bolometers. A fourth type is the superconducting bolometer. This bolometer operates on a conductivity transition in which the resistance changes

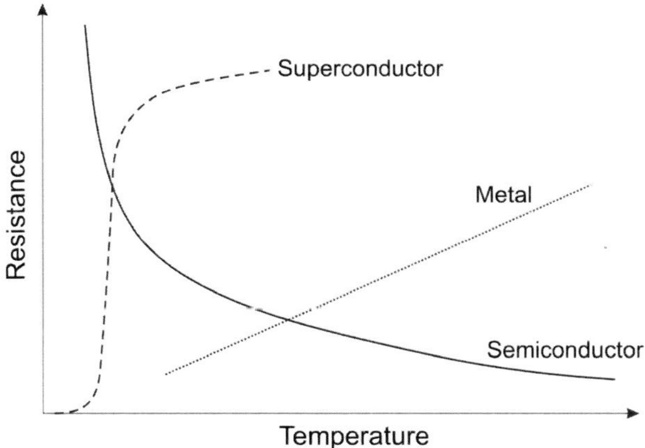

Figure 1.7 Temperature dependence of resistance of three bolometer material types.

dramatically over the transition temperature range. Figure 1.7 shows schematically the temperature dependence of resistance of different types of bolometers.

Many types of thermal detectors are operated in wide spectral range of electromagnetic radiation. The operation principles of thermal detectors are briefly described in Table 1.2.

Microbolometer detectors are now produced in larger volumes than all other IR array technologies together. At present, VO_x microbolometer arrays are clearly the most used technology for uncooled detectors. VO_x wins the battle between the amorphous silicon bolometers and barium strontium titanate (BST) ferroelectric detectors.

1.3 Detector Figures of Merit

It is difficult to measure the performance characteristics of infrared detectors because of the large number of experimental variables involved. A variety of environmental, electrical, and radiometric parameters must be taken into account and carefully controlled. With the advent of large 2D detector arrays, detector testing has become even more complex and demanding.

This section is intended to serve as an introductory reference for the testing of infrared detectors. Numerous texts and journals cover this issue, including: *Infrared System Engineering*[14] by R. D. Hudson; *The Infrared Handbook*,[15] edited by W. L. Wolfe and G. J. Zissis; *The Infrared and Electro-Optical Systems Handbook*,[16] edited by W. D. Rogatto; and *Fundamentals of Infrared Detector Operation and Testing*[17] by J. D. Vincent, and second edition of Vincent's book.[18] In this chapter we have restricted our consideration to detectors whose output consists of an electrical signal that is proportional to the radiant signal power.

Table 1.2 Thermal detectors.

Mode of operation	Schematic of detector	Operation and properties
Thermopile	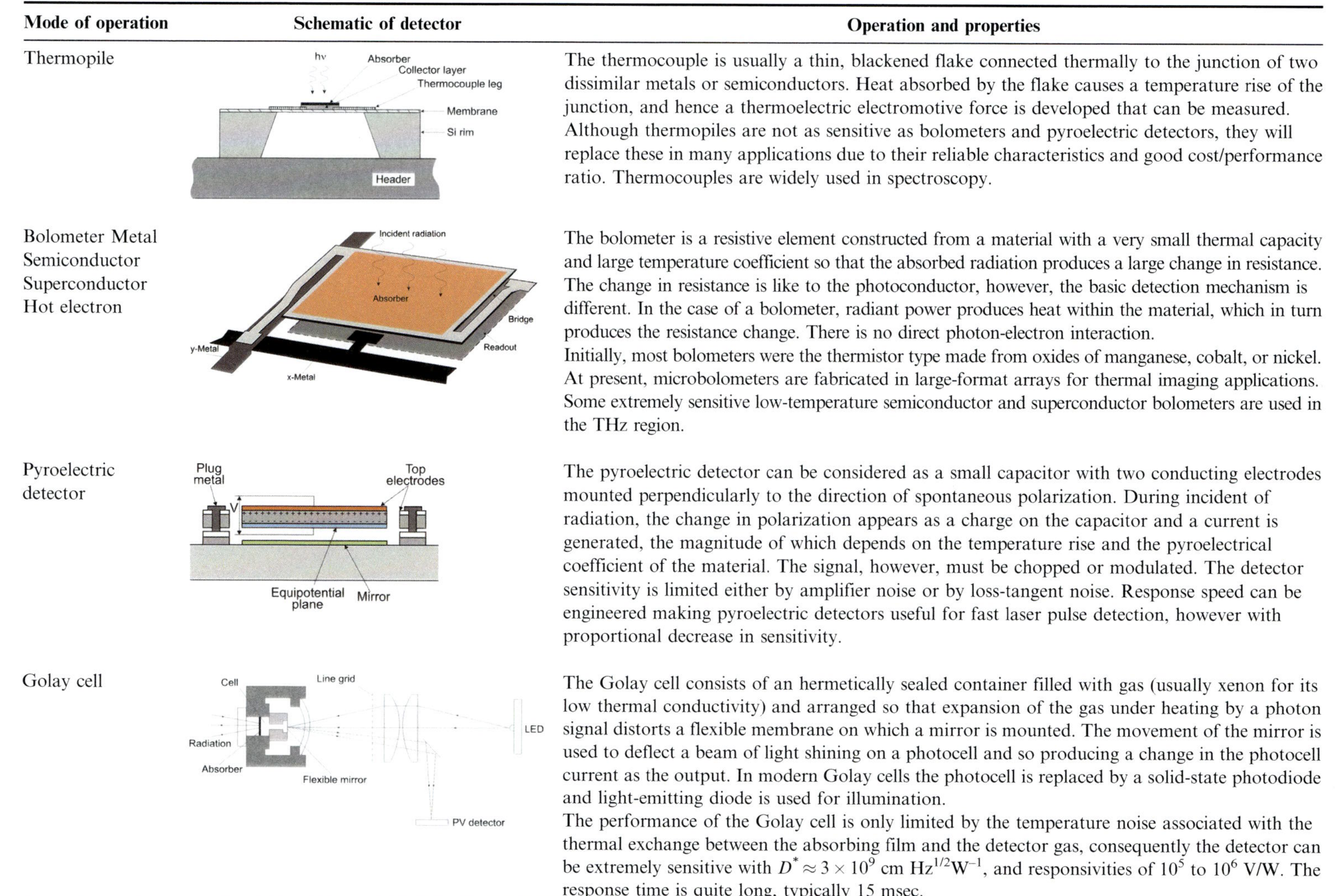 	The thermocouple is usually a thin, blackened flake connected thermally to the junction of two dissimilar metals or semiconductors. Heat absorbed by the flake causes a temperature rise of the junction, and hence a thermoelectric electromotive force is developed that can be measured. Although thermopiles are not as sensitive as bolometers and pyroelectric detectors, they will replace these in many applications due to their reliable characteristics and good cost/performance ratio. Thermocouples are widely used in spectroscopy.
Bolometer Metal Semiconductor Superconductor Hot electron		The bolometer is a resistive element constructed from a material with a very small thermal capacity and large temperature coefficient so that the absorbed radiation produces a large change in resistance. The change in resistance is like to the photoconductor, however, the basic detection mechanism is different. In the case of a bolometer, radiant power produces heat within the material, which in turn produces the resistance change. There is no direct photon-electron interaction. Initially, most bolometers were the thermistor type made from oxides of manganese, cobalt, or nickel. At present, microbolometers are fabricated in large-format arrays for thermal imaging applications. Some extremely sensitive low-temperature semiconductor and superconductor bolometers are used in the THz region.
Pyroelectric detector		The pyroelectric detector can be considered as a small capacitor with two conducting electrodes mounted perpendicularly to the direction of spontaneous polarization. During incident of radiation, the change in polarization appears as a charge on the capacitor and a current is generated, the magnitude of which depends on the temperature rise and the pyroelectrical coefficient of the material. The signal, however, must be chopped or modulated. The detector sensitivity is limited either by amplifier noise or by loss-tangent noise. Response speed can be engineered making pyroelectric detectors useful for fast laser pulse detection, however with proportional decrease in sensitivity.
Golay cell		The Golay cell consists of an hermetically sealed container filled with gas (usually xenon for its low thermal conductivity) and arranged so that expansion of the gas under heating by a photon signal distorts a flexible membrane on which a mirror is mounted. The movement of the mirror is used to deflect a beam of light shining on a photocell and so producing a change in the photocell current as the output. In modern Golay cells the photocell is replaced by a solid-state photodiode and light-emitting diode is used for illumination. The performance of the Golay cell is only limited by the temperature noise associated with the thermal exchange between the absorbing film and the detector gas, consequently the detector can be extremely sensitive with $D^* \approx 3 \times 10^9$ cm Hz$^{1/2}$W^{-1}, and responsivities of 10^5 to 10^6 V/W. The response time is quite long, typically 15 msec.

1.3.1 Responsivity

The responsivity of an infrared detector is defined as the ratio of the root mean square (rms) value of the fundamental component of the electrical output signal of the detector to the rms value of the fundamental component of the input radiation power. The units of responsivity are volts per watt (V/W) or amperes per watt (amp/W).

The voltage (or analogous current) spectral responsivity is given by

$$R_v(\lambda, f) = \frac{V_s}{\Phi_e(\lambda)}, \tag{1.2}$$

where V_s is the signal voltage due to Φ_e, and $\Phi_e(\lambda)$ is the spectral radiant incident power (in W).

An alternative to the above monochromatic quality, the blackbody responsivity, is defined by the equation

$$R_v(T, f) = \frac{V_s}{\int_0^\infty \Phi_e(\lambda) d\lambda}, \tag{1.3}$$

where the incident radiant power is the integral over all wavelengths of the spectral density of power distribution $\Phi_e(\lambda)$ from a blackbody. The responsivity is usually a function of the bias voltage, the operating electrical frequency, and the wavelength.

1.3.2 Noise equivalent power

The noise equivalent power (*NEP*) is the incident power on the detector generating a signal output equal to the rms noise output. Stated another way, the *NEP* is the signal level that produces a signal-to-noise ratio (SNR) of 1. It can be written in terms of responsivity:

$$NEP = \frac{V_n}{R_v} = \frac{I_n}{R_i}. \tag{1.4}$$

The unit of *NEP* is watts.

The *NEP* is also quoted for a fixed reference bandwidth, which is often assumed to be 1 Hz. This "*NEP* per unit bandwidth" has a unit of watts per square root hertz (W/Hz$^{1/2}$).

1.3.3 Detectivity

The detectivity D is the reciprocal of *NEP*:

$$D = \frac{1}{NEP}. \tag{1.5}$$

It was found by Jones[19] that for many detectors the *NEP* is proportional to the square root of the detector signal that is proportional to the detector area A_d. This means that both *NEP* and detectivity are functions of electrical bandwidth and detector area, so a normalized detectivity D^* (or *D*-star) suggested by Jones[19,20] is defined as

$$D^* = D(A_d \Delta f)^{1/2} = \frac{(A_d \Delta f)^{1/2}}{NEP}.$$ (1.6)

The importance of D^* is that this figure of merit permits comparison of detectors of the same type, but having different areas. Either a spectral or blackbody D^* can be defined in terms of the corresponding type of *NEP*.

Useful equivalent expressions to Eq. (1.6) include:

$$D^* = \frac{D(A_d \Delta f)^{1/2}}{V_n} R_v = \frac{D(A_d \Delta f)^{1/2}}{I_n} R_i = \frac{D(A_d \Delta f)^{1/2}}{\Phi_e} (SNR),$$ (1.7)

where D^* is defined as the rms SNR in a 1-Hz bandwidth per unit rms incident radiant power per square root of detector area. D^* is expressed in units of cm $Hz^{1/2}W^{-1}$, which recently has been referred to as "Jones."

Spectral detectivity curves for a number of commercially available IR detectors are shown in Fig. 1.8. Interest has focused mainly on the two atmospheric windows 3–5 μm (MWIR) and 8–14 μm (LWIR) (atmospheric transmission is the highest in these bands and the emissivity maximum of the objects at $T \approx 300$ K is at the wavelength $\lambda \approx 10$ μm), although in recent years there has been increasing interest in longer wavelengths stimulated by space applications. The spectral character of the background is influenced by the transmission of the atmosphere that controls the spectral ranges of the infrared for which the detector may be used when operating in the atmosphere.

1.3.4 Quantum efficiency

A signal whose photon energy is sufficient to generate photocarriers will continuously lose energy as the optical field propagates through the semiconductor. Inside the semiconductor, the field decays exponentially as energy is transferred to the photocarriers. The material can be characterized by an absorption length α and a penetration depth $1/\alpha$. Penetration depth is the point at which $1/e$ of the optical signal power remains.

The power absorbed in the semiconductor as a function of position within the material is then

$$P_a = P_i(1 - r)(1 - e^{-\alpha x}).$$ (1.8)

The number of photons absorbed is the power (in watts) divided by the photon energy ($E = h\nu$). If each absorbed photon generates a photocarrier,

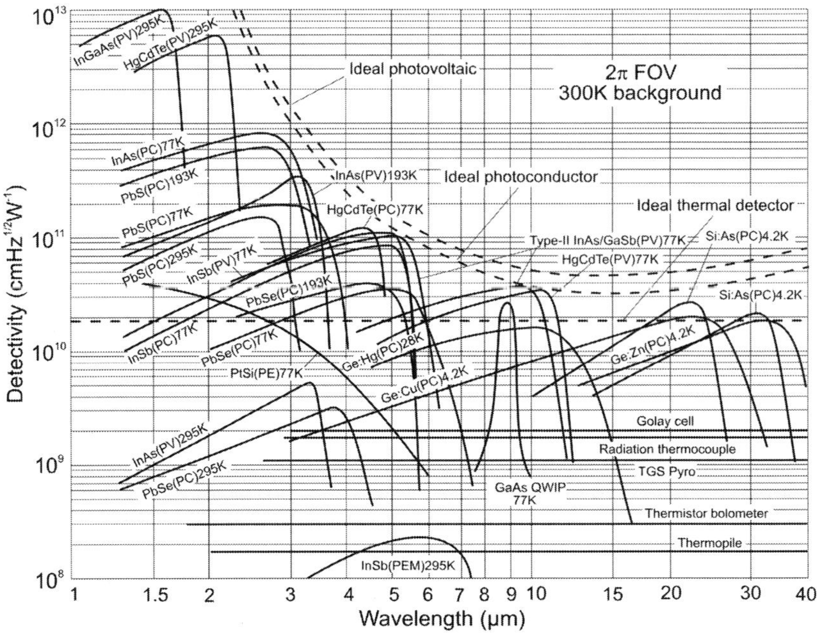

Figure 1.8 Comparison of the *D* of various commercially available infrared detectors when operated at the indicated temperature. The chopping frequency is 1000 Hz for all detectors except the thermopile (10 Hz), thermocouple (10 Hz), thermistor bolometer (10 Hz), Golay cell (10 Hz), and pyroelectric detector (10 Hz). Each detector is assumed to view a hemispherical surround at a temperature of 300 K. Theoretical curves for the background-limited *D* for ideal photovoltaic and photoconductive detectors and thermal detectors are also shown (adapted from Ref. 3).

the number of photocarriers generated per number of incident photons for a specific semiconductor with reflectivity *r* is given by

$$\eta(x) = (1 - r)(1 - e^{-\alpha x}), \tag{1.9}$$

where $0 \leq \eta \leq 1$ is a definition for the detector's quantum efficiency as the number of electron–hole pairs generated per incident photon.

Figure 1.9 shows the quantum efficiency of some of the detector materials used to fabricate arrays of ultraviolet (UV), visible, and infrared detectors. Photocathodes and AlGaN detectors are being developed in the UV region. Silicon p-i-n diodes are shown with and without antireflection coating. Lead salts (PbS and PbSe) have intermediate quantum efficiencies, while PtSi Schottky barrier types and quantum well infrared photodetectors (QWIPs) have low values. InSb can respond from the near UV out to 5.5 μm at 80 K. A suitable detector material for the near-IR (1.0–1.7 μm) spectral range is InGaAs lattice matched to the InP. Various HgCdTe alloys, in both photovoltaic and photoconductive configurations, cover from 0.7 μm to over 20 μm. InAs/GaSb strained layer superlattices have emerged as an alternative

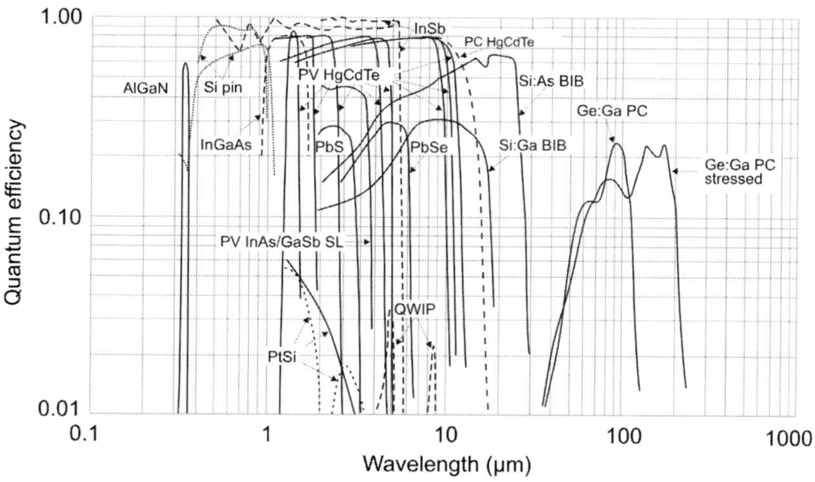

Figure 1.9 Quantum efficiency of different detectors.

to the HgCdTe. Impurity-doped (Sb, As, and Ga) silicon BIB detectors operating at 10 K have a spectral response cutoff in the range of 16– to 30– μm. Impurity-doped Ge detectors can extend the response out to 100–200 μm.

1.4 Fundamental Detector Performance Limits

In general, the detector can be considered as a slab of homogeneous semiconductor with actual "electrical" area A_e, thickness t, and volume $A_e t$ (see Fig. 1.10). Usually, the optical and electrical areas of the device are the same or similar. However, the use of some kind of optical concentrator can increase the A_o/A_e ratio by a large factor.

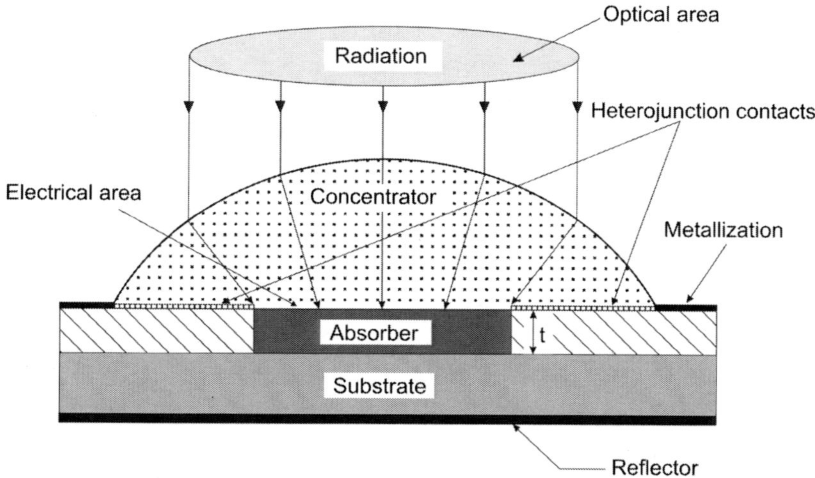

Figure 1.10 Model of a photodetector (adapted from Ref. 3).

The detectivity D^* of an infrared photodetector is limited by generation and recombination rates G and R (in $m^{-6}s^{-1}$) in the active region of the device.[21] It can be expressed as

$$D^* = \frac{\lambda}{2^{1/2}hc(G+R)^{1/2}} \left(\frac{A_o}{A_e}\right) \frac{\eta}{t^{1/2}}, \tag{1.10}$$

where λ is the wavelength, h is Planck's constant, c is the velocity of light, and η is the quantum efficiency.

For a given wavelength and operating temperature, the highest performance can be obtained by maximizing the ratio of the quantum efficiency to the square root of the sum of the sheet thermal generation and recombination rates $\eta/[(G+R)t]^{1/2}$. This means that high quantum efficiency must be obtained with a thin device.

A possible way to improve the performance of IR detectors is to reduce the physical volume of the semiconductor, thus reducing the amount of thermal generation. However, this must be achieved without decrease in quantum efficiency, optical area, and field of view (FOV) of the detector.

At equilibrium, the generation and recombination rates are equal. If we further assume $A_e = A_o$, the detectivity of an optimized infrared photodetector is limited by thermal processes in the active region of the device. It can be expressed as

$$D^* = 0.31 \frac{\lambda}{hc} k \left(\frac{\alpha}{G}\right)^{1/2}, \tag{1.11}$$

where $1 \leq k \leq 2$ and is dependent on the contribution of recombination and backside reflection. The k-coefficient can be modified by using more sophisticated coupling of the detector with IR radiation, e.g., using photonic crystals or surface plasmon-polaritons.

The ratio of the absorption coefficient to the thermal generation rate, α/G, is the fundamental figure of merit of any material intended for infrared photodetection. The α/G ratio versus temperature for various material systems capable of bandgap tuning is shown in Fig. 1.11 for a hypothetical energy gap equal to 0.25 eV ($\lambda = 5$ μm) [Fig. 1.11(a)] and 0.124 eV ($\lambda = 10$ μm) [Fig. 1.11(b)]. Procedures used in calculations of α/G for different material systems are given in Ref. 22. Analysis shows that narrow-gap semiconductors are more suitable for high-temperature photodetectors in comparison to competing technologies such as extrinsic devices, QWIP (quantum well IR photodetector) and QDIP (quantum dot IR photodetector) devices. The main reason for the high performance of intrinsic photodetectors is the high density of states in the valence and conduction bands, which results in strong absorption of infrared radiation. Figure 1.11(b) predicts that a recently emerging

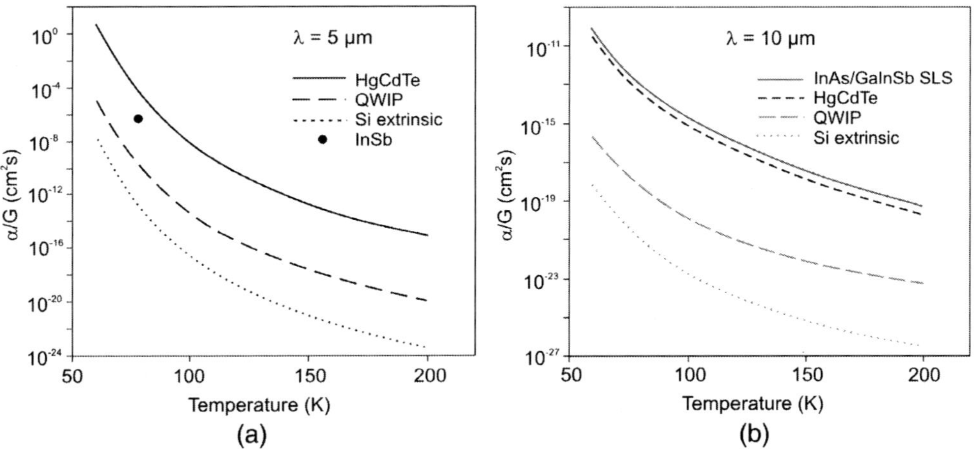

Figure 1.11 α/G ratio versus temperature for (a) MWIR ($\lambda = 5$ μm) and (b) LWIR ($\lambda = 10$ μm) photodetectors based on HgCdTe, QWIP, Si extrinsic, and type-II superlattice (for LWIR only) material technology (adapted from Ref. 3).

competing IR material, type-II superlattice, is the most efficient material technology for IR detection in the long-wavelength region, theoretically perhaps even better than HgCdTe if the influence of the Shockley–Read–Hall lifetime is not considered. It is characterized by a high absorption coefficient and relatively low fundamental (band-to-band) thermal generation rate. However, this theoretical prediction has not been confirmed by experimental data. It is also worth noting that theoretically AlGaAs/GaAs QWIP is also a better material than extrinsic silicon.

The ultimate performance of infrared detectors is reached when the detector and amplifier noise are low compared to the photon noise. The photon noise is fundamental in the sense that it arises not from any imperfection in the detector or its associated electronics but rather from the detection process itself, as a result of the discrete nature of the radiation field. The radiation falling on the detector is a composite of that from the target and that from the background. The practical operating limit for most infrared detectors is not the signal fluctuation limit but the background fluctuation limit, also known as the background-limited infrared photodetector (BLIP) limit.

The expression for shot noise can be used to derive the BLIP detectivity,

$$D^*_{BLIP}(\lambda, T) = \frac{\lambda}{hc} k \left(\frac{\eta}{2\Phi_B}\right)^{1/2}, \qquad (1.12)$$

where η is the quantum efficiency, and Φ_B is the total background photon flux density reaching the detector, denoted as

$$\Phi_B = \sin^2(\theta/2) \int_0^{\lambda_c} \Phi(\lambda, T_B) d\lambda, \tag{1.13}$$

where θ is the detector field of view angle.

Planck's photon emittance (in units of photons $cm^{-2}s^{-1}$ μm^{-1}) at temperature T_B is given by

$$\Phi(\lambda, T_B) = \frac{2\pi c}{\lambda^4 [\exp(hc/\lambda k T_B) - 1]} = \frac{1.885 \times 10^{23}}{\lambda^4 [\exp(14.388/\lambda k T_B) - 1]}. \tag{1.14}$$

Equation (1.12) holds for photovoltaic detectors, which are shot-noise limited. Photoconductive detectors that are generation-recombination noise limited have a lower D^*_{BLIP} by a factor of $2^{1/2}$:

$$D^*_{BLIP}(\lambda, f) = \frac{\lambda}{2\,hc} k \left(\frac{\eta}{\Phi_B}\right)^{1/2}. \tag{1.15}$$

Once background-limited performance is reached, quantum efficiency η is the only detector parameter that can influence detector's performance.

Figure 1.12 shows the peak spectral detectivity of a background-limited photodetector operating at 300, 230, and 200 K, versus the wavelength calculated for 300 K background radiation and hemispherical FOV ($\theta = 90$ deg). The minimum D^*_{BLIP} (300 K) occurs at 14 μm and equals 4.6×10^{10} cm $Hz^{1/2}$/W. For some photodetectors that operate at near

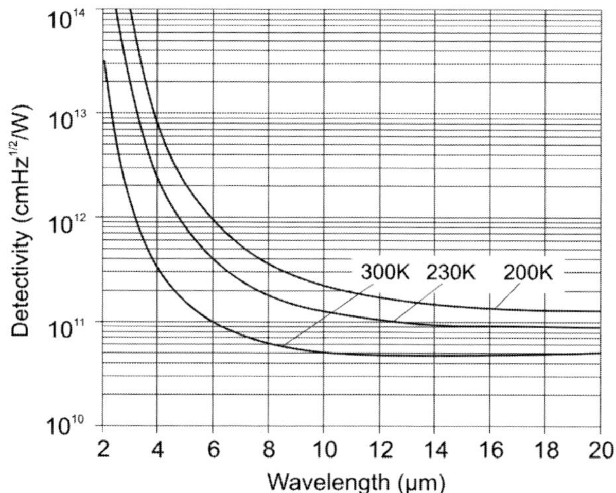

Figure 1.12 Calculated spectral detectivities of a photodetector limited by the hemispherical FOV background radiation of 300 K as a function of the peak wavelength for detector operating temperatures of 300, 230, and 200 K (reprinted from Ref. 23).

equilibrium conditions, such as non-sweep-out photoconductors, the recombination rate is equal to the generation rate. For these detectors the contribution of recombination to the noise will reduce D^*_{BLIP} by a factor of $2^{1/2}$. Note that D^*_{BLIP} does not depend on area and the A_o/A_e ratio. As a consequence, the background-limited performance cannot be improved by making A_o/A_e large.

The highest performance possible will be obtained by the ideal detector with unity quantum efficiency and ideal spectral responsivity [$R(\lambda)$ increases with wavelength to the cutoff wavelength λ_c at which the responsivity drops to zero]. This limiting performance is of interest for comparison with actual detectors.

The detectivity of BLIP detectors can be improved by reducing the background photon flux Φ_B. Practically, there are two ways to do this: a cooled or reflective spectral filter to limit the spectral band or a cooled shield to limit the angular field of view of the detector (as described above). The former eliminates background radiation from spectral regions in which the detector need not respond. The best detectors yield background-limited detectivities in quite narrow fields of view.

1.5 Performance of Focal Plane Arrays

This section discusses concepts associated with the performance of focal plane arrays (FPAs). For arrays the relevant figure of merit for determining the ultimate performance is not the detectivity D^*, but the noise equivalent difference temperature (*NEDT*) and the modulation transfer function (*MTF*). *NEDT* and *MTF* are considered as the primary performance metrics to thermal imaging systems: thermal sensitivity and spatial resolution. Thermal sensitivity is concerned with the minimum temperature difference that can be discerned above the noise level. The *MTF* concerns the spatial resolution and answers the question of how small an object can be and still be imaged by the system. The general approach of system performance is given by Lloyd in his fundamental monograph.[24]

1.5.1 Modulation transfer function

The modulation transfer function (*MTF*) expresses the ability of an imaging system to faithfully image a given object; it quantifies the ability of the system to resolve or transfer spatial frequencies.[25] Consider a bar pattern with a cross-section of each bar being a sine wave. Since the image of a sine wave light distribution is always a sine wave, the image is always a sine wave independent of the other effects in the imaging system, such as aberrations.

Usually, imaging systems have no difficulty in reproducing the bar pattern when the bar pattern is sparsely spaced. However, an imaging system reaches its limit when the features of the bar pattern get closer and closer together. When the imaging system reaches this limit, the contrast or the modulation M is defined as

$$M = \frac{E_{\max} - E_{\min}}{E_{\max} + E_{\min}}, \qquad (1.16)$$

where E is the irradiance. Once the modulation of an image is measured experimentally, the MTF of the imaging system can be calculated for that spatial frequency using

$$MTF = \frac{M_{\text{image}}}{M_{\text{object}}}. \qquad (1.17)$$

The system MTF is dominated by the optics, detector, and display MTFs and can be cascaded by simply multiplying the MTF components to obtain the MTF of the combination. In spatial frequency terms, the MTF of an imaging system at a particular operating wavelength is dominated by limits set by the size of the detector and the aperture of the optics. More details about this issue is given in section 9.2.

1.5.2 Noise equivalent difference temperature

Noise equivalent difference temperature ($NEDT$) is a figure of merit for thermal imagers that is commonly reported. In spite of its widespread use in infrared literature, it is applied to different systems, in different conditions, and with different meanings.[26]

$NEDT$ of a detector represents the temperature change, for incident radiation, that gives an output signal equal to the rms noise level. While normally thought of as a system parameter, detector $NEDT$ and system $NEDT$ are the same except for system losses. $NEDT$ is defined as

$$NEDT = \frac{V_n(\partial T/\partial \Phi)}{(\partial V_s/\partial \Phi)} = V_n \frac{\Delta T}{\Delta V_s}, \qquad (1.18)$$

where V_n is the rms noise, Φ is the spectral photon flux density (photons/cm^{2}s) incident on a focal plane, and ΔV_s is the signal measured for the temperature difference ΔT.

We follow Kinch[27] further to obtain useful equations for noise equivalent irradiance (NEI) and $NEDT$, used for estimation of detector performance (see e.g., section 6.6).

In modern IR FPAs the current generated in a biased photon detector is integrated onto a capacitive node with a carrier well capacity of N_w. For an ideal system, in absence of excess noise, the detection limit of the node is achieved when a minimum detectable signal flux $\Delta\Phi$ creates a signal equal shot noise on the node:

$$\Delta\Phi\eta A_d\tau_{\text{int}} = \sqrt{N_w} = \sqrt{\frac{(J_{dark} + J_\Phi)A_d\tau_{\text{int}}}{q}},\qquad(1.19)$$

where η is the detector collection efficiency, A_d is the detector area, τ_{int} is the integration time, J_{dark} is the detector dark current, and J_Φ is the flux current.

Associated with *NEDT* is the other critical parameter, the so-called noise equivalent flux (*NE$\Delta\Phi$*). This parameter is defined for spectral regions in which the thermal background flux does not dominate. By equating the minimum detectable signal to the integrated current noise, we have

$$\eta\Phi_s A_d\tau_{\text{int}} = \sqrt{\frac{(J_{dark} + J_\Phi)A_d\tau_{\text{int}}}{q}},\qquad(1.20)$$

giving

$$NE\Delta\Phi = \frac{1}{\eta}\sqrt{\frac{J_{dark} + J_\Phi}{qA_d\tau_{\text{int}}}}.\qquad(1.21)$$

This can be converted to a noise equivalent irradiance (*NEI*), which is defined as the minimum observable flux power incident on the system aperture, by renormalizing the incident flux density on the detector to the system aperture area A_{opt}. The *NEI* is given by

$$NEI = NE\Delta\Phi\frac{A_d h\nu}{A_{opt}},\qquad(1.22)$$

where monochromatic radiation of energy $h\nu$ is assumed.

NEI [photons/(cm^{2}sec)] is the signal flux level at which the signal produces the same output as the noise present in the detector. This unit is useful because it directly gives the photon flux above which the detector will be photon-noise limited.

For high-background-flux conditions, the signal flux can be defined as $\Delta\Phi = \Delta T(d\Phi_B/dT)$. Thus, for shot noise, substituting in Eq. (1.19), we have

$$\eta\Delta T\frac{d\Phi_B}{dT} = \sqrt{\frac{J_{dark} + J_\Phi}{qA_d\tau_{\text{int}}}}.\qquad(1.23)$$

Finally, after some re-arrangement,

$$NEDT = \frac{1 + (J_{dark}/J_\Phi)}{\sqrt{N_w C}}, \tag{1.24}$$

where $C = (d\Phi_B/dT)/\Phi_B$ is the scene contrast through the optics. In deriving Eq. (1.24) it was assumed that the optics transmission is unity, and that the cold shield of the detector is not contributing flux. This is reasonable at low detector temperatures but not at higher operating temperatures. At higher temperatures the scene contrast is defined in terms of the signal flux coming through the optics, whereas the flux current is defined by the total flux through the optics and the flux from the cold shield.

1.5.3 Other issues

Infrared photodetectors are typically operated at cryogenic temperatures to decrease the noise of the detector arising from various mechanisms associated with the narrow bandgap. There are considerable efforts to decrease system cost, size, weight, and power consumption, to increase the operating temperature in so-called high-operating-temperature (HOT) detectors. Increasing the operating temperature of the detector reduces the cooling load, allowing more compact cooling systems with higher efficiency. Because the cost of the optics made from Ge (the standard material for IR optics) rises approximately with the square of the lens diameter, the reduction of the pixel size results in a significantly reduced cost of the optics. In addition, the reduction in pixel size allows for a larger number of FPAs to be fabricated on each wafer.

Pixel reduction is also needed to increase the detection and identification range of infrared imaging systems. It appears that, e.g., the detection range of many uncooled IR imaging systems is limited by pixel resolution rather than sensitivity. Figure 1.13 presents a trade-off analysis of the detection range and

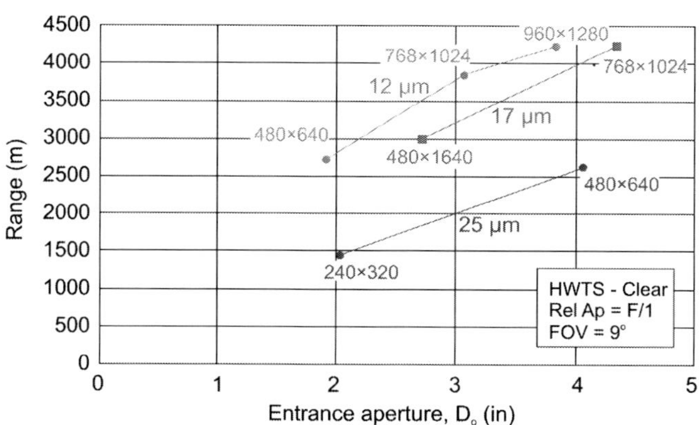

Figure 1.13 Calculated detection range as a function of sensor optics for increasing detector pixel size and format using NVESD NVTherm IP modeling, assuming a 35-mK *NEDT* (*F*/1, 30Hz) for all detectors (reprinted from Ref. 28).

sensor optics for a thermal weapon sight using the NVESD NVTherm IP model, assuming a detector sensitivity of 35-mK *NETD* (*F*/1, 30 Hz) for the 25-, 17-, and 12-μm pitch pixel of uncooled FPAs. The advantages of small pixel pitch and large-format FPAs are obvious. By switching to smaller pitch and larger format detectors, the detection range of a weapon sight increases significantly with a fixed optical entrance aperture.

Key challenges in realizing ultimate pixel dimensions in FPA design including dark current, pixel hybridization, pixel delineation, and unit cell readout capacity are considered in Refs. 27 and 29, and Section 9.2 of this book.

It is interesting to consider the performance requirements of near-room-temperature photodetectors for thermal cameras. It can be shown[3] that the thermal resolution of infrared thermal systems is characterized by the equation

$$NEDT = \frac{4(F/\#)^2 \Delta f^{1/2}}{A_d^{1/2} t_{opt}} \left[\int_{\lambda_a}^{\lambda_b} \frac{dM}{dT} D^*(\lambda) d\lambda \right], \qquad (1.25)$$

where $F/\#$ is the optics f-number, Δf is the frequency band, A_d is the detector area, t_{opt} is the optics transmission, and M is the spectral emittance of the blackbody described by Planck's law.

As Eq. (1.25) shows, the thermal resolution improves with an increase in detector area. However, increasing detector area results in reduced spatial resolution. Hence, a reasonable compromise between the requirement of high thermal resolution and spatial resolution is necessary. Improvement of thermal resolution without detrimental effects on spatial resolution may be achieved by:

- an decrease of detector area combined with a corresponding decrease of the optics f-number,
- improved detector performance, and
- an increase in the number of detectors.

As was mentioned before, increasing the aperture is undesirable because it increases the size, mass, and price of an IR system. It is more appropriate to use a detector with higher detectivity. This can be achieved by better coupling of the detector with the incident radiation. Another possibility is the application of a multi-elemental sensor, which reduces each element bandwidth proportionally to the number of elements for the same frame rate and other parameters.

Figure 1.14 shows the dependence of detectivity on the cutoff wavelength for a photodetector thermal imager with a resolution of 0.1 K. Detectivities of 1.9×10^8 cm Hz$^{1/2}$/W, 2.3×10^8 cm Hz$^{1/2}$/W, and 2×10^9 cm Hz$^{1/2}$/W are necessary to obtain *NEDT* = 0.1 K for 10-μm, 9-μm, and 5-μm cutoff wavelengths, respectively. The above estimations indicate that the ultimate

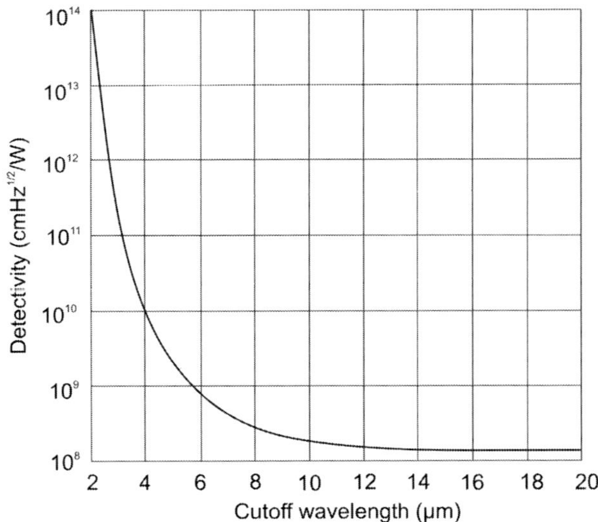

Figure 1.14 Detectivity needed to obtain *NEDT* = 0.1 K in a photon-counter detector thermal imager as a function of cutoff wavelength (reprinted from Ref. 23).

performance of the uncooled photodetectors is not sufficient to achieve a thermal resolution of 0.1 K. Thermal resolution below 0.1 K is achieved for staring thermal imagers containing thermal FPAs.

The previous considerations are valid assuming that the temporal noise of the detector is the main source of noise. However, this assertion is not true for staring arrays, where the nonuniformity of the detector response is a significant source of noise. This nonuniformity appears as a fixed-pattern noise (spatial noise) and is defined in various ways in the literature. The most common definition is that it is the dark signal nonuniformity arising from an electronic source (i.e., other than thermal generation of the dark current); e.g., clock breakthrough or from offset variations in row, column, or pixel amplifiers/switches. So, estimation of IR sensor performance must include a treatment of spatial noise that occurs when FPA nonuniformities cannot be compensated correctly.

Mooney et al.[30] have given a comprehensive discussion of the origin of spatial noise. The total noise of a staring array is the composite of the temporal noise and the spatial noise. The spatial noise is the residual nonuniformity *u* after application of nonuniformity compensation, multiplied by the signal electrons *N*. Photon noise, which equals $N^{1/2}$, is the dominant temporal noise for the high IR background signals for which spatial noise is significant. Then, the total *NEDT* is

$$NEDT_{total} = \frac{(N + u^2 N^2)^{1/2}}{\partial N/\partial T} = \frac{(1/N + u^2)^{1/2}}{(1/N)(\partial N/\partial T)},\qquad (1.26)$$

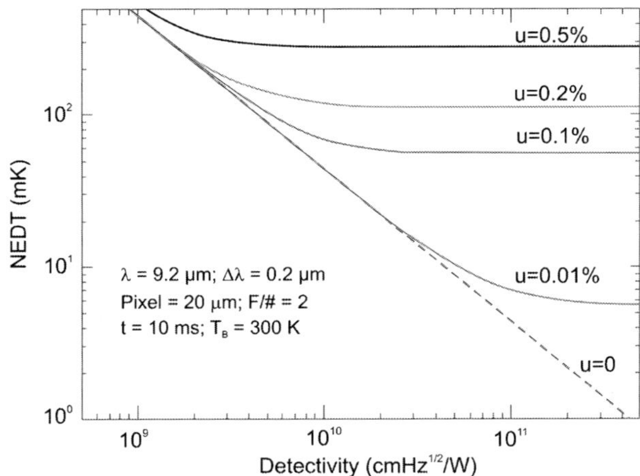

Figure 1.15 *NEDT* as a function of detectivity. The effects of nonuniformity are included for $u = 0.01\%, 0.1\%, 0.2\%$, and 0.5%. Note that for $D^* > 10^{10}$ cm Hz$^{1/2}$/W, detectivity is not the relevant figure of merit 9 (adapted from Ref. 3).

where $\partial N / \partial T$ is the signal change for a 1-K source temperature change. The denominator, $(\partial N / \partial T)/N$, is the fractional signal change for a 1-K source temperature change. This is the relative scene contrast.

The dependence of the total *NEDT* on detectivity for different residual nonuniformity is plotted in Fig. 1.15 for 300 K scene temperature and the set of parameters shown in the figure insert. When detectivity approaches a value above 10^{10} cm Hz$^{1/2}$/W, the FPA performance is uniformity-limited prior to correction and thus essentially independent of the detectivity. An improvement in nonuniformity from 0.1% to 0.01% after correction could lower the *NEDT* from 63 to 6.3 mK.

References

1. R. J. Cashman, "Film-type infrared photoconductors," *Proc. IRE* **47**, 1471–1475 (1959).
2. E. Burstein, G. Pines, and N. Sclar, "Optical and photoconductive properties of silicon and germanium," in *Photoconductivity Conference at Atlantic City*, edited by R. Breckenbridge, B. Russell, and E. Hahn, Wiley, New York, pp. 353–413 (1956).
3. A. Rogalski, *Infrared Detectors*, 2nd edition, CRC Press, Boca Raton, Florida (2010).
4. R. A. Soref, "Extrinsic IR photoconductivity of Si dped with B, Al, Ga, P, As or Sb," *J. Appl. Phys.* **38**, 5201–5209 (1967).
5. W. S. Boyle and G. E. Smith, "Charge-coupled semiconductor devices," *Bell Syst. Tech. J.* **49**, 587–593 (1970).

6. F. Shepherd and A. Yang, "Silicon Schottky retinas for infrared imaging," *IEDM Tech. Dig.*, 310–313 (1973).

7. C. Hilsum and A. C. Rose-Innes, *Semiconducting III-V Compounds*, Pergamon Press, Oxford (1961).

8. J. Melngailis and T. C. Harman, "Single-crystal lead-tin chalcogenides," in *Semiconductors and Semimetals*, Vol. **5**, edited by R. K. Willardson and A. C. Beer, Academic Press, New York, pp. 111–174 (1970).

9. T. C. Harman and J. Melngailis, "Narrow gap semiconductors," in *Applied Solid State Science*, Vol. **4**, edited by R. Wolfe, Academic Press, New York, pp. 1–94 (1974).

10. W. D. Lawson, S. Nielson, E. H. Putley, and A. S. Young, "Preparation and properties of HgTe and mixed crystals of HgTe-CdTe," *J. Phys. Chem. Solids* **9**, 325–329 (1959).

11. S. Krishna, "The infrared retina," *J. Phys. D: Appl. Phys.* **42**, 234005 (2009).

12. A. Rogalski and F. Sizov, "Terahertz detectors and focal plane arrays," *Opto-Electr. Rev.* **19**(3), 346–404 (2011).

13. D. Long, "Photovoltaic and photoconductive infrared detectors," in *Optical and Infrared Detectors*, edited by R. J. Keyes, Springer, Berlin, pp. 101–147 (1980).

14. R. D. Hudson, *Infrared System Engineering*, Wiley, New York (1969).

15. *The Infrared Handbook*, edited by W. I. Wolfe and G. J. Zissis, Office of Naval Research, Washington, D.C. (1985).

16. *The Infrared and Electro-Optical Systems Handbook*, edited by W. D. Rogatto, Infrared Information Analysis Center, Ann Arbor and SPIE Press, Bellingham, Washington (1993).

17. J. D. Vincent, *Fundamentals of Infrared Detector Operation and Testing*, Wiley, New York (1990).

18. J. D. Vincent, S. E. Hodges, J. Vampola, M. Stegall, and G. Pierce, *Fundamentals of Infrared and Visible Detector Operation and Testing*, Wiley, Hoboken, New Jersey (2016).

19. R. C. Jones, "Performance of detectors for visible and infrared radiation," in *Advances in Electronics*, Vol. **5**, edited by L. Morton, Academic Press, New York, pp. 27–30 (1952).

20. R. C. Jones, "Phenomenological description of the response and detecting ability of radiation detectors," *Proc. IRE* **47**, 1495–1502 (1959).

21. J. Piotrowski and A. Rogalski, Comment on "Temperature limits on infrared detectivities of InAs/In$_x$Ga$_{1-x}$Sb superlattices and bulk Hg$_{1-x}$Cd$_x$Te" [*J. Appl. Phys.* 74, 4774 (1993)], *J. Appl. Phys.* **80**(4), 2542–2544 (1996).

22. A. Rogalski, "Quantum well photoconductors in infrared detector technology," *J. Appl. Phys.* **93**, 4355 (2003).

23. J. Piotrowski and A. Rogalski, *High-Operating Temperature Infrared Photodetectors*, SPIE Press, Bellingham, Washington (2007) [doi: 10.1117/3.717228].

24. J. M. Lloyd, *Thermal Imaging Systems*, Plenum Press, New York (1975).
25. G. C. Holst, "Infrared imaging testing," in *The Infrared & Electro-Optical Systems Handbook*, Vol. **4** *Electro-Optical Systems Design, Analysis, and Testing*, edited by M. C. Dudzik, SPIE Press, Bellingham, Washinton (1993).
26. J. M. Lopez-Alonso, "Noise equivalent temperature difference (NETD)," in *Encyclopedia of Optical Engineering*, edited by R. Driggers, Marcel Dekker Inc., New York, pp. 1466–1474 (2003).
27. M. A. Kinch, *State-of-the-Art Infrared Detector Technology*, SPIE Press, Bellingham, Washington (2014) [doi: 10.1117/3.1002766].
28. C. Li, G. Skidmore, C. Howard, E. Clarke, and C. J. Han, "Advancement in 17 micron pixel pitch uncooled focal plane arrays," *Proc. SPIE* **7298**, 72980S (2009) [doi: 10.1117/12.818189].
29. A. Rogalski, P. Martyniuk, and M. Kopytko, "Challenges of small-pixel infrared detectors: a review," *Rep. Prog. Phys.* **79**, 046501 (2016).
30. J. M. Mooney, F. D. Shepherd, W. S. Ewing, and J. Silverman, "Responsivity nonuniformity limited performance of infrared staring cameras," *Opt. Eng.* **28**, 1151 (1989) [doi: 10.1117/12.7977112].

Chapter 2
Antimonide-based Materials

The wide body of information that is now available concerning different methods of antimonide-based crystal growth and physical properties of materials used for IR photodetectors makes it difficult to review all aspects in detail. As a result, only general topics are reviewed in this chapter to give the basic information about crystal growth and physical properties. More information can be found in many comprehensive reviews and monographs (see, for example, Refs. 1–5).

Here we describe crystal growth methods and some of the physical properties of III-V semiconductors of interest to the preparation of infrared detectors. More attention is placed on the development of large InSb and GaSb single crystals. InSb is a well-established sensor material in the manufacture of FPAs and is suitable for detection in the MWIR region. Instead, rapid development of GaSb epi-ready wafers are designed to meet the needs of the upcoming type-II superlattice market.

InSb semiconductor properties were first reported in the early 1950s,[6,7] and it was soon discovered that InSb had the smallest energy gap among any semiconductors known at that time, and naturally its application as an infrared detector became obvious. The bandgap energy of 0.18 eV at room temperature indicated that it would have a long wavelength limit of approximately 7 μm, and when cooled with liquid nitrogen the gap increased to 0.23 eV, enabling it to cover the entire MWIR region up to 5.5 μm (see Fig. 2.1).

GaSb is an intermediate-gap semiconductor, since its gap of 0.8 eV is neither as wide as that of GaAs and InP nor as narrow as that of InAs and InSb. GaSb is particularly interesting as a substrate material because it is lattice matched to solid solutions of various ternary and quaternary III-V compounds whose bandgaps cover a wide spectral range from 0.3 to 1.6 eV (i.e., 0.8–4.3 μm), as shown in Fig. 2.1. GaSb (lattice constant $a_o = 6.0954$ Å), AlSb ($a_o = 6.1355$ Å), and InAs ($a_o = 6.0584$ Å), and their related compounds with small x-composition ($In_xGa_{1-x}Sb$, $AlAs_xSb_{1-x}$, and $GaAs_xSb_{1-x}$) form a nearly lattice-matched family of semiconductors known as the 6.1 Å family because the lattice constants of these materials are about 6.1 Å.

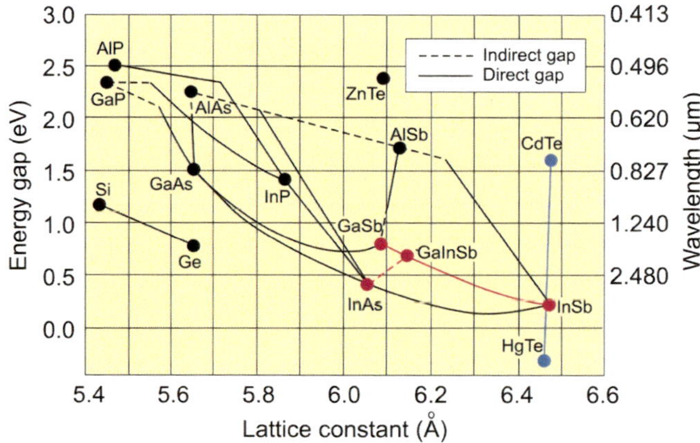

Figure 2.1 Composition and wavelength diagram of semiconductor material systems (adapted from Ref. 8).

2.1 Bulk Materials

In order to grow homogeneous and stoichiometric bulk and epitaxial single crystals of binaries and ternaries, it is essential to understand the phase diagrams of these materials. Phase diagrams of InSb and GaSb are reproduced in Fig. 2.2. A minor deviation from a 1:1 In(Ga) to Sb ratio in the melt still produces very nearly stoichiometric single crystals. When trying to grow single crystals too far from a stoichiometric melt, several complications can arise, including increased occurrence of twins, increased defect density, and second phase precipitations.

The phase diagram of InSb is characterized by the presence of two eutectics occurring at 0.8% and 68.2 at. % Sb. At the extreme left there exists a phase consisting of pure In (α-phase) with a melting point of 154 °C. At the extreme right is elemental Sb (γ-phase) with a melting point of 630 °C.

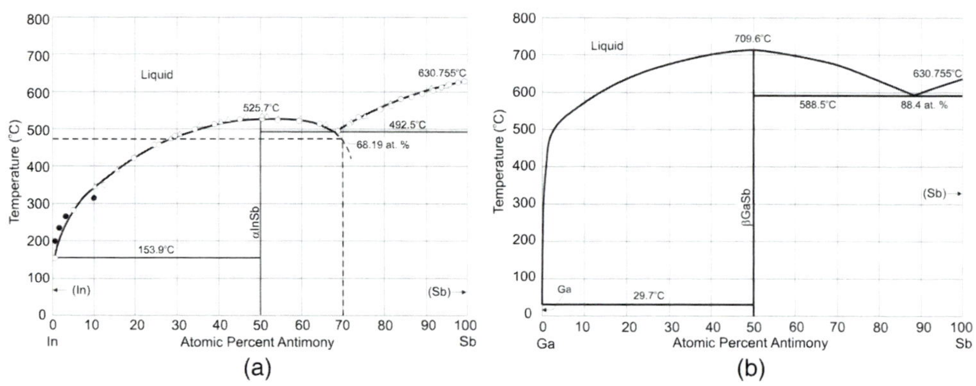

Figure 2.2 Equilibrium phase diagrams of (a) InSb and (b) GaSb.

As early as 1926, Goldschmidt synthesized GaSb and determined its lattice constant.[9] The phase diagram of this compound was determined simultaneously by Koster and Thoma[10] and Greenfield and Smith.[11]

Development of crystal growth techniques in the early 1950s led to bulk single-crystal detectors.[11] Since then, the quality of single-crystal growth has improved immensely.[5,12] Several methods have been developed, among which the Czochralski, Bridgman, and vertical gradient freeze (VGF) methods are the most popularly used.[13] Every crystal growth method has advantages and disadvantages. For commercial production, trade-offs need to be made to maintain balance between crystal quality, desirable electrical and optical properties, infrastructure investments and operational costs, etc. While the quantity of III-V compound semiconductor materials produced by the Czochralski method remains low when compared with more volume-oriented crystal growth methods such as VGF, the substrate industry has met the challenge to develop more volume-production-oriented approaches to crystal growth and expanded the road map of advanced III-V device technologies that are reliant on Czochralski-grown materials. For small-scale fundamental research, much wider crystal growth methods can be suitably adopted based on available infrastructure.

Currently, a modified Czochralski method is commercially used for InSb and GaSb growth.[14,15] Due to the low vapor pressure of Sb, there is no need for a high-pressure vessel. Growth is carried out under dry, pure, inert gas that helps in reducing the oxide scum formed on the top of the melt.

InSb single crystals are grown with relatively high purity and low dislocation density; their ingot sizes have increased from small 2-inch ingots pulled in the early 1980s through to today's 4- to 5-inch boules that support routine InSb substrate production, suitable for convenient handling and photolithography. Recently, larger format ($\geq$6 inch) InSb boule growth has been successfully demonstrated.[16]

While smaller crystals that are low in weight (below 10 kg) can be grown with relative ease, more-sophisticated and automated growth systems are required to deliver large ingots. By precisely controlling the temperature gradients, the rate of pulling, and the speed of rotation, it is possible to extract a large, single-crystal, cylindrical boule from the melt.

Currently, IQE Infrared is a global market leader in the supply of antimonide materials to the IR detector industry, with IQE's US (Galaxy Compound Semiconductors) and UK (Wafer Technology) operations. IQE has the World's largest antimonide wafer production capacity in the industry, using multiple production tools (pullers), volume double-side polishing platforms, and state-of-the-art product metrology.

Galaxy elaborated on the modified version of the Czochralski technique. The growth starts with the highest-purity metals (Ga, In, and Sb), which are placed in the growth crucible of the puller system. A surrounding inert atmosphere (H_2 or N_2) is typically implemented. At the beginning, the quartz crucible is heated until melting and compound synthesis of the raw elements occurs. To support the stoichiometric melt, fume-off of the more volatile Sb element from the melt requires a steady addition of Sb to the melt as the boule is pulled. If desired, tiny amounts of Te are added to the melt for n-type or p-type extrinsic crystal doping. In the Czochralski method, the seed crystal of either GaSb or InSb is rotated and slowly withdrawn from the melt, forming the initial boule growth. To reduce boule dislocations, a crystal necking is implemented. The final boule diameter and length are a function of the starting crucible size, melt weight, and total time of the pulling process. Finally, the boule must be slowly cooled in order to prevent slip or crystal cracking.

For commercial production of InSb, the majority of ingots are grown in the (211) direction. Production of (100) InSb is possible, but difficulties are encountered in the growth of this orientation type, and applications of (100) InSb remain limited. Grown wafers are processed into epi-ready substrates and characterized for bulk crystal quality, electrical properties, and uniformity. The key surface quality characteristics of roughness (<0.5 nm rms), oxide thickness (<100 Å), and flatness (total thickness variation <7 μm) have been maintained across the production process that scale 4- to 6-inch wafer formats.[16] At present the ultrahigh-purity InSb bulk crystal's carrier concentration can be less than 10^{13} cm^{-3}. Typical etch pit densities below 10^2 cm^{-2} are found for InSb, which is considered one of the lowest-defect-type compound semiconductor materials available commercially.

Also, in the case of growth of GaSb single crystals from a melt, several challenges must be overcome that require modifications of conventional Czochralski equipment. Problems include controlling oxide scum on the melt surface and evaporation of Sb from the melt surface. Initially, researchers used liquid-encapsulated Czochralski growth, using molten B_2O_3 to prevent water from reaching the surface as oxidation scum, and preventing Sb vapor from escaping.[17] This method is still in use today but presents its own set of complications (the molten B_2O_3 causes melt contamination, changes in surface tension and viscosity, significant alterations in the growth process, and modification of the heat and energy flow at the melt meniscus). In addition, since B_2O_3 is hygroscopic, extra care must be taken to keep it dry by vacuum baking or bubbling with dry N_2. Growth without encapsulant has been revisited a number of times. GaSb crystals grown without encapsulant in a hydrogen environment had reduced oxide formation, higher crystal quality, and less twinning probability than encapsulated runs. Historical challenges with Czochralski GaSb crystal growth are described in more detail in Ref. 18.

The majority of GaSb detector production relies on epitaxial growth using 3-inch-diameter substrates, with a small volume of 4-inch material consumed, too. However, to bring GaSb substrate production technologies to the same level of maturity as InSb, interests in 6-inch GaSb substrates for very

Figure 2.3 Standard 2- to 3-inch (left) and large-format 4- to 7-inch (right) GaSb boules manufactured by IQE Infrared (reprinted from Ref. 18).

large-area detector applications has recently emerged. Figure 2.3 shows the evolution in GaSb boule size from 2- to 3-inch ingot up to the 4- to 7-inch crystal that exceed 30 kg in charge size. From the largest size of ingot grown, an average of 4×7 inch and 10×6 inch GaSb substrates can be cut from each boule.

GaSb is intrinsically p-type, with hole carrier concentrations of $\sim10^{17}$ cm^{-3} at room temperature. The residual hole concentration of GaSb bulk crystals is about 2×10^{16} cm^{-3}. The intrinsic defect is mostly Ga antisite defects or complexes of $V_{Ga}Ga_{Sb}$ providing acceptor sites. High resistivity or n-type GaSb can be obtained by compensation doping with group VI elements such as Te—the resulting crystal is characterized by a lower absorption coefficient in comparison with undoped crystal.

The growth of type-II superlattice structures demands the flattest possible substrates, especially where high strain contribution from the epitaxy adversely affects flatness. A low level of bow and total thickness variation must be maintained to secure effective photolithography processes, yielding well-defined pixel features. In addition, array hybridization steps require flat substrate, too, so that the connections between the pixels and readout integrated circuits (ROICs) are effective with several numbers of outputs. Typically, bow and total thickness variation parameters for bare 6-inch GaSb substrate diameters are <10 μm. Additional wafer polishing is required to deliver sub-5-μm specification. Common GaSb substrates show threading dislocation densities of $\sim10^3$ to 10^4 cm^{-2}.

Recently, a research group of 5N Plus Semiconductors has achieved a rapid-development cycle in fabrication of epi-ready 6-inch GaSb wafers, leveraging their learning from the development of InSb and Ge crystal growth projects.[19] Figure 2.4 demonstrates a timeline for GaSb crystal growth using an encapsulant-free Czochralski method and a modified puller, starting from initial growth attempt, across images of several crystals, to the manufactured wafers.

InAs single crystals of up to 100-mm diameter are being commercially grown using a high-pressure, liquid-encapsulated Czochralski technique. pBN crucibles and ultralow-water-content B_2O_3 encapsulation are being used. For research purposes, InAs single crystals can be grown in vacuum-sealed crucibles using a vertical Bridgman/vertical gradient freeze method. In the latter case, an appropriately placed As reservoir inside the crucible and a pressure vessel for providing counteracting pressure outside the crucible are required during crystal growth.

Compared with other antimonide-based III-V compounds, little work has been done on the investigation of AlSb. The large, high-quality AlSb single crystals are rarely fabricated, and their surfaces react rapidly with air. AlSb single crystals are not commercially available.

Due to the immiscible gap of multi-element antimonides, their growth process is very immature and the ternary and quaternary antimonide bulk crystal materials are rarely used.

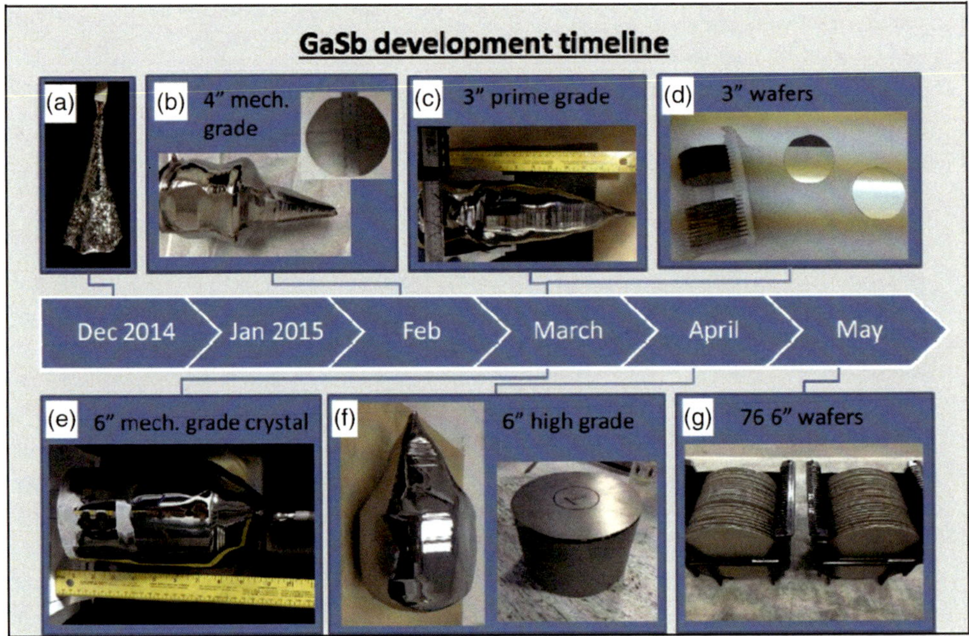

Figure 2.4 Development timeline for GaSb crystal growth by 5N Plus Semiconductors (reprinted from Ref. 19).

The substitution of a fraction of antimony sites in InSb with isovalent arsenic reduces the energy gap of InSb-InAs ($InAs_{1-x}Sb_x$) to a value lower than the energy gap of either of the parent binary compounds. Consequently, $InAs_{1-x}Sb_x$ ternary alloy has the lowest energy gap among the III-V semiconductors. A room-temperature energy gap in both 3- to 5-μm and 8- to 14-μm atmospheric wavelength windows can be achieved. However, progress in this ternary system has been limited by crystal synthesis problems. The large separation between the solidus and liquidus [see Fig. 2.5(a)] and the lattice mismatch (6.9% between InAs and InSb) place stringent demands on the method of crystal growth. These difficulties are being overcome systematically using molecular beam epitaxy (MBE) and metal organic chemical vapor deposition (MOCVD) growth methods.

Considering the $Ga_{1-x}In_xSb$ (GaInSb) pseudobinary phase diagram shown in Fig. 2.5(b), the separation between the solidus and liquidus curves leads to alloy segregation. The vertical Bridgman or vertical gradient freezing process is the most suitable method for growing large-diameter GaInSb bulk crystals. The established process has been successfully demonstrated in laboratory-scale experiments for growing GaInSb crystals (up to 50-mm diameter) with a wide range of alloy compositions.

A wide spectrum of topics in materials that today's engineers, material scientists, and physicists need is included in a comprehensive treatise on electronic and photonic materials gathered in the *Springer Handbook of Electronic and Photonic Materials*.[4]

2.2 Epitaxial Layers

Until the early 1990s, difficulties in the preparation of single crystals and high-quality epitaxial layers were the main obstacle in the rapid development of

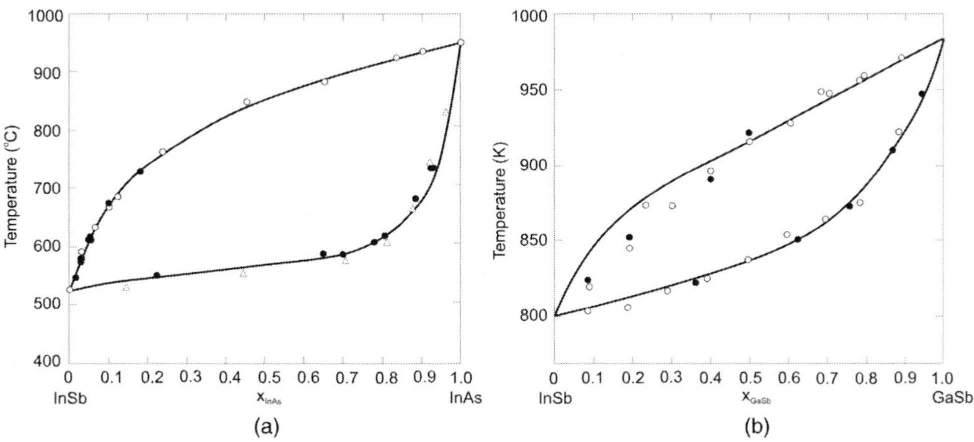

Figure 2.5 Pseudobinary phase diagram for the (a) InAs-InSb and (b) GaSb-InSb systems.

antimonide-based devices. The commonly used methods for preparation of antimonide-based epitaxial layers are liquid phase epitaxy (LPE), MBE, and MOCVD.

LPE is a relatively simple, high-quality technique, with less-expensive epitaxial equipment (in comparison with MOCVD and MBE), and a high utilization rate of the source material. It is particularly suitable for the preparation of thick-film layers. Since the LPE technique is a near-thermodynamic equilibrium growth method, it cannot be used for the growth of metastable ternary and quaternary antimonide compounds with miscibility gaps. The growth rate of LPE is generally higher than that of MOCVD and MBE, and it changes from different substrate crystalline phases with the typical growth rate of 100 nm/min to a few μm/min. The disadvantage of LPE is that it cannot be used for precision-controlled growth of very thin epilayers, especially superlattices, quantum-well devices, and other complex micro-structure materials. Moreover, the morphology of LPE layers is usually worse than that grown by MOCVD or MBE.

The era of MBE and MOCVD growth of III-V semiconductor materials began in the early 1970s. It is not a simple matter to discern which epitaxial growth technique is better, MBE or MOCVD. Each has specific merits in specific device applications. Some generalizations about different classes of techniques for III-V compounds are given in Table 2.1.[20]

MBE epilayer structures are grown on a heated substrate in an ultrahigh vacuum (UHV) environment (base pressure $\sim 10^{-10}$ Torr), typically using elemental sources. Materials are vaporized from high-temperature sources (effusion cells) and subsequently transported toward the substrate in the form of vapor without any chemical change. The temperature of the substrate is independently controlled to facilitate layer-by-layer material incorporation to

Table 2.1   Comparison of MOCVD and MBE techniques (from Ref. 20).

Category	Characteristics of MOCVD	Characteristics of MBE
Technical	High growth rate for bulk layers Growth near thermodynamic equilibrium; excellent quality/crystallinity Ability to explicitly control background doping	Fast switching for superior interfaces Able to grow thermodynamically forbidden materials No hydrogen passivation; no burn in inherent to MOCVD Uniformity easier to tune, largely set by reactor geometry
Commercial	Shorter maintenance periods More flexibility for source and reactor configuration changes Higher safety risk, increasing scrutiny of legislative bodies worldwide Economic to idle; overhead cost scales with run rate.	Longer individual campaigns, less setup variability Lower material cost/wafer Overhead does not scale with run rate. Contribution per wafer increases with wafer volume.

the substrate. The UHV environment ensures high material purity; in addition, the inherently long mean-free paths result in directional elemental beams—no carrier gas is required. More sophisticated layer structures (quantum wells and superlattices) are deposited by shuttering. Due to relatively high vapor pressures for group V materials, the valved sources are typically used. Also, decomposition sources (e.g., GaTe for Te doping), gas sources (e.g., CBr_4 for C doping), and plasma sources (e.g., nitrogen plasma for nitride applications) are used.

The growth rate in MBE is ~1.0 monolayer per second; this slow rate coupled with shutters placed in front of the crucibles allows one to switch the composition of the growing crystal with monolayer control. The low background pressure in MBE allows one to use electron beams to monitor and calibrate the growing crystal. The reflection high-energy electron diffraction (RHEED) technique relies on electron diffraction to monitor both the quality of the growing substrate, the layer-by-layer growth mode, and the III/V flux ratio determination. Because the growth occurs far from thermodynamic equilibrium, MBE allows the ability to physically control the interfaces by shutter sequence actuation with a precision of 0.1 s. However, since the growth involves high vacuum, leaks can be a major problem. The growth chamber walls are usually cooled by liquid N_2 to ensure high vacuum and to prevent atoms/molecules from dislodging from the chamber walls.

The MBE system is divided into three different temperature zones [see Fig. 2.6(b)]: the effusion cells (where the species material is evaporated), the gaseous region (close to the sample surface where molecular beams overlap and stabilize), and the sample surface (where the epitaxial growth occurs). As a result, the MBE system allows a low growth temperature since the substrate temperature has influence only on the diffusion of atoms on the growth surface, not on cracking the molecules.

MOCVD is another important growth technique widely used for heteroepitaxy, quantum well structures, and superlattices. Like MBE, it is

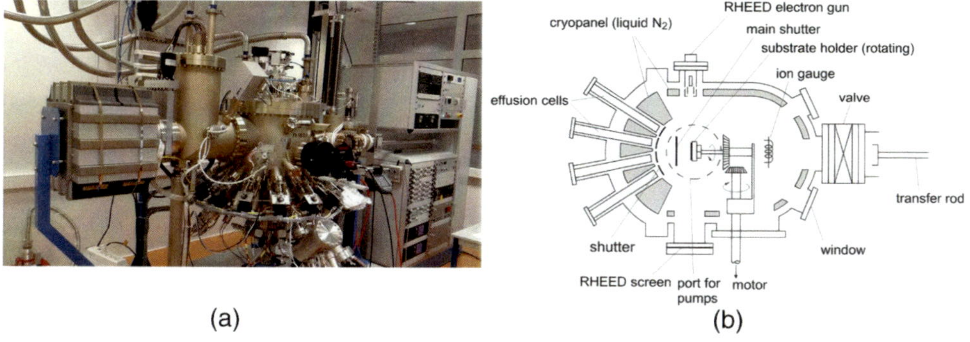

Figure 2.6 MBE system: (a) a photograph of a Riber III-V system and (b) a schematic of MBE growth.

also capable of producing monolayer abrupt interfaces between semiconductors. The MOCVD epitaxial layer also grows on a heated substrate but in a much different pressure regime than MBE (typically 15 to 750 Torr). A typical MOCVD system is shown in Fig. 2.7.

There are several varieties of MOCVD reactors. In atmospheric MOCVD, the growth chamber is essentially at atmospheric pressure. One needs a large amount of gas for growth in this case. In low-pressure MOCVD, the growth chamber pressure is kept low. The growth rate is then slowed down, as in the MBE case.

The use of MOCVD equipment requires very serious safety precautions. The gases used are highly toxic and a great many safety features have to be incorporated to avoid any deadly accidents. Safety and environmental concerns are important issues in almost all semiconductor manufacturing since quite often one has to deal with toxic and hazardous materials.

Usually, the MOCVD technique uses more-complex compound sources, namely metal-organic sources (e.g., tri-methyl or tri-ethyl Ga, In, Al, etc.), hydrides (e.g., AsH_3, etc.), and other gas sources (e.g., disilane). In contrast to MBE, the growth of crystals is by chemical reaction and not physical deposition. Mass flow controllers control the species deposited. In MOCVD, the reactants are flowed across the substrate where they react, resulting in epitaxial growth. In contrast to MBE, MOCVD requires the use of a carrier gas (typically H) to transport reagent materials across the substrate surface. Layered structures are achieved by valve actuation for differing injection ports of a gas manifold. Both techniques (MBE and MOCVD) are preferred for the formation of devices incorporating thermodynamically metastable alloys and have become important processes in the manufacture of optoelectronics.

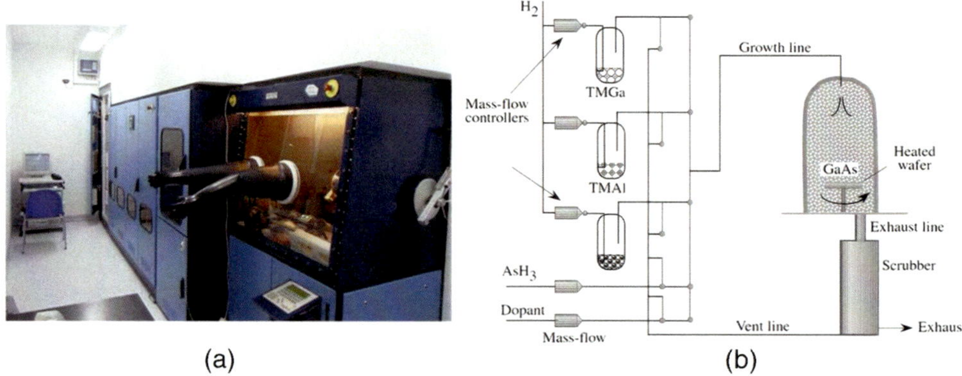

Figure 2.7 MOCVD system: (a) Aixtron III-V system and (b) a schematic of the MOCVD growth employing alkyds—trimethyl gallium (TMGa) and trimethyl aluminium (TMAl)—and metal hydride (arsine) material sources with hydrogen as a carrier gas. Chemical reactions at the heated substrate deposits GaAs or AlAs.

From the commercial side of consideration (see Table 2.1 and Ref. 20),

- the cost profiles for each technique are very different due to the specific utility/overhead requirements for each:
  - MOCVD overhead costs tend to scale with production volume, while
  - MBE overhead is relatively fixed and does not scale with volume; therefore,
- MOCVD wins in a situation of significant overcapacity (significant idle time), and the opposite is true for MBE, which excels on a cost basis when fully loaded;
- in terms of manufacturing uptime, the two techniques are very similar; however,
  - MBE downtime is concentrated in multiple-month periods when the reactor is down, while
  - MOCVD maintenance is a much more frequent but less time-consuming occurrence; therefore,
- for MBE, long bake times are required each time the growth chamber must be brought to atmospheric pressure for repairs;
- in contrast, MOCVD does not require such significant bake times (MOCVD is able to recover more quickly from equipment failures).

The first epitaxial growth of antimonide thin film materials using MOCVD was done in 1969 by Manasevit and Simpson, who used TMGa and SbH_3 (stibine) sources for growing GaSb films.[21] At present, the commonly used III-group metal-organic sources for antimonide-based compounds are 3-methyl compound and 3-ethyl compound, such as TMGa, TMIn, TMAl, TEGa, TEIn, etc.[22,23] The commonly used V-group sources are TMSb, AsH_3, PH_3, TMBi, RF-N_2, etc.

Generally, antimonides are low-melting-point materials, and their growth temperature is about 500 °C. It appears that the majority of III-group metal-organic sources cannot be completely decomposed below 500 °C. Therefore, new organic source materials with lower decomposition temperatures are introduced, including TDMASb (trisdimethylaminoantimony), TASb (triallyantimony), TMAA (trimethylamine alane), TTBAl (tritertiarybutylaluminum), EDMAA (ethyldimethylaminealane), etc. In the case of Al-containing antimonide materials, carbon and oxygen contamination problems exist. It is expected that this phenomenon is related to the lack of active hydrogen atoms on the surface of epitaxial layers. Carbon is typically doped with p-type impurities, which causes certain difficulties in the growing of n-type-doping, Al-containing antimonide epitaxial layers.

The growth of semiconductor antimonides by MBE was first reported in the late 1970s.[24,25] In comparison with GaAs and other arsenides, the growth of GaSb-based compounds is characterized by relatively low vapor pressures of Sb, or, equivalently, by its high sublimation energy. Much of the published papers focus on GaSb and AlSb because they are nearly lattice matched to each other and to InAs. The substrate temperature during the growth of GaSb and AlSb is usually between 550 and 600 °C. MBE avoids the C-pollution problem in Al-containing materials grown by MOCVD and greatly reduces the concentration of O-doping. Most devices having complex fine structures and low-dimensional structures (quantum wells, quantum wires, and quantum dots) were first achieved using materials grown by MBE. It appears that substrates whose surface orientation have a small angle offset (i.e., low-density atomic step on the surface of the substrate) result in accessible high-quality epitaxial layers.[26]

The growth of antimonide-based III-V epitaxial layers is usually performed on InSb, InAs, and GaSb low-defect substrates. To overcome the challenge of antimonides including on semi-insulating substrate materials, the use of GaAs, and GaAs-coated Si substrates and other heterogeneous substrate materials for epitaxy have attracted great attention. A variety of substrate structures can be realized either by effecting gradual, continuous compositional grading of thick epilayers, or by growing multilayers with abrupt but incremental compositional changes between adjacent layers. In infrared detector fabrications, both approaches can be combined with selective removal of the seeding substrate and wafer-bonding techniques. Low-defect-alloy substrates—with increased functionality and lattice constants, and bandgaps significantly different from those available with binary compound wafers (e.g., InAs or GaSb)—appear to be feasible.[27]

2.3 Physical Properties

Table 2.2 presents some physical properties of semiconducting families, including narrow-gap semiconductors, used in fabrication of infrared photodetectors. All compounds have diamond (D) or zincblende (ZB) crystal structure. Moving across the table from the left to the right, there is a trend in change of chemical bond from the covalent group IV-semiconductors to more ionic II-VI semiconductors with increasing of the lattice constant. The chemical bonds become weaker and the materials become softer as reflected by the values of the bulk. The materials with larger contribution of covalent bonds are more mechanically robust, which leads to better manufacturability. This is evidenced in the dominant position of silicon in electronic materials and GaAs in optoelectronics ones. On the other hand, the band gap energy of semiconductors on the right side of the table tends to have smaller values. Due to their direct band gap structure, strong band-to-band absorption leading to high quantum efficiency is observed (e.g, in InSb and HgCdTe).

The properties of narrow-gap semiconductors that are used as the material systems for IR detectors result from the direct energy bandgap structure: a high

Table 2.2 Selected properties of common families of semiconductors used in fabrication of infrared photodetectors (D – diamond, ZB – zinc blende, id – indirect, d – direct, L – light hole, H – heavy hole).

	Si	Ge	GaAs	AlAs	InP	InGaAs	AlInAs	InAs	GaSb	AlSb	InSb	HgT	CdTe
Group	IV	IV	III-V	III-V	III-V	III-V	III-V	III-V	III-V	III-V	III-V	II-V	II-V
Lattice constant (A)/structure	5.431 (D)	5.658 (D)	5.653 (ZB)	5.661 (ZB)	5.870 (ZB)	5.870 (ZB)	5.870 (ZB)	6.058 (ZB)	6.096 (ZB)	6.136 (ZB)	6.479 (ZB)	6.453 (ZB)	6.476 (ZB)
Bulk modulus (Gpa)	98	75	75	74	71	69	66	58	56	55	47	43	42
Band gap (eV)	1.124 (id)	0.660 (id)	1.426 (d)	2.153 (id)	1.350 (d)	0.735 (d)		0.354 (d)	0.730 (d)	1.615 (id)	0.175 (d)	−0.141 (d)	1.475 (d)
Electron effective mass	0.26	0.39	0.067	0.29	0.077	0.041		0.024	0.042	0.14	0.014	0.028	0.090
Hole effective mass	0.19	0.12	0.082(L) 0.45(H)	0.11(L) 0.40(H)	0.12(L) 0.55(H)	0.05(L) 0.60(H)		0.025(L) 0.37(H)	0.4	0.98	0.018(L) 0.4(H)	0.40	0.66
Electron mobility (cm^2/Vs)	1450	3900	8500	294	5400	13800		3×10^4	5000	200	8×10^4	26500	1050
Hole mobility (cm^2/Vs)	505	1900	400	105	180			500	880	420	800	320	104
Electron saturation velocity (10^7 cm/s)	1.0	0.70	1.0	0.85	1.0			4.0			4.0		
Thermal cond. (W/cmK)	1.31	0.31	0.5		0.7			0.27	0.4	0.7	0.15		0.06
Relative dielectric constant	11.9	16.0	12.8	10.0	12.5			15.1	15.7	12.0	17.9	21	10.2
Substrate	Si,Ge		GaAs		InP			InAs,GaSb			InSb	CdZnTe,GaAs,Si	
MW/LW detection mechanism	Heterojunction internal		QWIP,QDIP		QWIP			Bulk (MW) Superllatice (MW/LW) Band-to-band (BBT)			Bulk BBT	Bulk BBT	

density of states in the valence and conduction bands, which results in strong absorption of IR radiation and a relatively low rate of thermal generation. From the viewpoint of producibility, III-V materials offer much stronger chemical bonds and thus higher chemical stability compared to HgCdTe.

The shape of the electron band and the light-mass hole band at the Brillouin zone center is determined by the $\mathbf{k \cdot p}$ perturbation theory. The momentum matrix element varies only slightly for different materials and has an approximate value of 9.0×10^{-8} eVcm. Then, the electron effective masses and conduction band densities of states are similar for materials with the same energy gap.

These materials have a conventional negative temperature coefficient of the energy gap, which is well described by the Varshni relation,[28]

$$E_g(T) = E_0 + \frac{\alpha T^2}{T + \beta},\qquad (2.1)$$

where α and β are fitting parameter characteristics of a given material.

Figure 2.8 shows the composition dependence of the energy gap and the electron effective mass at the Γ-conduction bands of $Ga_xIn_{1-x}As$, $InAs_xSb_{1-x}$, and $Ga_xIn_{1-x}Sb$ ternaries.

Table 2.3 contains some physical parameters of the InAs, InSb, GaSb, and $InAs_{0.35}Sb_{0.65}$ semiconductors.[12,29] Among these, InSb has been investigated most broadly. The temperature-independent portions of the Hall curves indicate that most of the electrically active impurity atoms in InSb have shallow activation energies and above 77 K are thermally ionized. The Hall coefficient for p-type samples is positive in the low-temperature extrinsic

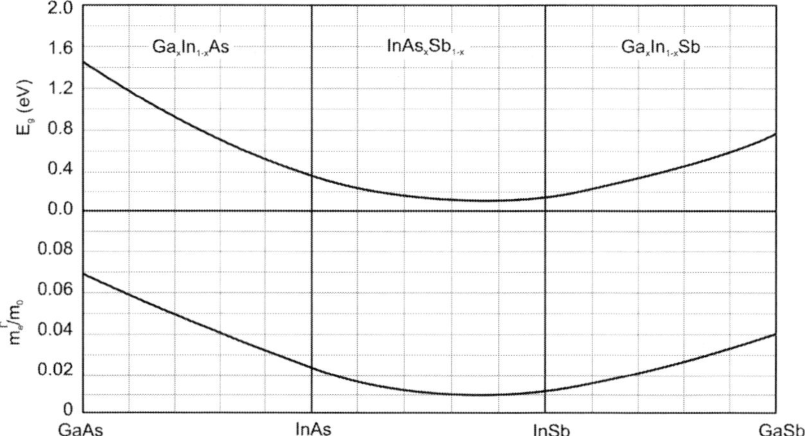

Figure 2.8 Variation of the bandgap energy and electron effective mass at the Γ-conduction bands of $Ga_xIn_{1-x}As$, $InAs_xSb_{1-x}$, and $Ga_xIn_{1-x}Sb$ ternary alloys at room temperature.

Table 2.3 Physical properties of narrow-gap III-V alloys.

	T(K)	InAs	InSb	GaSb	InAs$_{0.35}$Sb$_{0.65}$
Lattice structure		cub. (ZnS)	cub. (ZnS)	cub. (ZnS)	cub. (ZnS)
Lattice constant a (nm)	300	0.60584	0.647877	0.6094	0.636
Thermal expansion coefficient α (10^{-6}K^{-1})	300	5.02	5.04	6.02	
	80		6.50		
Density ρ (g/cm^3)	300	5.68	5.7751	5.61	
Melting point T_m (K)		1210	803	985	
Energy gap E_g (eV)	4.2	0.42	0.2357	0.822	0.138
	80	0.414	0.228	0.725	0.136
	300	0.359	0.180		0.100
Thermal coefficient of E_g	100–300	-2.8×10^{-4}	-2.8×10^{-4}		
Effective masses:					
$\quad m_e^*/m$	4.2	0.023	0.0145	0.042	
	300	0.022	0.0116		0.0101
$\quad m_{lh}^*/m$	4.2	0.026	0.0149		
$\quad m_{hh}^*/m$	4.2	0.43	0.41	0.28	0.41
Momentum matrix element P (eVcm)		9.2×10^{-8}	9.4×10^{-8}		
Mobilities:					
$\quad \mu_e$ (cm^2/Vs)	77	8×10^4	10^6		5×10^5
	300	3×10^4	8×10^4	5×10^3	5×10^4
$\quad \mu_h$ (cm^2/Vs)	77		1×10^4	2.4×10^3	
	300	500	800	880	
Intrinsic carrier concentration n_i (cm^{-3})	77	6.5×10^3	2.6×10^9		2.0×10^{12}
	200	7.8×10^{12}	9.1×10^{14}		8.6×10^{15}
	300	9.3×10^{14}	1.9×10^{16}		4.1×10^{16}
Refractive index n_r		3.44	3.96	3.8	
Static dielectric constant ε_s		14.5	17.9	15.7	
High frequency dielectric constant ε_∞		11.6	16.8	14.4	
Optical phonons:					
$\quad$ Longitudinal optical (LO) (cm^{-1})		242	193		$\approx$210
$\quad$ Transverse optical (TO) (cm^{-1})		220	185		$\approx$200

range and reverses sign to become negative in the intrinsic range because of the higher mobility of the electrons (the mobility ratio $b = \mu_e/\mu_h$ on the order of 10^2 is observed). The transition temperature for the p-type samples, at which R_H changes sign, depends on the purity. The samples become intrinsic above certain temperature (above 150 K for pure n-type samples); below these temperatures (below 100 K for pure n-type samples), there is little variation of Hall coefficients.

There are various carrier-scattering mechanisms in semiconductors, as shown in Fig. 2.9 for InSb.[30] Reasonably pure n-type and p-type InSb samples exhibit an increase in mobility up to approximately 20–60 K, after which the mobility decreases due to polar and electron–hole scattering. Carrier mobility systematically increases with a decrease in impurity concentration both in temperature 77 K as well as in 300 K.

In alloy semiconductors, the charged carriers see potential fluctuations as a result of the composition disorder. This kind of scattering mechanism,

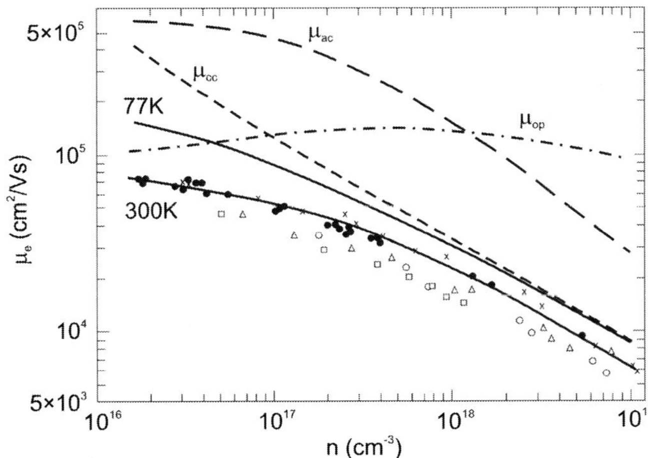

Figure 2.9 Electron mobility in n-type InSb at 300 K and 77 K versus free electron concentration. The dashed lines denote the theoretical mobilities at 300 K for charged-center, polar-optical, and acoustic-scattering modes. The experimental data are taken at 300 K (adapted from Ref. 30).

so-called alloy scattering, is important in some III-V ternaries and quaternaries. Let us simply express the total carrier mobility μ_{tot} in alloy $A_xB_{1-x}C$ as[4]

$$\frac{1}{\mu_{tot}(x)} = \frac{1}{x\mu_{tot}(AC) + (1-x)\mu_{tot}(BC)} + \frac{1}{\mu_{al,0}/[x/(1-x)]}. \qquad (2.2)$$

The first term on the right side of Eq. (2.2) comes from the linear interpolation scheme, and the second term accounts for the effects of alloying.

Because of the very small effective mass of electrons, the conduction band density of states is small, and it is possible to fill the available band states by doping, thereby appreciably shifting the absorption edge to shorter wavelengths. This is referred to as the Burstein–Moss effect, which is shown, for example, in Fig. 2.10 for n-type InSb with different electron concentrations.[31]

The development of $InAs_{1-x}Sb_x$ ternary alloys has a long history. InAsSb ternary alloy as an alternative to HgCdTe for IR applications was demonstrated in the mid-1970s.[32] A III-V detector technology would benefit from superior bond strengths and material stability (compared to HgCdTe), well-behaved dopants, and high-quality III-V substrates.

The properties of InAsSb were first investigated in the 1960s by Woolley and co-workers, who established the InAs-InSb miscibility,[33] the pseudobinary phase diagram,[34] scattering mechanisms,[35] and the dependence of fundamental properties such as bandgap[36] and effective mass on composition.[36,37] All of the above measurements were performed on polycrystalline

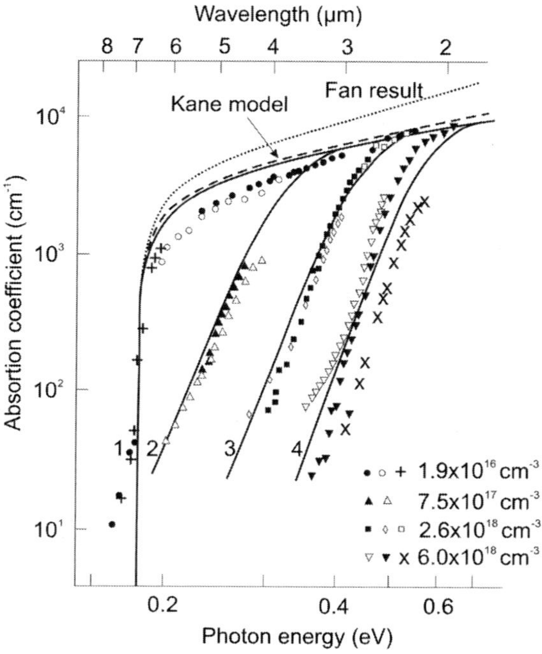

Figure 2.10 Dependence of the optical absorption coefficient of InSb on photon energy at 300 K. Carrier concentration: 1.9×10^{16} cm^{-3} (1), 7.5×10^{17} cm^{-3} (2), 2.6×10^{18} cm^{-3} (3), and 6.0×10^{18} cm^{-3} (4) (adapted from Ref. 31).

samples prepared by various freezing and annealing techniques. A review of the early-stage development of InAsSb crystal growth techniques, physical properties, and detector fabrication procedures is presented in Rogalski's papers.[38,39]

Recently, the rapid development of InAs$_{1-x}$Sb$_x$ crystal growth techniques and detector fabrication procedure has been observed due to a new development in infrared detector design, called a barrier detector. The electronic properties of this ternary alloy have been reconsidered in a wide range of alloy compositions.[40–44]

The electronic properties of ternary alloys are commonly described with the virtual crystal approximation (VCA) approach.[45] In this model, the disordered alloy is considered as an ideal crystal with an average potential modelled by linear interpolation of the potentials of corresponding binary compounds. The alloy bandgap depends nonlinearly on the composition and is lower than the bandgap of binary compounds.

The nonlinearity of the composition dependence of the InAs$_{1-x}$Sb$_x$ ternary alloy bandgap is described by the bowing parameter C as

$$E_g(x) = E_{gInSb}x + E_{gInAs}(1 - x) - Cx(1 - x). \tag{2.3}$$

Initial reports based on experimental data at temperatures above or near 100 K put the direct-gap bowing parameter in InAsSb at 0.58–0.6 eV.[46] Theoretical considerations led to a higher projected bowing parameter of 0.7 eV, which was recommended by Rogalski and Jóźwikowski.[47] More-recent photoluminescence studies on unrelaxed MBE-grown $InAs_{1-x}Sb_x$ in a wide range of compositions gave bowing parameters of 0.83 to 0.87.[42,43] Figure 2.11 summarizes the experimental data and theoretical predictions of the energy gap in the temperature range 4–77 K as was published in different papers. Discrepancies in the $E_g(x,T)$-dependence can be caused by several reasons, including structural quality of samples and CuPt-type ordering effect. It is possible that the low-energy-gap data in earlier reports were masked by electron filling of the conduction band, which would be due to background doping as these samples were grown with various degrees of residual strain and relaxation. High-quality unstrained, unrelaxed InAsSb epilayers have been developed by using specially graded buffer layers, which accommodate the large difference between the lattice constant of the substrate and alloy. Electron diffraction patterns of unstrained InAsSb alloys, described in Refs. 43 and 44, showed an ordering-free distribution of group V elements, indicating that the observed energy gaps of ternary alloys are inherent (both ordering and residual strain effects were eliminated). The described $E_g(x,T)$-dependence indicates that conventional InAsSb has a sufficiently small gap at 77 K for operation in the 8- to 14-μm wavelength range and differs from that previously described by Wieder and Clawson:[48]

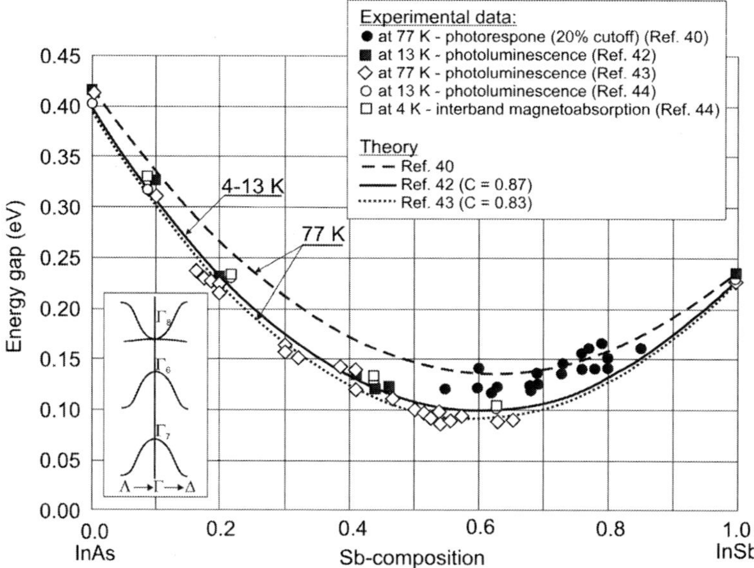

Figure 2.11 Bandgap energy of $InAs_{1-x}Sb_x$ as a function of the Sb composition. The experimental data are taken with different papers as indicated in the key.

$$E_g(x,T) = 0.411 - \frac{3.4 \times 10^{-4}T^2}{210 + T} - 0.876x + 0.70x^2 + 3.4 \times 10^{-4}xT(1 - x).$$

$$(2.4)$$

It was found that the bandgap energy of the InAsSb ternary compound is generally a square function of the composition and indicates a weak dependence of the band edge on composition in comparison with HgCdTe (see Fig. 2.1).

To obtain a good agreement between experimental room-temperature effective masses and calculations, Rogalski and Jóźwikowski[47] have taken into account the conduction-valence-band mixing theory[49] (see Fig. 2.12). More recently, published low-temperature data, especially for the mid-composition range, indicate lower electron effective masses. The estimated negative bowing parameter for the electron effective mass is $C_m = 0.038$ and is slightly less than expected from the Kane model ($C_m = 0.045$), reaching the lowest effective mass ($0.0082m_0$ at $x = 0.63$ and 4 K) ever reported for III-V semiconductors. A possible reason for the effective mass value discrepancy, shown in Fig. 2.12, is mixing of the conduction and valence band states caused by the random potential due to alloy disorder.

Figure 2.13 summarizes the composition dependence of the spin–orbit splitting energy Δ of $InAs_{1-x}Sb_x$ ternary alloy at 10 K. More-recently-measured results of $\Delta(x)$ dependence by Cripps et al.[50] are in strong disagreement with the paper of Van Vechten et al. published in 1972.[51]

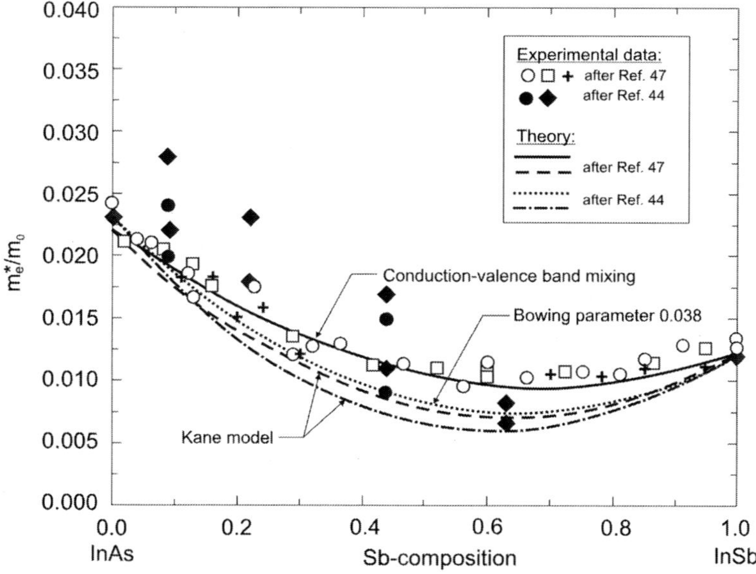

Figure 2.12 Dependence of electron effective mass on composition for the $InAs_{1-x}Sb_x$ alloy system (data from Refs. 44 and 47).

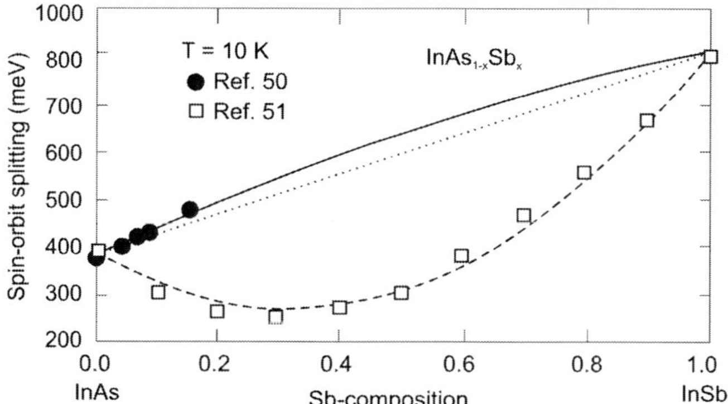

Figure 2.13   The composition dependence of the spin–orbit splitting energy of InAs$_{1-x}$Sb$_x$ ternary alloy at $T = 10$ K (data from Refs. 50 and 51). The dotted line indicates the zero bowing behavior, which is consistent with the virtual crystal approximation.

Almost no bowing for the Δ parameter as a function of x (Sb fraction) was observed. A good approximation gives

$$\Delta(x) = 0.81x + 0.373(1 - x) + 0.165x(1 - x) \text{ in eV.} \tag{2.5}$$

This spin–orbit splitting bandgap energy much more satisfactorily agrees with measurements and is independent of temperature. However, the recently measured composition and temperature dependence of the energy gap is consistent with that predicted from the literature.

The intrinsic carrier concentration in InAsSb as a function of composition x for various temperatures can be approximated by the following relation:[47]

$$n_i = (1.35 + 8.50x + 4.22 \times 10^{-3}T - 1.53 \times 10^{-3}xT$$

$$- 6.73x^2) \times 10^{14}T^{3/2}E_g^{3/4} \exp\left(-\frac{E_g}{2kT}\right). \tag{2.6}$$

For a given temperature, the maximum of n_i appears at $x \approx 0.63$, which corresponds to the minimum energy gap.

The first measurements of transport properties of n-type InAsSb alloys were performed on samples prepared in the late 1960s by various freezing and annealing techniques.[35,52] The properties of high-quality InAsSb epitaxial layers with $x < 0.35$ manufactured by LPE are similar to those of pure InAs (when $n = 2 \times 10^{16}$ cm^{-3}, typical mobilities are 30,000 cm^2/Vs at 300 K and 50,000 cm^2/Vs at 77 K). For InSb-rich alloys with $x \geq 0.90$, typical mobilities are 60,000 cm^2/Vs at 300. When As is added to InAs$_{1-x}$Sb$_x$ alloys, the background carrier concentration increases to a low 10^{17}cm^{-3}; instead, the mobility first increases and then decreases by a factor of 1.5 to 2 for temperatures decreasing from 300 to 77 K. At the current stage of

MBE-growth development, the background electron concentration in $InAs_{1-x}Sb_x$ alloys with 40% Sb at 77 K is as low as $1.5 \times 10^{15} cm^{-3}$.[43]

Chin et al. have calculated the electron mobility of InAsSb by considering all of the possible scattering mechanisms: impurities, acoustic phonons, optical phonons, alloy scattering, and dislocations.[53,54] Comparison with experimental results confirms that dislocation scattering has a strong effect on transport, while alloy scattering limits mobility in ternary samples grown with a minimum of defects (see Fig. 2.14).

The ternary alloy $Ga_xIn_{1-x}Sb$ is an important material for the fabrication of detectors designed for MWIR applications. The long-wavelength limit of $Ga_xIn_{1-x}Sb$ detectors has been tuned compositionally from 1.52 μm ($x = 1.0$) at 77 K to 6.8 μm ($x = 0.0$) at room temperature. The bandgap energy of $In_xGa_{1-x}As_ySb_{1-y}$ at room temperature can be fitted by the relationship[55]

$$E_g(x) = 0.726 - 0.961x - 0.501y + 0.08xy + 0.451x^2$$
$$-1.2y^2 + 0.021x^2y + 0.62xy^2. \tag{2.7}$$

The lattice-matching condition for GaSb imposes the additional constraint that x and y are related as $y = 0.867/(1-0.048x)$.

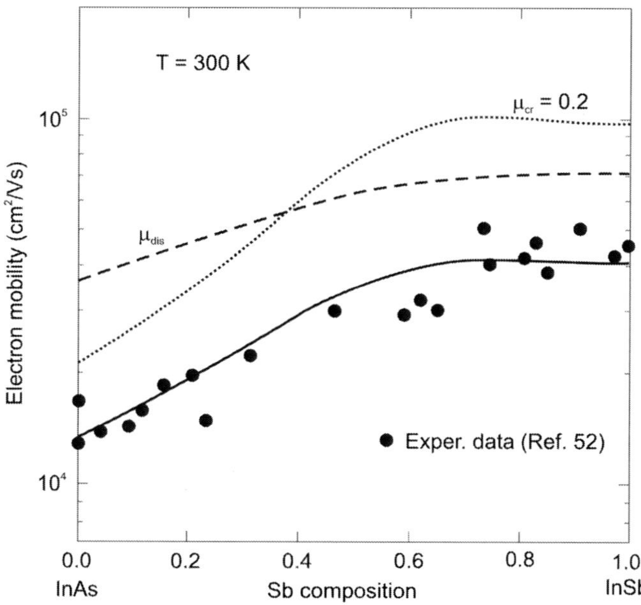

Figure 2.14 Compositional dependence of room-temperature electron mobility for a carrier concentration of 10^{17} cm^{-3} with a compensation ratio of 0.2 and the effects of a dislocation density of 3.8×10^8 cm^{-3} (reprinted from Ref. 54 with permission from AIP).

2.4 Thermal Generation–Recombination Processes

The generation processes that compete against the recombination processes directly affect the performance of photodetectors, setting up a steady-state concentration of carriers in semiconductors subjected to thermal and optical excitation and, frequently, determining the kinetics of photogenerated signals. Generation–recombination (GR) processes in semiconductors are widely discussed in the literature (see, for example, Refs. 56–58). We present here only some carrier lifetime data directly related to the performance of photodetectors. Assuming bulk processes only, there are three main thermal generation–recombination processes to be considered with narrow-bandgap semiconductors, namely, Shockley–Read, radiative, and Auger.

The statistical theory for the GR processes via intermediate centers was developed first by Shockley and Read,[59] and Hall.[60] This type of recombination is often called Shockley–Read–Hall (SRH) recombination. It can be reduced by lowering concentrations of native defects and foreign impurities, which can be achieved by low-temperature growth and by progress in the purification of materials. Though a considerable research effort is still necessary, the SRH process does not represent a fundamental limit to photodetector performance; it can be reduced with progress toward purer and higher-quality material, which is viable in the case of narrow-gap semiconductors.

In the SRH mechanism, generation and recombination occur via energy levels introduced into the forbidden energy gap by impurities or lattice defects. As recombination centers they trap electrons and holes; as generation centers they successively emit them. The rates of generation and recombination depend on (1) the individual nature of the center and on its predominant occupation state of charge carriers, and (2) the local densities of those carriers in the bands of the semiconductor. In general, the transition rates via energy levels for electrons and holes are quite different, thus causing different lifetimes.

Radiative generation of charge carriers is a result of absorption of internally generated photons. The radiative recombination is an inverse process of annihilation of electron–hole pairs with emission of photons. For a long time, internal radiative processes have been considered to be the main fundamental limit to detector performance, and the performance of practical devices has been compared to that limit. The role of the radiative mechanism in the detection of IR radiation has been critically re-examined.[61–63] Humpreys[61] indicated that most of the photons emitted in photodetectors as a result of radiative decay are immediately reabsorbed so that the observed radiative lifetime is only a measure of how well photons can escape from the body of the detector. Due to reabsorption, the radiative lifetime is highly extended and dependent on the semiconductor geometry. Therefore, internal combined recombination–generation processes in one detector are essentially noiseless. In contrast, the recombination act with cognate escape of a photon from the detector, or generation of photons by thermal radiation from outside the active body of the detector, are noise-producing processes. This may readily occur for a case of a detector array, where an element may absorb photons emitted by another detector or by a passive part of the structure.[64] Deposition of the reflective layers (mirrors) on the back and side of the detector may significantly improve optical insulation, preventing noisy emission and absorption of thermal photons. It should be noted that internal radiative generation can be suppressed in detectors operating under reverse bias, where the electron density in the active layer is reduced to well below its equilibrium level.[65]

As follows from the above considerations, the internal radiative processes, although of fundamental nature, do not essentially limit the ultimate performance of infrared detectors, especially LWIR devices.[66]

Auger mechanisms dominate generation and recombination processes in high-quality narrow-gap semiconductors such as $Hg_{1-x}Cd_xTe$ and InSb at near-room temperatures. The Auger generation is essentially the impact ionization by electrons of holes in the high-energy tail of Fermi–Dirac distribution.

The band-to-band Auger effects are classified in several processes according to related bands. In Fig. 2.15 we show the three most important mechanisms in the case of this type of band structure. The three mechanisms have the smallest threshold ($E_T \approx E_g$) and the largest combined density of states. The CHCC recombination mechanism (also labeled Auger 1) involves two electrons and a heavy hole and is dominant in n-type material. The CHLH process (labeled Auger 7) is dominant in p-type material if the spin split-off band can be ignored. For materials such as InSb and HgCdTe where the spin split-off energy Δ is much larger than the bandgap energy E_g, the probability of the Auger transition through the conduction

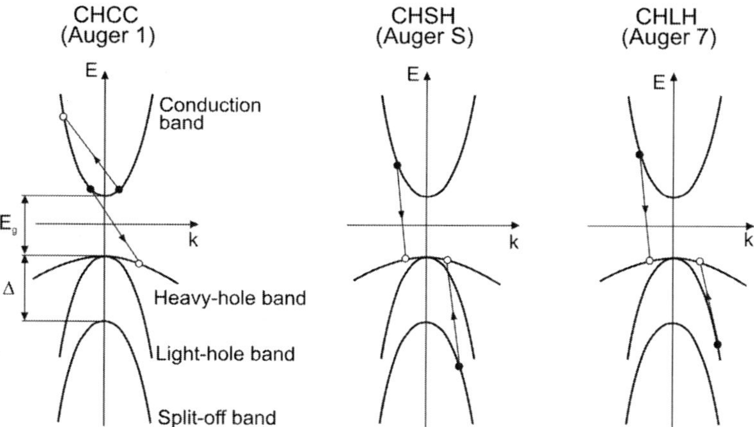

Figure 2.15 The three band-to-band Auger recombination processes. Arrows indicate electron transitions; solid circle – occupied state; and circle – unoccupied state.

band/heavy-hole band/spin split-off band mechanism (called CHSH or Auger S hereafter) may be negligibly small in comparison with that of the CHLH Auger transition. The spin split-off band plays a far more important role than the light-hole band for direct-bandgap materials, especially when the bandgap energy E_g approaches the spin–orbit splitting Δ (as in the case of InAs and InAsSb).

By the late 1950s and early 1960s several papers had been published on the photoconductive lifetime of InSb and InAs. Since then, the quality of single-crystal growth and fabrication technology has improved immensely. Pines and Stafsudd[67] have presented the photoconductive lifetime of a high-quality state-of-the-art n-type InSb ingot. The measured photoconductive lifetime is the parallel combination of the SRH photoconductive lifetime $\tau'_{pc} = (\mu_e \tau_e + \mu_h \tau_h)/(\mu_e + \mu_h)$ (when the lifetimes of the electron and hole are not equal), the Auger lifetime τ_{A1}, and the radiative lifetime τ_R, and is given by $\tau_{pc} = (1/\tau_{A1} + 1/\tau_R + 1/\tau'_{pc})^{-1}$.

Figure 2.16(a) shows the photoconductive lifetime as a function of inverse temperature for various impurity electron concentrations and for the density of the recombination centers: 8×10^{13} cm^{-3}. The SRH photoconductive lifetime is predominant in extrinsic temperatures up to 250 K, but above that temperature, the Auger 1 lifetime becomes the limiting lifetime. At higher temperatures ($T > 200$ K), the parallel combination of the SRH photoconductive lifetime and the Auger lifetime is presented by a broken line. Measured photoconductive lifetime data from work by Hollis et al.[68] are also presented in Fig. 2.16(a). At temperatures below 130 K the electron and hole lifetimes obey the expression for the SRH recombination process:

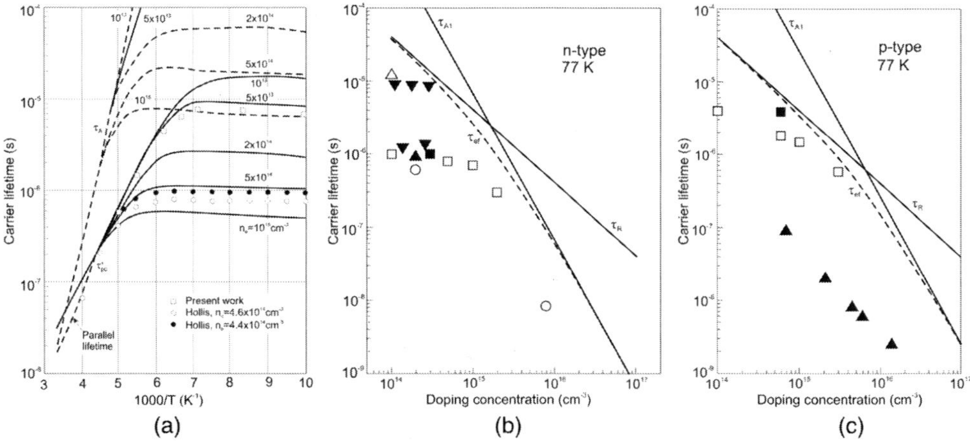

Figure 2.16 Carrier lifetime in InSb: (a) photoconductive and Auger lifetime in n-type material for various impurity concentration (adapted from Ref. 67) and dependence of the carrier lifetime on doping concentration for (b) n-type and (c) p-type materials (reprinted from Ref. 69).

$$\tau_h = \frac{4.4 \times 10^8}{N_d}, \tag{2.8}$$

where N_d is the doping concentration. This lifetime dependence on doping has been observed in a number of different samples fabricated at different laboratories.[67-74] The electron and hole lifetimes are equal because in n-type material there is a single set of recombination centers. To explain the recombination process in the extrinsic temperature range, Pines and Stafsudd[65] have presented a two-step recombination model in which a dipole moment interaction potential and a Coulomb interaction potential are involved.

Figures 2.16(a) and 2.16(b) give additional insights into carrier lifetime recombination mechanisms in InSb. As seen from Fig. 2.16(b), the Auger 1 process is dominant at the electron concentration $n > 4 \times 10^{15}$ cm^{-3}, whereas at $n < 1 \times 10^{15}$ cm^{-3} the carrier lifetime is determined by the radiative mechanism. In p-type InSb material the radiative recombination is dominant at the concentration $p < 5 \times 10^{15}$ cm^{-3} [see Fig. 2.16(c)]. The observed scatter of experimental data in samples with approximately the same concentration of carriers can be attributed to SRH recombination, as is especially observed for p-type material.[70] In p-type InSb at 77 K, hole and electron lifetimes are far different, τ_e being less than 10^{-9} s, and τ_h being equal to about $10^9/p_o$, where p_o is the equilibrium hole concentration. This behavior is attributed to the presence of donor-like recombination centers situated at about 0.05 eV above the valence band and possibly a second deeper level.[70,71] The values of the carrier lifetimes in samples investigated earlier (see, e.g., closed triangle according to Ref. 70) were lower than those obtained today. Thus, the relation between the improvement in technology of InSb and the increase in the carrier lifetime is clearly seen from experimental data shown in Figs. 2.16(b,c).

InSb detectors have usually been restricted to operating temperatures ranging from 77 K to 90 K. This limitation is mainly due to the small generation lifetimes that result from impurities and defects in the crystal lattice of available bulk material. It has been found that these impurities and defects can be greatly reduced by growing epitaxial layers of InSb on InSb wafers by LPE[72] and MBE.[73] Despite the fact that a significant reduction in impurities and defects in InSb has been achieved by the use of epitaxial growth techniques, the SRH lifetime of InSb has remained at $\sim$1 μs for 50 years.

Figure 2.17 shows the experimental and calculated dependences of the carrier lifetime on doping concentration in n-type and p-type InAs binary compound. The contribution of the radiative mechanism increases with an energy gap increase and a temperature decrease. As a result, at a low temperature ($T = 77$ K) the experimental data for n-type InAs are well described by the radiative recombination mechanism. As the temperature increases, the Auger recombination comes to the fore [see Figs. 2.17(b) and 2.17(c)].

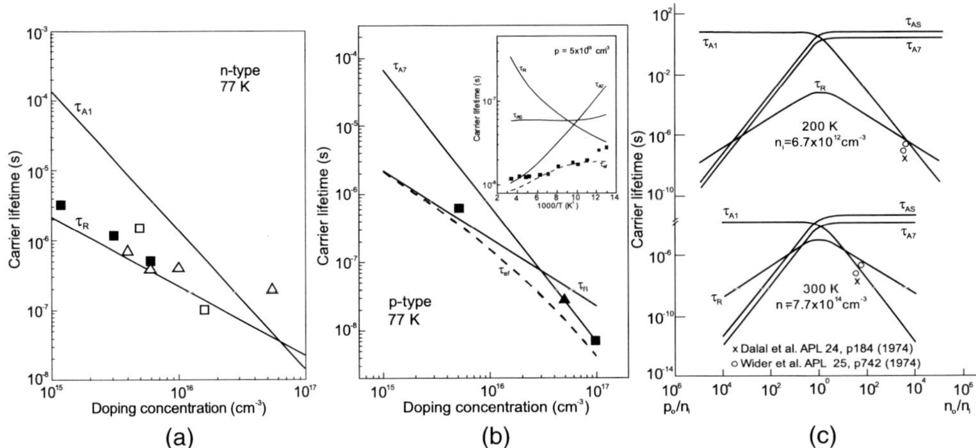

Figure 2.17 Dependence of carrier lifetime versus doping level at 77 K for (a) n-type and (b) p-type InAs (reprinted from Ref. 74). Insert of figure (b) shows carrier lifetime versus temperature in p-InAs with $p = 5 \times 10^{16}$ cm^{-3}. (c) Dependence of carrier lifetime on normalized doping concentration for InAs at 200 K and 300 K (reprinted from Ref. 75).

In contrast to InSb, for p-type InAs the influence of the Auger S process is apparent, so three recombination mechanisms—radiative, Auger 7 and Auger S—are taken into account. The contribution of the radiative mechanism at 77 K is essential at a concentration of holes below 10^{16} cm^{-3}. The Auger S recombination mechanism is weakly dependent on temperature, and its contribution to the lifetime is less important in comparison with the dominant contribution of the Auger 7 process. The Auger S mechanism is strongly dependent on the difference between Δ and E_g.[29]

Careful analysis of experimental values of the carrier lifetime obtained from the early 1960s to the present day clearly indicate a gradual increase to the theoretical limit determined by the radiative and Auger recombination processes. Obviously, these results are caused by improvements in technology of InAs single crystals. On the other hand, the SRH recombination process still remains insufficiently investigated.

Rogalski and Orman[75] calculated the carrier lifetime in InAs$_{1-x}$Sb$_x$ for radiative and Auger recombination for the temperature range 77–300 K and the composition range $0 \leq x \leq 1$. In the low-temperature range, the carrier lifetime is determined by radiative recombination. At higher temperatures the Auger 1 process is dominant in n-type InAs$_{1-x}$Sb$_x$, but in p-type material, for the composition range $0 \leq x \leq 0.3$, competition takes place between Auger 7 and Auger S processes; for the composition range $x > 0.3$ the Auger 7 process is dominant. Various recombination channels in InAs$_{0.35}$Sb$_{0.65}$, which is potentially capable of operating at the longest cutoff wavelengths, with E_g close to 0.1 eV at room temperature, are also discussed in Ref. 75.

In the calculations of the influence of the Auger S process, the spin–orbit splitting energy $\Delta(x)$ (after Van Vechten et al. published in 1972) has been assumed. A recently published paper indicates, however,[50] that the influence of spin-off bandgap energy $\Delta(x)$ on carrier lifetime in p-type $InAs_{1-x}Sb_x$ ternary alloy should be re-examined due to new insights into the composition dependence of $\Delta(x)$. Assuming new experimental data for $\Delta(x)$, the predicted influence of the Auger S process is important in $InAs_{1-x}Sb_x$ alloys close to InAs—in the composition range $0 \leq x \leq 0.15$. This means that InAsSb devices have lower Auger S nonradiative losses than were predicted previously.[76] This correction has important implications for interpretation of the performance of p-type InAsSb-based photodiodes.

References

1. W. F. M. Micklethwaite and A. J. Johnson, "InSb: Materials and devices," in *Infrared Detectors and Emitters: Materials and Devices*, edited by P. Capper and C. T. Elliott, Kluwer Academic Publishers, Boston, pp. 178–204 (2001).
2. P. S. Dutta, H. L. Bhat, and V. Kumar, "The physics and technology of gallium antimonide: An emerging optoelectronic material," *J. Appl. Phys.* **81**, 5821–5879 (1997).
3. W. Zhang and M. Razeghi, "Antimony-based materials for electro-optics," in *Semiconductor Nanostructures for Optoelectronic Applications*, edited by T. Steiner, Artech House, Norwood, Massachusetts, pp. 229–288 (2004).
4. *Springer Handbook of Electronic and Photonic Materials*, edited by S. Kasap and P. Capper, Springer, Heidelberg (2006).
5. *Springer Handbook of Crystal Growth*, edited by G. Dhanaraj, K. Byrappa, V. Prasad, and M. Dudley, Springer, Heidelberg (2010).
6. T. S. Liu and E. A. Peretti, "The indium-antimony system," *Trans. Am. Soc. Met.* **44**, 539–548 (1951).
7. H. Walker, "Über Neue Halbleitende Verbindungen," *Z. Naturf.* **7A**, 744 (1952).
8. A. Rogalski, J. Antoszewski, and L. Faraone, "Third-generation infrared photodetector arrays," *J. Appl. Phys.* **105**, 091101 (2009).
9. V. M. Goldschmidt, *Skrifter Norshe Vide. Akads. Oslo I: Mat. Naturv. Kl. VIII* (1926).
10. W. Koster and B. Thoma, "Building of the gallium-antimony, gallium-arsenic and alluminium-arsenic systems," *Z. Metallkd.* **46**, 291–293 (1955).
11. I. G. Greenfield and R. L. Smith, "Gallium-antimony system," *Trans. AIME* **203**, 351–353 (1955).
12. A. Rogalski, *Infrared Detectors*, 2nd edition, CRC Press, Boca Raton, Florida (2011).

13. R. Fornari, "Bulk crystal growth of semiconductors: An Overview," in *Comprehensive Semiconductor Science and Technology. Six-Volume Set*, Vol. **3**, edited by P. Bhattacharya, R. Fornari, and H. Kamimura, Elsevier, Amsterdam, Chapter 3.01 (2011).

14. M. J. Furlong, R. Martinez, S. Amirhaghi, D. Small, B. Smith, and A. Mowbray, "Scaling up antimonide wafer production: Innovation and challenges for epitaxy ready GaSb and InSb substrates," *Proc. SPIE* **8012**, 801211 (2011) [doi: 10.1117/12.884450].

15. M. J. Furlong, R. Martinez, S. Amirhaghi, B. Smith, A. P. Mowbray, J. P. Flint, G. Dallas, G. Meshew, and J. Trevethan, "Towards the production of very low defect GaSb and InSb substrates: bulk crystal growth, defect analysis and scaling challenges," *Proc. SPIE* **8631**, 86311N (2013) [doi: 10.1117/12.2005130].

16. M. J. Furlong, G. Dallas, G. Meshew, J. P. Flint, D. Small, B. Martinez, and A. P. Mowbray, "Growth and characterization of 6" InSb substrates for use in large area infrared imaging applications," *Proc. SPIE* **9070**, 907016 (2014) [doi: 10.1117/12.2042393].

17. J. B. Mulin, "Melt-growth of III-V compounds by the liquid encapsulation and horizontal growth techniques," in *III-V Semiconductor Materials and Devices*, edited by R. J. Malik, North-Holland, Amsterdam, pp. 1–72 (1989).

18. M. J. Furlong, B. Martinez, M. Tybjerg, B. Smith, and A. Mowbray, "Growth and characterization of ≥6" epitaxy-ready GaSb substrates for use in large area infrared imaging applications," *Proc. SPIE* **9451**, 94510S (2015) [doi: 10.1117/12.2178040].

19. N. W. Gray, A. Prax, D. Johnson, J. Demke, J. G. Bolke, and W. A. Brock, "Rapid development of high-volume manufacturing methods for epi-ready GaSb wafers up to 6" diameter for IR imaging applications," *Proc. SPIE* **9819**, 981914 (2016) [doi: 10.1117/12.2223998].

20. R. Pelzel, "A comparison of MOVPE and MBE growth technologies for III-V epitaxial structures," *CS MANTECH Conference*, May 13th–16th, 2013, New Orleans.

21. H. M. Mansevit and W. I. Simpson, "The use of metal-organics in the preparation of semiconductor materials," *J. Electrochem. Soc.* **116**(12), 1725–1732, pp. 105–108 (1969).

22. R. M. Biefeld, "The metal-organic chemical vapor deposition and properties of III-V antimony-based semiconductor materials," *Materials Science and Engineering* **R36**, 105–142 (2002).

23. C. A. Wang, "Progress and continuing challenges in GaSb-based III-V alloys and heterostructures grown by organometallic vapor-phase epitaxy," *Journal of Crystal Growth* **272**, 664–681 (2004).

24. C.-A. Chang, R. Ludeke, L. L. Chang, and L. Esaki, "Molecular beam epitaxy of InGaAs and GaSbAs," *Appl. Phys. Lett.* **31**, 759 (1977).

25. R. Leduke, "Electronic properties of the (100) surfaces of GaSb and InAs and their alloys with GaAs," *IBM J. Res. Dev.* **22**, 304 (1978).

26. B. R. Bennett and B. V. Shanabrook, "Molecular beam epitaxy of Sb-based semiconductors," in *Thin Films: Heteroepitaxial Systems*, edited by W. K. Liu and M. B. Santos, World Scientific Publishing Co., Singapore, pp. 401–452 (1999).

27. M. G. Mauk and V. M. Andreev, "GaSb-related materials for TPV cells," *Semicond. Sci. Technol.* **18**, S191–S201 (2003).

28. Y. P. Varshni, "Temperature dependence of the energy gap in semiconductors," *Physica* **34**, 149–154 (1967).

29. A. Rogalski, K. Adamiec, and J. Rutkowski, *Narrow-Gap Semiconductor Photodiodes*, SPIE Press, Bellingham, Washington (2000).

30. W. Zawadzki, "Electron transport phenomena in small-gap semiconductors," *Advances in Physics* **23**, 435–522 (1974).

31. F. P. Kesamanli, Yu. V. Malcev, D. N. Nasledov, Yu. I. Uhanov, and A. S. Filipczenko, "Magnetooptical investigations of InSb conduction band," *Fiz. Tverd. Tela* **8**, 1176–1181 (1966).

32. D. T. Cheung, A. M. Andrews, E. R. Gertner, G. M. Williams, J. E. Clarke, J. L. Pasko, and J. T. Longo, "Backside-illuminated $InAs_{1-x}Sb_x$-InAs narrow-band photodetectors," *Appl. Phys. Lett.* **30**, 587–598 (1977).

33. J. C. Woolley and B. A. Smith, "Solid solution in III-V compounds," *Proc. Phys. Soc.* **72**, 214–223 (1958).

34. J. C. Woolley and J. Warner, "Preparation of InAs-InSb alloys," *J. Electrochem. Soc.* **111**, 1142–1145 (1964).

35. M. J. Aubin and J. C. Woolley, "Electron scattering in InAsSb alloys," *Can. J. Phys.* **46**, 1191–1198 (1968).

36. J. C. Woolley and J. Warner, "Optical energy-gap variation in InAs-InSb alloys," *Can. J. Phys.* **42**, 1879–1885 (1964).

37. E. H. Van Tongerloo and J. C. Woolley, "Free-carrier Faraday rotation in $InAs_{1-x}Sb_x$ alloys," *Can. J. Phys.* **46**, 1199–1206 (1968).

38. A. Rogalski, "InAsSb infrared detectors," *Prog. Quant. Electr.* **13**, 191–231 (1989).

39. A. Rogalski, *New Ternary Alloy Systems for Infrared Detectors*, SPIE Press, Bellingham, Washington (1994).

40. M. Razeghi, "Overview of antimonide based III-V semiconductor epitaxial layers and their applications at the center for quantum devices," *Eur. Phys. J. AP* **23**, 149–205 (2003).

41. W. Zhang and M. Razeghi, "Antimony-based materials for electro-optics," in *Semiconductor Nanostructures for Optoelectronic Applications*, edited by T. Steiner, Artech House, Norwood, Massachusetts, pp. 229–288 (2004).

42. S. P. Svensson, W. L. Sarney, H. Hier, Y. Lin, D. Wang, D. Donetsky, L. Shterengas, G. Kipshidze, and G. Belenky, "Band gap of $InAs_{1-x}Sb_x$ with native lattice constant," *Phys. Rev. B* **86**, 245205 (2012).

43. Y. Lin, D. Donetsky, D. Wang, D. Westerfeld, G. Kipshidze, L. Shterengas, W. L. Sarney, S. P. Svensson, and G. Belenky, "Development of bulk InAsSb alloys and barrier heterostructures for long-wavelength infrared detectors," *J. Electron. Mater.* **44**(10), 3360–3066 (2015).

44. S. Suchalkin, L. Ludwig, G. Belenky, B. Laikhtman, G. Kipshidze, Y. Lin, L. Shterengas, D. Smirnov, S. Luryi, W. L. Sarney, and S. P. Svensson, "Electronic properties of unstrained narrow gap $InAs_{1-x}Sb_x$ alloys," *J. Phys. D: Appl. Phys.* **49**(10) 105101 (2016).

45. J. M. Schoen, "Augmented-plane-wave virtual-crystal approximation," *Phys. Rev.* **184**(3), 858–863 (1969).

46. I. Vurgaftman, J. R. Meyer, and L. R. Ram-Mohan, "Band parameters for III-V compound semiconductors and their alloys," *J. Appl. Phys.* **89** (11), 5815–5875 (2001).

47. A. Rogalski and K. Jóźwikowski, "Intrinsic carrier concentration and effective masses in $InAs_{1-x}Sb_x$," *Infrared Phys.* **29**, 35–42 (1989).

48. H. H. Wieder and A. R. Clawson, "Photo-electronic properties of $InAs_{0.07}Sb_{0.93}$ films," *Thin Solid Films* **15**, 217–221 (1973).

49. O. Berolo, J. C. Woolley, and J. A. Van Vechten, "Effect of disorder on the conduction-band effective mass, valence-band spin-orbit splitting, and the direct band gap in III-V alloys," *Phys. Rev. B* **8**, 3794 (1973).

50. S. A. Cripps, T. J. C. Hosea, A. Krier, V. Smirnov, P. J. Batty, Q. D. Zhuang, H. H. Lin, P. W. Liu, and G. Tsai, "Determination of the fundamental and spin-orbit-splitting band gap energies of InAsSb-based ternary and pentenary alloys using mid-infrared photoreflectance," *Thin Solid Films* **516**, 8049–8058 (2008).

51. J. A. VanVechten, O. Borolo, and J. C. Woolley, "Spin-orbit splitting in compositionally disordered semiconductors," *Phys. Rev. Lett.* **29**, 1400 (1972).

52. W. M. Coderre and J. C. Woolley, "Electrical properties of electron effective mass in III-V alloys," *Can. J. Phys.* **46**, 1207 (1968).

53. V. W. L. Chin, R. J. Egan, and T. L. Tansley, "Electron mobility in $InAs_{1-x}Sb_x$ and the effect of alloy scattering," *J. Appl. Phys.* **69**, 3571 (1991).

54. R. J. Egan, V. W. L. Chin, and T. L. Tansley, "Dislocation scattering effects on electron mobility in InAsSb," *J. Appl. Phys.* **69**, 2473–2478 (1994).

55. A. Joullie, F. Jia Hua, F. Karouta, H. Mani, and C. Alibert, "III-V alloys based on GaSb for optical communications at 2.0–4.5 μm," *Proc. SPIE* **587**, 46 (1985) [doi: 10.1117/12.952100].

56. J. S. Blakemore, *Semiconductor Statistics*, Pergamon Press, Oxford (1962).

57. G. Nimtz, "Recombination in narrow-gap semiconductors," *Phys. Rep.* **63**, 265–300 (1980).

58. P. Landsberg, *Recombination in Semiconductors*, Cambridge University Press, Cambridge (1991).
59. W. Shockley and W. T. Read, "Statistics of the recombination of holes and electrons," *Phys. Rev.* **87**, 835–842 (1952).
60. R. N. Hall, "Electron-hole recombination in germanium," *Phys. Rev.* **87**, 387 (1952).
61. R. G. Humpreys, "Radiative lifetime in semiconductors for infrared detections," *Infrared Phys.* **23**(3), 171–175 (1983).
62. R. G. Humpreys, "Radiative lifetime in semiconductors for infrared detection," *Infrared Phys.* **26**(6), 337–342 (1986).
63. T. Elliott, N. T. Gordon, and A. M. White, "Towards background-limited, room-temperature, infrared photon detectors in the 3-13 μm wavelength range," *Appl. Phys. Lett.* **74**, 2881–2883 (1999).
64. N. T. Gordon, C. D. Maxey, C. L. Jones, R. Catchpole, and L. Hipwood, "Suppression of radiatively generated currents in infrared detectors," *J. Appl. Phys.* **91**, 565–568 (2002).
65. C. T. Elliott and C. L. Jones, "Non-equilibrium devices in HgCdTe," in *Narrow-gap II-VI Compounds for Optoelectronic and Electromagnetic Applications*, edited by P. Capper, Chapman & Hall, London, pp. 474–485 (1997).
66. K. Jóźwikowski, M. Kopytko, and A. Rogalski, "Numerical estimations of carrier generation-recombination processes and the photon recycling effect in HgCdTe heterostructure photodiodes," *J. Electron. Mater.* **41**, 2766–2774 (2012).
67. M. Y. Pines and O. M. Stafsudd, "Recombination processes in intrinsic semiconductors using impact ionization capture cross sections in indium antimonide and mercury cadmium telluride," *Infrared Phys.* **20**, 73–91 (1980).
68. J. E. Hollis, S. C. Choo, and E. L. Heasell, "Recombination centers in InSb," *J. Appl. Phys.* **38**, 1626–1636 (1967).
69. V. Tetyorkin, A. Sukach, and A. Tkachuk, "Infrared Photodiodes on II-VI and III-V Narrow-Gap Semiconductors," licensee InTech. 2012, 10.5772/52930.
70. R. N. Zitter, A. J. Strauss, and A. E. Attard, "Recombination processes in p-type indium antimonide," *Phys. Rev.* **115**, 266–273 (1959).
71. P. W. Kruse, "Indium antimonide photoconductive and photoelectro-magnetic detectors," in *Semiconductors and Semimetals*, Vol. **5**, edited by R. K. Willardson and A. C. Beer, Academic Press, New York, pp. 15–83 (1970).
72. S. R. Jost, V. F. Meikleham, and T. H. Myers, "InSb: A key for IR detector applications," *Mater. Res. Soc. Sym. Proc.* **90**, 429–435 (1987).
73. I. Shtrichman, D. Aronov, M Ben Ezra, I. Barkai, E. Berkowicz, M. Brumer, R. Fraenkel, A. Glozman, S. Grossman, E. Jacobsohn, O. Klin,

P. Klipstein, I. Lukomsky, L. Shkedy, N. Snapi, M. Yassen, and E. Weiss, "High operating temperature epi-InSb and XBn-InAsSb photodetectors," *Proc. SPIE* **8353**, 83532Y (2012) [doi: 10.1117/12.918324].

74. V. Tetyorkin, A. Sukach, and A. Tkachuk, "InAs infrared photodiodes," in *Advances in Photodiodes*, edited by G. F. D. Betta, InTech, 2011, http://www.intechopen.com/books/advances-in-photodiodes/inas-infrared-photodiodes

75. A. Rogalski and Z. Orman, "Band-to-band recombination in $InAs_{1-x}Sb_x$," *Infrared Phys.* **25**, 551–560 (1985).

76. J. Wróbel, R. Ciupa, and A. Rogalski, "Performance limits of room-temperature InAsSb photodiodes," *Proc. SPIE* **7660**, 766033 (2010) [doi: 10.1117/12.855196].

Chapter 3
Type-II Superlattices

Since their initial proposal by Esaki and Tsu[1] and the advent of MBE, the interest in semiconductor superlattices (SLs) and quantum well (QW) structures has continuously increased over the years, driven by technological challenges, new physical concepts and phenomena, as well as promising applications. A new class of materials and heterojunctions with unique electronic and optical properties has been developed. Here we focus on antimonide-based type-II superlattices, which involve infrared excitation of carriers and can be realized in several material systems.

The physical properties of the respective quantum well structures are strongly determined by the band discontinuities at the interface, i.e., the band alignment. An abrupt discontinuity in the local band structure is usually associated with a gradual band bending in its neighborhood, which reflects space-charge effects. The conduction- and valence-band discontinuities determine the character of carrier transport across the interfaces; therefore, they are the most important quantities that determine the suitability of present SLs or QWs for IR detector purposes. The presence of an additional SL periodic potential changes the electronic spectrum of a semiconductor in such a manner that the Brillouin zone is divided into a series of mini-zones, giving rise to narrow subbands separated by minigaps. Thus, SLs possess new properties not exhibited by homogeneous semiconductors. Surprisingly, the corresponding values for the band discontinuities of the conduction band ΔE_c and the valence band ΔE_v cannot be obtained by simple considerations. Band lineups based on electron affinity do not work in most cases when two semiconductors form a heterostructure. This is because of subtle charge-sharing effects that occur across atoms on the interface. There have been a number of theoretical studies that predict general trends in how bands line up.[2-4] However, the techniques are quite complex and heterostructure designs usually depend on experiments to provide line-up information.[5-7] It must be taken into account that electrical and optical methods do not measure the band offsets themselves but instead measure the quantities associated with the electronic structure of the heterostructure. The band offset determination from such experiments requires an appropriate theoretical model.

In dependence on the values for the band discontinuities, known heterointerfaces can be classified into four groups: type I, type II-staggered, type II-misaligned, and type III, as shown in Fig. 3.1.

Type I occurs for systems such as GaAs/AlAs, GaSb/AlSb, strained-layer structure GaAs/GaP, and most II-VI and IV-VI semiconductor structures with a nonzero bandgap. The sum of ΔE_c and ΔE_v is seen equal the bandgap difference $E_{g2} - E_{g1}$ of the two semiconductors. The electrons and holes are confined in one of the semiconductors that are in contact. Such types of SLs and multiple quantum wells (MQWs) are preferentially used as effective

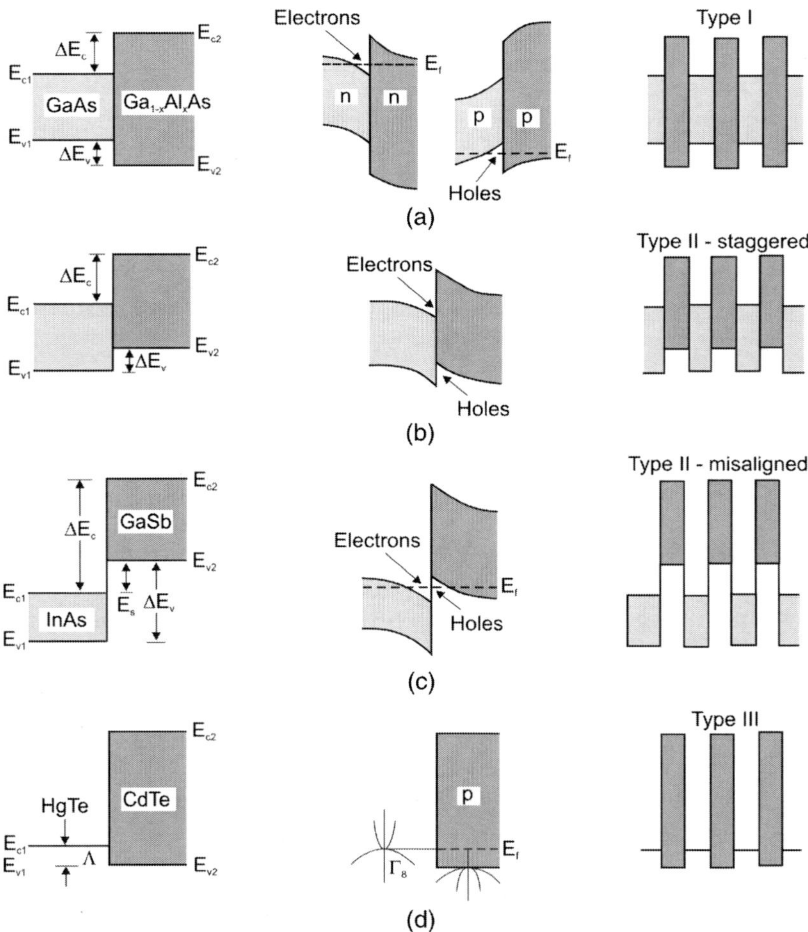

Figure 3.1 Various types of semiconductor SL and multiple-quantum-well structures: (a) type-I structure; (b) type-II-staggered structure; (c) type-II-misaligned structure; and (d) type-III structure (adapted from Ref. 8).

injection lasers, in which the threshold currents can be made much lower than those of heterolasers.

Type-II structures can be divided into two groups: staggered [Fig. 3.1(b)] and misaligned, also called broken-gap [Fig. 3.1(c)] structures. Here it is seen that $\Delta E_c - \Delta E_v$ equals the bandgap difference $E_{g2} - E_{g1}$. The type-II-staggered structure is found in certain superlattices of ternary and quaternary III-Vs, where the bottom of the conduction band and the top of the valence band of one of the semiconductors are below the corresponding values of the other (e.g., as in the case of $InAs_xSb_{1-x}/InSb$, $In_{1-x}Ga_xAs/GaSb_{1-y}As_y$ structures). As a consequence, the bottom of the conduction bands and the top of the valence bands are located in opposite layers of SLs and MQWs, so the spatial separation of confined electrons and holes takes place.

The type-II-misaligned structure is an extension of this, in which the conduction band states of semiconductor 1 overlap the valence band states of semiconductor 2. This has been established as occurring, for example, in InAs/GaSb, PbTe/PbS, and PbTe/SnTe systems. Electrons from the GaSb valence band enter the InAs conduction band and produce a dipole layer of electron and hole gas, as shown in Fig. 3.1(c). With smaller periods of SLs or MQWs it is possible to observe the semimetalic-to-semiconductor transition and to use such systems as photosensitive structures, in which the spectral responsivity range can be changed by the thickness of the components.

Type-III structures are formed from one semiconductor with a positive band gap, e.g., $E_g = E_{\Gamma6} - E_{\Gamma8} > 0$ (such as CdTe or ZnTe), and one semiconductor with a negative bandgap, e.g., $E_g = E_{\Gamma6} - E_{\Gamma8} < 0$ (such as HgTe-type semiconductors). At all temperatures the HgTe-type semiconductors behave like semimetals since there are no activation energies between the light- and heavy-hole states in the Γ_8 band [see Fig. 3.1(d)]. This type of superlattice can not be formed with III-V compounds.

3.1 Bandgap-Engineered Infrared Detectors

Figure 3.2 shows the bandgap diagrams of three basic types of superlattices used in infrared detector fabrications: type I, type II, and type III.

Type-I superlattices consist of alternating thin wider-bandgap layers of AlGaAs and GaAs. Their bandgaps are approximately aligned—the valence band (with symmetry of Γ_8) of one does not overlap the conduction band (with symmetry of Γ_6) of the other. Various forms exist, but generally the device is a majority photoconductor with infrared absorption achieved by transitions between the energy levels induced in the conduction band by dimensional quantization. The AlGaAs layers are very thick barriers that inhibit excess current, such as tunneling through the superlattice. Absorption coefficients of quantum well infrared photodetectors (QWIPs) are typically very small, and ingenious tricks must be employed to efficiently couple incident normal

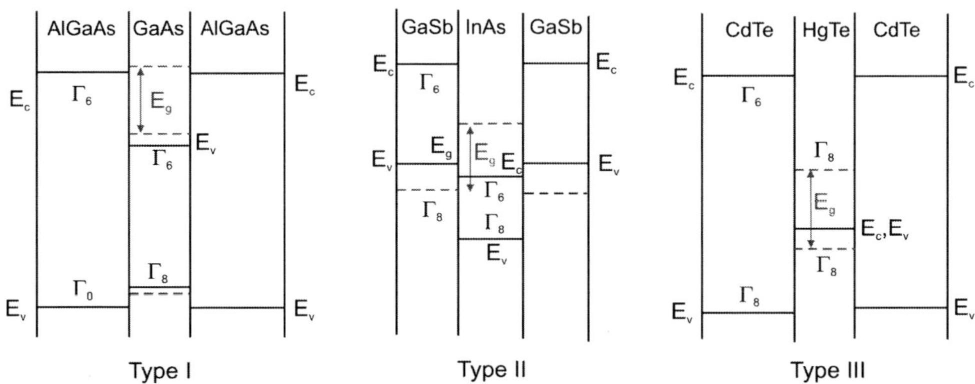

Figure 3.2 Bandgap diagrams of three basic types of superlattices designated for infrared detector applications (adapted from Ref. 9).

radiation into the structure due to selection rules for the optical transitions. The advantages of AlGaAs/GaAs QWIP architecture are: foundry materials technology, design complexity capability, and low $1/f$ noise. The disadvantages are: high dark current, low quantum efficiency, and low operating temperatures.

Type-II superlattices, a representative of which is the InAs/GaSb superlattice, are similar to type I with the exception of overlapping conduction and valence bands in adjacent bands. A type-II superlattice utilizes the quantized levels associated with the conduction band of one layer and the valence band of the adjacent layer. The electron and hole levels are separated in real space, and transitions only occur in spatial regions in which the wave functions of the carriers overlap. To provide suitable absorption, the extremely thin layers are used. The enhancement of absorption can be achieved by the additional introduction of lattice misfit and strain between alternating layers.

In a type-III superlattice the alternating layers are of different conduction and valence band symmetry. The architecture is essentially that of the type I except for the use of a semimetal instead of a semiconductor alternated with a semiconductor barrier layer. However, in this case the thickness of the semimetal layer determines a system of 2D quantized levels in both the conduction and valence bands. Conduction of electrons and holes occurs via tunneling through the thin barrier layers of the superlattice.

Type-II and type-III superlattices are essentially minority-carrier intrinsic semiconductor materials. Their absorption coefficients of IR radiation are similar to direct-bandgap alloys, and their effective masses are larger than those associated with a direct-bandgap alloy of the same bandgap.

Summarizing, the advantages of type-II and type-III superlattices are direct-bandgap absorption, large effective masses, and reduced Auger generation (this is especially the case for type-II InAs/GaSb superlattices due to space-charge separation). The disadvantages of type-III superlattices are surface passivation and layer interdiffusion at typical processing temperatures. Also, surface passivation is a serious issue for type II superlattices. However, the main drawback of type-II superlattices is short SRH lifetimes.

3.2 Growth of Type-II Superlattices

When the constituent materials are rather closely lattice matched, it is possible to design an electronic type-II SL or MQW band structure by controlling only the layer thickness and the height of the barriers. But it is also possible to grow high-quality III-V type-II SL (T2SL) devices with reduced conduction–valence bandgap for IR detector applications, in which a QW layer can be controlled on the atomic scale, too, but with a significantly different lattice constant of the well material compared to the barrier material, which gives additional opportunities to design the electronic band structure by deformation potential effects.[10–12] A schematic of the typical strained-layer superlattice (SLS) structure is given in Fig. 3.3. The thin SLS layers are alternatively in compression and tension so that the in-plane lattice constants of the individual strained layers are equal. The entire lattice mismatch is accommodated by layer strains without the generation of misfit dislocations if the individual layers are below the critical thickness for dislocation generation. Since misfit defects are not generated in SLS structures, the SLS layer can be of sufficiently high crystalline quality for a variety of scientific and device applications. Strain can change the bandgaps of the constituents

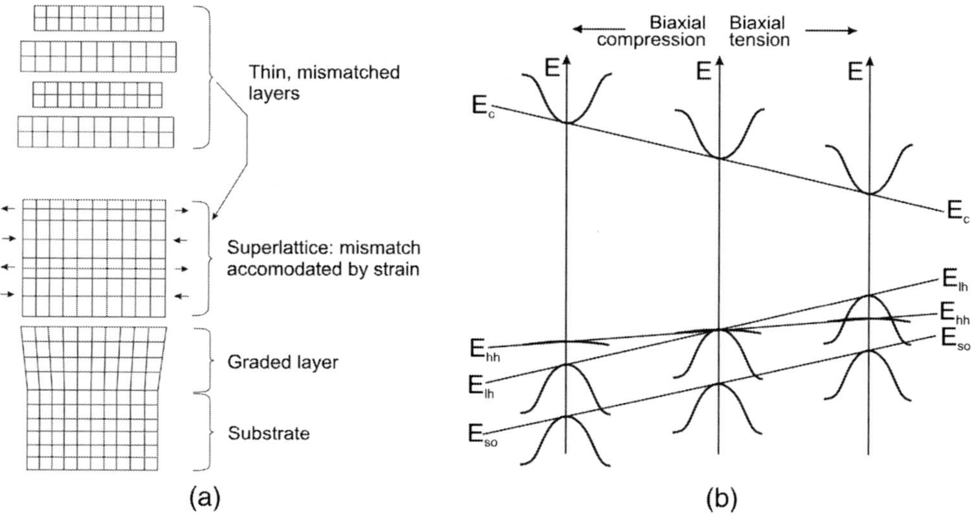

Figure 3.3 Strained-layer superlattice: (a) schematic of the fabrication and (b) the biaxial strain-induced shifts of the out-of-growth-plane energies of the valence and conduction bands.

and split the degeneracy of heavy- and light-hole bands in such a way that these changes and band splitting can lead not only to energy level reversals in the SL electronic band structure but also to appreciable suppression of recombination rates of photoexcited carriers.[13] In such systems the conduction–valence bandgap can be made much smaller than that in any III-V alloy bulk crystals.[14] For example, Fig. 3.3(b) shows the effect of biaxial strain in a tetrahedrally coordinated, direct-bandgap semiconductor. The out-of-plane conduction, light-hole, heavy-hole, and split-off band energies are shown for different biaxial strain components. As a result, e.g., in a InAsSb/InSb SLS system, the bandgap of small bandgap component (InAsSb) is decreased, and the bandgap of the InSb layers is increased. Therefore, from the effects of strain alone, InAsSb can potentially absorb at longer wavelengths than InAsSb alloys.

Type-II band alignment and some of its interesting physical behavior were originally suggested by Sai-Halasz, Tsu, and Esaki in 1977.[15] Soon after that, they reported the optical absorption of type-II superlattices[16] and their semimetal behavior,[17] and the potential use of this system in IR optoelectronics was recognized.

Progress in the growth of InAs/InAsSb SLSs by both MBE and MOCVD has been observed since 1984, when Osbourn showed in theoretical work that strain effects were sufficient to achieve wavelength cutoffs above 12 μm at 77 K.[10] The first decade's worth of efforts in development of epitaxial layers are presented in Rogalski's monograph.[18] Difficulties have been encountered in finding the proper growth conditions, especially for SLSs in the middle region of composition.[19–21] This ternary alloy tends to be unstable at low temperatures, exhibiting miscibility gaps, and this can generate phase separation or clustering. Control of alloy composition has been problematic, especially for MBE. Due to the spontaneous nature of CuPt orderings, which result in substantial bandgap shrinkage, it is difficult to accurately and reproducibly control the desired bandgap for optoelectronic device applications.[22]

Although InAsSb T2SL structures were successfully demonstrated in the 1990s,[23] they were set aside as potential infrared detector materials in favor of the InAs/InGaSb SL. Recently, new impact on their development have been observed[24] due to limits of InAs/InGaSb detector performance (short carrier lifetime and reduced quantum efficiency).

MBE is the best technique for the growth of antimonide-based superlattices due to its unique advantages. These advantages are described in Section 2.2. In the case of the InAs/GaSb superlattice, these structures contain InSb interfaces, which have weak bonds and a low melting point. For these reasons, growth is restricted to a temperature range between 390 and 450 °C. This growth condition is not possible in MOCVD since the substrate susceptor requires much higher temperatures to crack metalorganic sources.

Antimonide-based superlattices are grown by MBE using standard metal effusion cells for Ga and In, and valved cracker cells for As and Sb. According to Ref. 24, in order to minimize cross contamination of the anion fluxes, the V/III flux ratio is set at a minimum of three for both the InAs and GaSb depositions. Growths are performed primarily on 2-inch GaSb (100) epiready wafers (both n-type and p-type) with rates typically about 1 Å/s for GaSb and InGaSb and lower for InAs. To strain balance the lattice-mismatched InAs, controlled InSb-like interfaces are used. The SL stack has a thickness of several microns, and the GaSb buffer layer is on the order of one micron thick.

The 6.1-Å materials can be epitaxially grown on GaSb and GaAs substrates. In particular, 4-inch diameter GaSb substrates became commercially available in 2009, offering improved economy of scale for fabrication of large-format FPA arrays. Recently, interest in 6-inch GaSb substrates for very large-area detector applications has emerged.

The structural parameters of the samples, such as SL period, residual strain, and individual layer thickness, can be confirmed by high-resolution transmission electron microscopy (HRTEM) and high-resolution x-ray diffraction (HRXRD) measurements. For example, Fig. 3.4 shows a HRTEM image of the individual layers in the SL near the substrate region, where the $In_{0.25}Ga_{0.75}Sb$ and InAs layers appear dark and bright, respectively. From these images the average SL period was determined to be 67.6 ± 0.3 Å, with an average layer thickness of 24.1 ± 1.6 Å for InAs and 44.1 ± 1.2 Å for $In_{0.25}Ga_{0.75}Sb$.

The residual background carrier concentration of SL structures has influence on the photodiode performance (the depletion width and the

Figure 3.4 A HRTEM image of the first few layers in the InGaSb/InAs SL near the GaSb substrate (reprinted from Ref. 24).

minority-carrier response). Therefore, reduction of the background carrier concentration is a major task in the optimization of the growth conditions.

InAs/InAsSb SLs have been grown both by MBE and MOCVD on GaSb substrates. MOCVD growth of these SLs are more suitable than growth of InAs/GaSb SLs. It appears that InAsSb-based SLs are strain balanced simply by the layer thicknesses without any interfacial control.[25] The substrate growth temperature is similar to that used in fabrication of InAs/GaSb SLs.

3.3 Physical Properties

The 6.1-Å III-V semiconductor family plays a decisive role in offering new concepts for high-performance IR detectors connected with high design flexibility, direct energy gaps, and strong optical absorption. This family is formed by three semiconductors of an approximately matched lattice constant of around 6.1 Å—InAs, GaSb, and AlSb—with the low-temperature energy gaps ranging from 0.417 eV (InAs) to 1.696 eV (AlSb).[26] Like other semiconductor alloys, they are of interest principally for their heterostructures, especially combining InAs with the two antimonides (GaSb and AlSb) and their alloys. This combination offers a band alignment that is drastically different from that of the more widely studied AlGaAs system; in fact, the flexibility in the band alignment that forms is one of the principal reasons for the interest in the 6.1-Å family. The most exotic band alignment is that of InAs/GaSb heterojunctions, which is identified as broken-gap alignment. At the interface, the bottom of the conduction band of InAs is located below the top of the valence band of GaSb by about 150 meV. In such a heterostructure, with partial overlapping of the InAs conduction band with the GaSb-rich solid solution valence band, electrons and holes are spatially separated and localized in self-consistent quantum wells formed on both sides of the heterointerface. This leads to unusual tunneling-assisted radiative recombination transitions and transport properties. As illustrated in Fig. 3.5, with the availability of type-I (nested, or straddling), type-II-staggered, and type-II-broken-gap (misaligned) band offsets between the GaSb/AlSb, InAs/AlSb, and InAs/GaSb material pairs, respectively, there is considerable flexibility in forming a rich variety of alloys and superlattices.

The band alignment of the T2SL shown in Fig. 3.6(a) creates a situation in which the energy bandgap of the superlattice can be adjusted to form either a semimetal (for wide InAs and GaInSb layers) or a narrow-bandgap (for narrow layers) semiconductor material. The resulting energy gaps depend on the layer thicknesses and interface compositions. In reciprocal **k**-space, the superlattice is a direct-bandgap material that enables optical coupling, as shown in Fig. 3.6(b). Unlike an AlGaAs/GaAs superlattice, the electrons in an InAs/GaSb T2SL are confined in InAs, while holes are confined in GaSb. Since the layers are thin, the overlap of electron and hole wave functions in

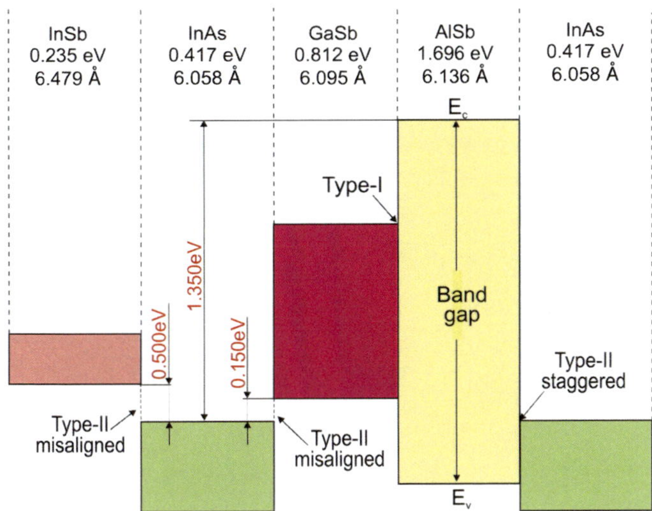

Figure 3.5 Schematic illustration of the low-temperature energy band alignment in the nearly 6.1-Å lattice-matched InAs/GaSb/AlSb material system. Three types of band alignment are available in this material system: type-I (nested) band alignment between GaSb and AlSb, type-II-staggered alignment between InAs and AlSb, and type-II-misaligned (or broken-gap) alignment between InAs and GaSb. The approximate values of band offsets are marked.

InAs and GaSb, respectively, forms electron and hole minibands. The bandgap of the SL is determined by the energy difference between the electron miniband C1 and the first heavy-hole state HH1 at the Brillouin zone center, and can be varied continuously in a range between 0 and about 400 meV. One advantage of using a T2SL is the ability to fix one component of the material and vary the other to tune the wavelength. An example of the wide tunability of SLs is shown in Figs. 3.6(c) and (d).

Remarkable theoretical modeling efforts have also been reported in calculating the T2SL band structure.[29] Several methods such as the **k·p** method,[30–33] effective bond-orbital method,[34] the empirical tight-binding method,[27] and the empirical pseudopotential method (EPM),[25–40] have been considered, and a reasonable agreement between the predictions from each method has been achieved. From theoretical modeling results it can be concluded that:

- the bandgap is defined as the gap between the bottom of the lowest electron miniband (C1) and the top of the highest hole miniband (HH1), as is shown in Fig. 3.6(a);
- depending on the layer thickness, C1 can be located anywhere between conduction bands of InAs and GaSb, while HH1 can be anywhere between the valence bands of GaSb and InAs;
- theoretically, the bandgap can be varied continuously in a range from 0 to about 400 meV;

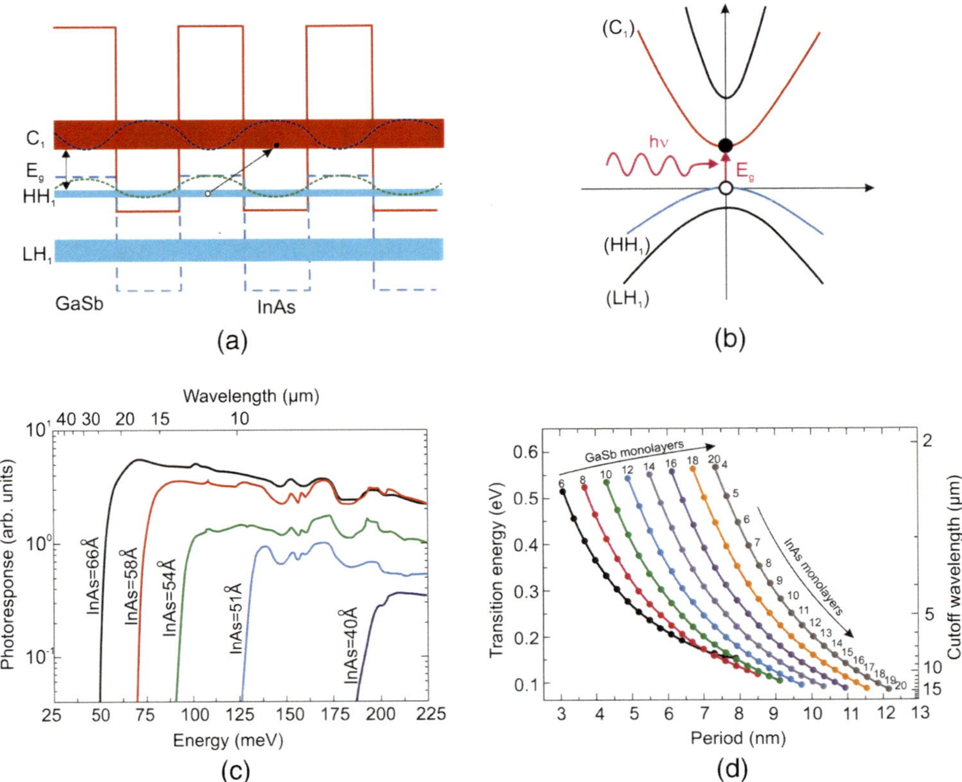

Figure 3.6 InAs/InGaSb strained-layer superlattice: (a) band edge diagram illustrating the confined electron and hole minibands that form the energy bandgap; (b) band structure with direct bandgap and absorption process in **k**-space; (c) experimental data of type-II InAs/GaSb SLS-cutoff-wavelength change with InAs thickness while GaSb is fixed at 40 Å (reprinted from Ref. 27); and (d) change in cutoff wavelength with change of InAs and GaSb monolayers for InAs/GaSb SLSs (reprinted from Ref. 28).

- transitions between electron and hole bands are indirect;
- the hole effective mass is extremely high in the growth direction, while the electron effective mass is slightly heavier than that of InAs and weakly dependent on the T2SL design; and
- the hole mobility in the growth direction is extremely low.

Figure 3.7 shows the EPM-calculated variation of the C1 and HH1 band positions of $(InAs)_N GaSb_N$ T2SLs, where N is the layer thickness in monolayers (MLs). The InAs conduction band set at 0 meV as the reference level. It is clearly shown that the C1 band is more sensitive to layer thickness than the HH1 band. The thickness of the GaSb layers has a minimal effect on the T2SL bandgap because of the large value of the GaSb heavy-hole mass ($\sim$0.41 m_o); however, the thickness of the GaSb has a significant effect on the conduction band dispersion due to tunneling of the InAs electron wave

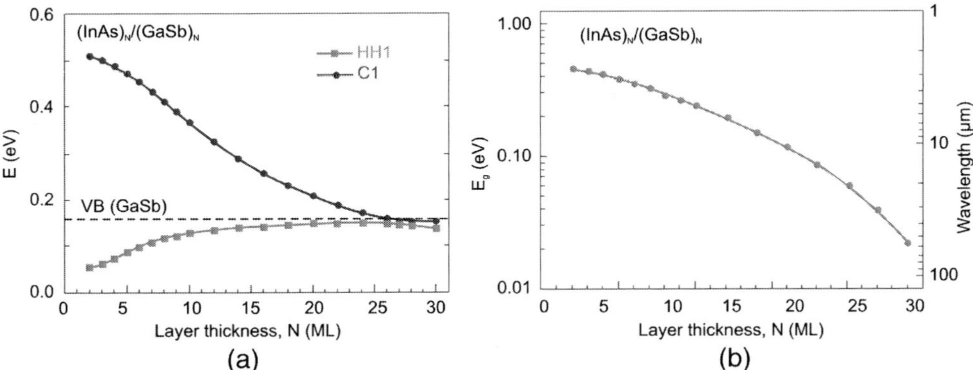

Figure 3.7 (a) Calculated HH1 and C1 band positions for InAs/GaSb SLs with equally thick InAs and GaSb layers (VB – valence band). (b) Variation of the bandgap with layer thickness (adapted from Ref. 40).

functions through the GaSb barriers. Similar GaSb layer thickness creates similar values for the conduction band effective mass in the superlattice direction. It should be noted that changing the layer thickness requires a good understanding of the effects of strain on the material quality since InAs is not lattice matched to GaSb.

An additional important observation on the InAs/GaSb T2SL is the blue-shift of the bandgap,[30,37,38] which can be explained by much higher broadening of the C1 band compared to the HH1 band. The HH1 band shifts very slowly when the layer thickness is changed. The calculation also indicates that the T2SLs designated for long-wavelength absorption exhibit a higher HH1–LH1 gap compared to their bangaps.

The basic properties of artificial InAs/GaSb T2SL material supported by simple theoretical considerations are given by Ting et al.[41] These properties may be superior to those of the HgCdTe alloys and are completely different from those of constituent layers.

The SL band structure reveals important information about carrier transport properties. The C1 band shows strong dispersion along both the growth z and in-plane direction x, whereas the HH1 band is highly anisotropic and appears nearly dispersionless along the growth (transport) direction. The electron effective mass along the growth direction is quite small and even slightly smaller than the in-plain electron effective mass. The values estimated by Ting et al.[41] for LWIR SL material (22 ML InAs/6 ML GaSb) are as follows: $m_e^{x*} = 0.023\,m_0$, $m_e^{z*} = 0.022\,m_0$, $m_{HH1}^{x*} = 0.04\,m_0$, and $m_{HH1}^{z*} = 1055\,m_0$. The SL conduction band structure near the zone center is approximately isotropic, as opposed to the highly anisotropic valence band structure. For this reason we would expect very low hole mobility along the growth direction, which is unfavorable in detector design for LWIR FPAs.[41] The estimation of effective masses for MWIR SL material

(6 ML InAs/34 ML GaSb) is:[40] $m_e^{x*} = 0.173m_0$, $m_e^{z*} = 0.179m_0$, $m_{HH1}^{x*} = 0.0462m_0$, and $m_{HH1}^{z*} = 6.8m_0$.

The effective masses are not directly dependent on the bandgap energy, as is the case in a bulk semiconductor. The electron effective mass of InAs/GaSb SL is larger compared to $m_e^* = 0.009m_0$ in HgCdTe alloy with the same bandgap ($E_g \approx 0.1$ eV). Thus, diode tunneling currents in the SL can be reduced compared to the HgCdTe alloy.

Electronic transport properties of T2SLs are anisotropic. Although in-plane mobilities drop precipitously for thin wells, electron mobilities approaching 10^4 cm^2/Vs have been observed in InAs/GaSb SLs with layers less than 40-Å thick. Among the possible low-temperature scattering mechanisms, two are particularly important for superlattices: interface roughness scattering and alloy scattering. Alloy scattering of carriers is shown not to be a factor in electronic transport—it can be concluded that there are no disadvantages to incorporating indium into the GaSb layers. Theoretical modeling indicates that the electron mobility perpendicular to the SL layer is nearly equal to the in-plane mobility in the low-temperature range and decreases to a greater degree with temperature increase (see Fig. 3.8).[24] It should be mentioned that vertical transport in the photodiode structure has an important influence on IR detector performance.[42]

Since the hole mobility is much lower than the electron mobility, by keeping electrons as minority carriers, higher photodetector performance can be expected compared to the performance when holes are minority carriers.

InAs/InAs$_{1-x}$Sb$_x$ SLs are a viable alternative to the well-studied InAs/Ga$_{1-x}$In$_x$Sb SLSs; however, InAs/InAs$_{1-x}$Sb$_x$ SLs have been less studied. Steenbergen et al.[43] have reviewed the band-edge alignment models for

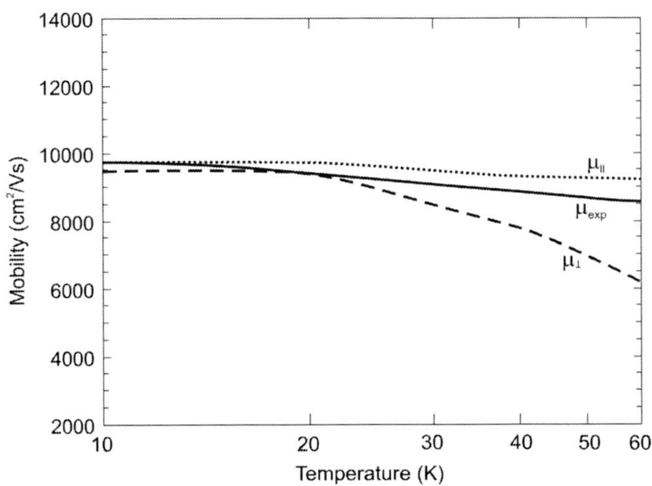

Figure 3.8 Comparison of the calculated horizontal and vertical electron mobilities with the measured horizontal mobility as a function of temperature for the 48.1-Å InAs/20.4-Å GaSb SL (reprinted from Ref. 24).

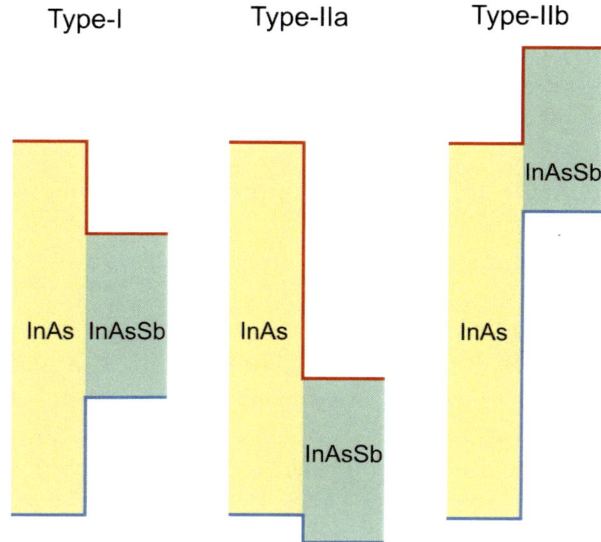

Figure 3.9 Three possible band alignments between InAs and InAs$_{1-x}$Sb$_x$.

InAs-InAs$_{1-x}$Sb$_x$ systems and considered different types of heterojunctions. Figure 3.9 shows three possible band alignments between InAs and InAs$_{1-x}$Sb$_x$, including two type-II band alignments—one with the InAs conduction band higher in energy than the InAsSb conduction band, and the other with the InAsSb conduction band higher in energy than the InAs conduction band.

Klipstein et al. have demonstrated accurate simulation of experimental absorption spectra for both InAs/GaSb and InAs/InAsSb T2SLs.[31–33] Results are gathered in Fig. 3.10 for both mid-wavelength and long-wavelength

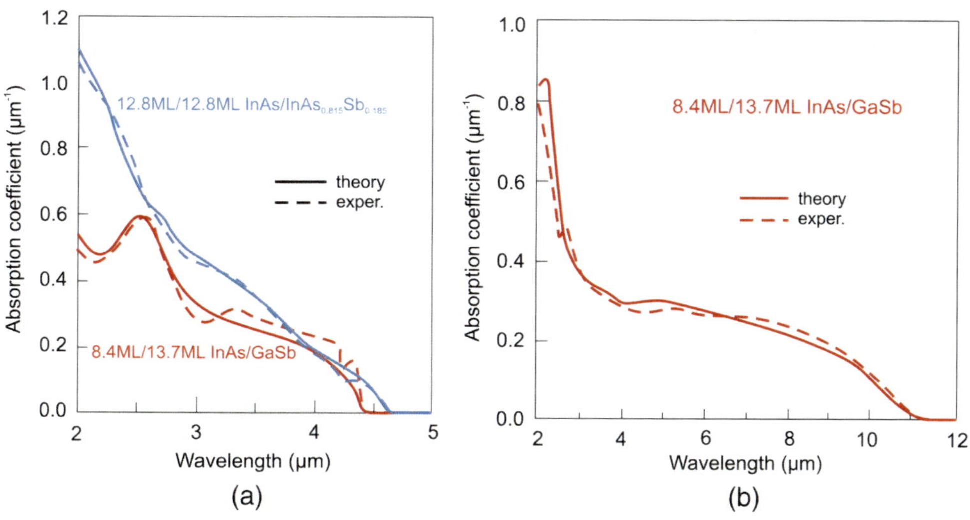

Figure 3.10 Measured and calculated absorption spectra for (a) MWIR InAs/GaSb and InAs/InAsSb T2SLs and (b) LWIR InAs/GaSb T2SL (reprinted from Ref. 33).

material systems. The strong peak below 3 µm is due to the zone boundary HH2 → C1 transitions. Note that this peak is much stronger than the longer-wavelength zone center LH1 → C1 transition (at ∼3.4 µm). The main features of the experimental spectrum of a 12.8 ML/12.8 ML InAs/InAsSb SL are reproduced by theoretical calculations shown in Fig. 3.10(a). We can see that the absorption coefficient of the InAs/InAsSb SL near the cutoff wavelength is weaker than that of the InAs/GaSb SL.

Recently, Vurgaftman et al.[44] have calculated the absorption spectra and compared these data to the measured data for LWIR T2SLs and bulk materials (HgCdTe and InAsSb) with the same energy gap (see Fig. 3.11). Absorption coefficients for bulk HgCdTe and InAsSb are very similar, reflecting that relatively minor differences between the optical matrix elements and joint densities of states exist in bulk semiconductors with the same energy gap. The SLs with an average lattice constant matched to GaSb have significantly lower absorption. This behavior is explained by the electron–hole overlap in the strain-balanced T2SLs, which occurs primarily in the hole well, having a relatively small fraction of the total thickness. It has been shown, however, that the absorption strength in small-period metamorphic $InAs_{1-x}Sb_x/InAs_{1-y}S_y$ SLs is similar to that of bulk materials.

3.4 Carrier Lifetimes

The competition between generation processes and recombination processes directly affects the performance of photodetectors. The decay of optically generated carriers due to recombination reduces the quantum efficiency of the

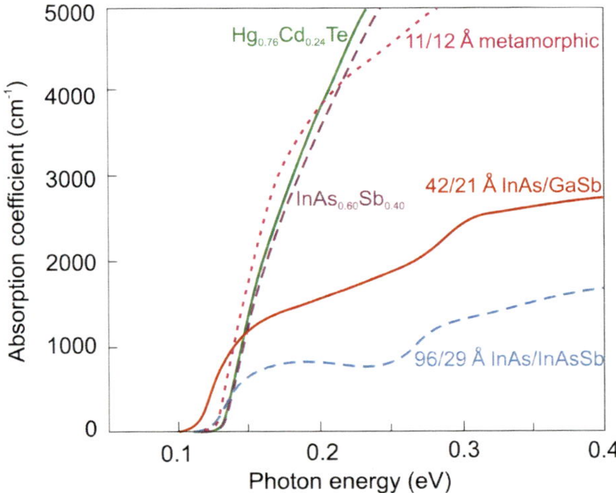

Figure 3.11 Calculated interband absorption coefficients as a function of photon energy at 80 K for bulk InAs$_{0.60}$Sb$_{0.4}$ and Hg$_{0.76}$Cd$_{0.24}$Te, and T2SLs: 42-Å InAs/21-Å GaSb, 96-Å InAs/29-Å InAs$_{0.61}$Sb$_{0.39}$, and 11-Å InAs$_{0.66}$Sb$_{0.34}$/12-Å InAs$_{0.36}$Sb$_{0.64}$ metamorphic.

device and photoelectric gain. Worse, the statistical nature of the generation–recombination processes results in fluctuation of the carrier concentration, causing noise that limits the performance of photodetectors. Here, we compare the influence of different generation–recombination mechanisms on carrier lifetimes in HgCdTe and T2SL material systems.

There are three main thermal generation–recombination (GR) processes to be considered in narrow-bandgap semiconductors, namely, SRH, radiative, and Auger. In this section, the radiative process is ignored since its contribution is small enough to be neglected due to the photon recycling effect.[45]

Theoretical analysis of band-to-band Auger and radiative recombination lifetimes for InAs/GaInSb SLSs shows that in these objects the p-type Auger recombination rates are suppressed by several orders compared to those of bulk HgCdTe with a similar bandgap;[46,47] however, n-type materials are less advantageous. In a p-type superlattice, Auger rates are suppressed due to lattice-mismatch-induced strain that splits the highest two valence bands (the highest light band lies significantly below the heavy-hole band and thus limits available phase space for Auger transitions). In the n-type superlattice, Auger rates are suppressed by increasing the InGaSb layer widths, thereby flattening the lowest conduction band and thus limiting available phase space for Auger transition.

The GR mechanisms have been generally well established in HgCdTe ternary alloys.[48] The carrier lifetime measurements of MWIR materials with a cutoff wavelength of $\lambda_c = 5$ μm and LWIR materials ($\lambda_c = 10$ μm) for both n-type and p-type HgCdTe at 77 K are summarized in Figs. 3.12 and 3.13, respectively. The trend lines of carrier lifetimes are given according to Kinch et al.[49] The experimental data are taken from many sources.

The SRH mechanism is responsible for lifetimes in lightly doped n- and p-type HgCdTe. The possible factors are SRH centers associated with native defects and residual impurities. Measured values for n-type LWIR HgCdTe at 77 K lie in a broad range from 2–20 μs and are independent of doping concentration for values below 10^{15} cm^{-3}. The MWIR values are typically somewhat longer, in the range of 2–60 μs.

Minority-carrier recombination in p-type material is associated with Auger 7 at higher extrinsic doping levels. The highest lifetime was measured in high-quality undoped and extrinsically doped materials grown by low-temperature epitaxial techniques from Hg-rich LPE and MOCVD. It is believed that the increase in lifetime of impurity-doped HgCdTe arises from a reduction of SRH centers. One other very common recombination mechanism in p-type material is of the SRH type associated with Hg-vacancy-doped HgCdTe observed in both MWIR and LWIR materials. Vacancy-doped material exhibits SRH-center densities roughly proportional to the vacancy concentration.

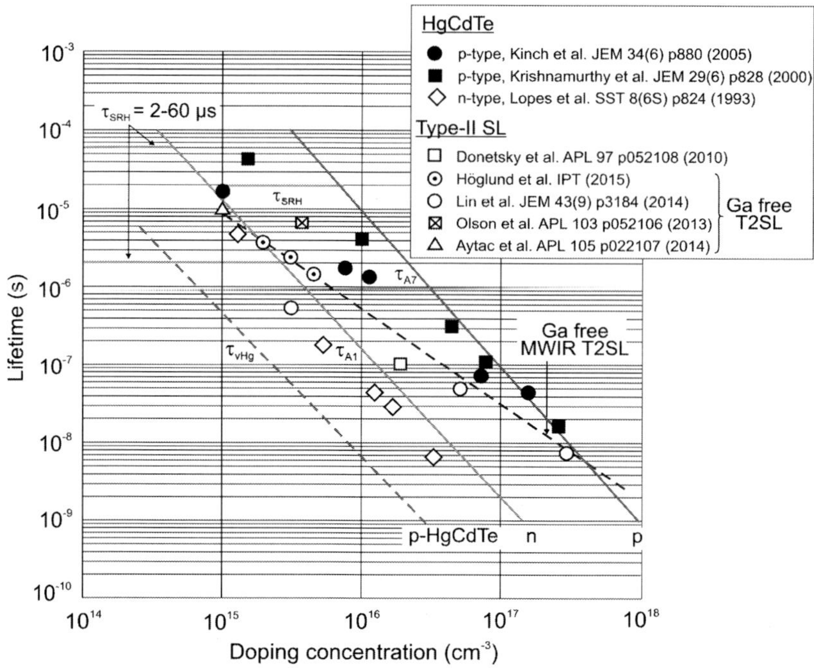

Figure 3.12   Carrier lifetimes for MWIR HgCdTe and type-II superlattices at 77 K as a function of doping concentration. Theoretical trend lines for n-type and p-type HgCdTe ternary alloys are taken from Ref. 49. The dashed line for Ga-free T2SLs follows experimental data.

The well-established GR mechanism in n-type HgCdTe at higher doping concentration is Auger 1 mechanisms.

In comparison with HgCdTe ternary alloys, III-V semiconductor bulk materials exhibit more active SRH centers, resulting in lower lifetime. The situation is more complex in the case of T2SLs.

In the type-II InAs/GaSb SL, separation of electrons (mainly located in the InAs layers) and holes (confined in the GaSb layers) suppresses Auger recombination mechanisms and thereby enhances carrier lifetime. Optical transitions occur spatially indirectly; thus, the optical matrix element for such transitions is relatively small. Theoretical analysis of band-to-band Auger recombination lifetimes for InAs/GaSb SLs shows that Auger recombination rates are suppressed by several orders of magnitude compared to those of bulk HgCdTe with a similar bandgap.[46]

The agreement between theory and experiment for LWIR T2SLs is good for carrier densities above 2×10^{17} cm^{-3}. The discrepancy between both types of results for lower carrier densities is due to SRH recombination processes having $\tau \approx 6 \times 10^{-9}$ s, which has not been taken into account in the calculations. For higher carrier densities, the SL carrier lifetime is longer than in HgCdTe; however, in low-doping regions (below 10^{15} cm^{-3}, as is necessary

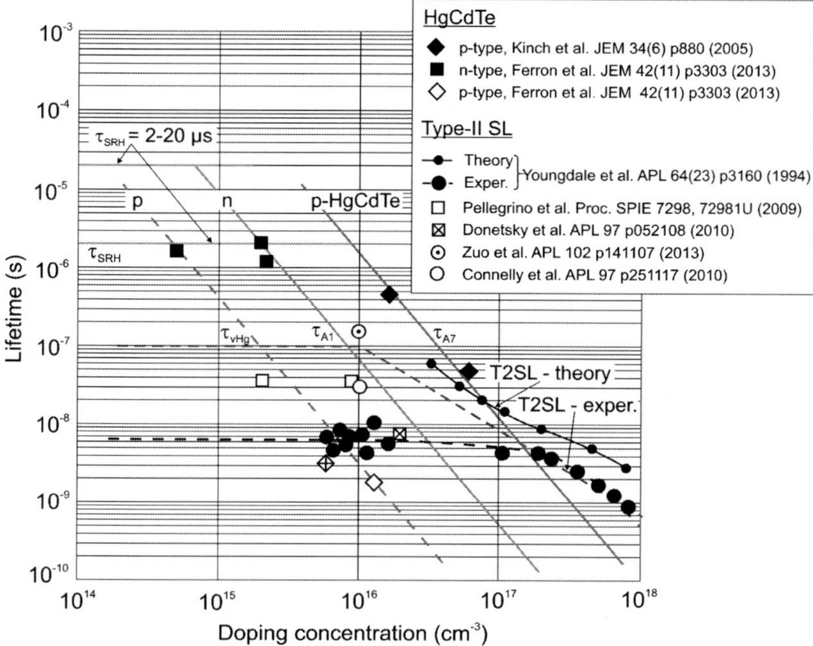

Figure 3.13 Carrier lifetimes for LWIR HgCdTe and type-II superlattices at 77 K as a function of doping concentration. Theoretical trend lines for n-type and p-type HgCdTe ternary alloys are taken from Ref. 49. The dashed line for T2SLs follows experimental data.

in the fabrication of high-performance p-on-n HgCdTe photodiodes) the experimentally measured carrier lifetime in HgCdTe is more than two orders of magnitude longer than in SL.

The promise of Auger suppression has not yet been observed in practical device material. At the present time, the measured carrier lifetime is typically below 100 ns and is limited by the SRH mechanism in both MWIR and LWIR compositions. More recently, an increase in the minority-carrier lifetime to 157 ns has been observed due to incorporation of an InSb interfacial layer in the InAs/GaSb T2SL.[50] There is no clear understanding of why the minority-carrier lifetime varies within the device structure.[51] It is interesting to note that InSb, a member of III-V compound family, has had a similar SRH lifetime issue since its inspection in the 1950s.

According to the statistical theory of the SRH process, the SRH rate approaches a maximum as the energy level of the trap center approaches midgap. Analysis of the defect formation energy of native defects is dependent on the location of the Fermi level stabilization energy. In bulk GaSb the stabilized Fermi level is located near either the valence band or the midgap, whereas in bulk InAs the stabilized Fermi level is located above the conduction-band edge. This observation results in the midgap trap levels in GaSb being available for SRH recombination, whereas in InAs they are inactive for the SRH process, suggesting a longer carrier lifetime in bulk InAs than in bulk GaSb material. It may then be hypothesized that native defects associated with GaSb are responsible for the SRH-limited minority-carrier lifetimes observed in InAs/GaSb T2SLs.

The origin of the abovementioned recombination centers has been attributed to the presence of gallium, as the gallium-free InAs/InAsSb superlattices possess much longer lifetimes, up to 10 μs for undoped material in MWIR region,[52] comparable to those obtained for HgCdTe alloys (see Fig. 3.12). It has been observed that the minority-carrier lifetime increases with increasing antimony content and decreasing layer thickness. In this system, the electrons and holes are spatially separated in InAs and InAsSb layers so that the recombination process is drastically reduced. However, it appears that the carrier separation (due to electrons and holes being spatially separated in InAs and InAsSb layers) reduces the optical absorption of the material, leading to considerably weak quantum efficiency. As a result, the performance of photodiodes manufactured from gallium-free InAs/InAsSb superlattices is worse than that of InAs/GaSb T2SL competitors.

Generally, in comparison with HgCdTe ternary alloys, III-V semiconductor bulk materials exhibit more active SRH centers, resulting in lower lifetime. In good-quality HgCdTe photodiodes fabricated on MBE and LPE materials, the absence of a measurable depletion current component is observed. The values of τ_{SRH}, gathered by Kinch[53] and given in Table 3.1, are larger by approximately three orders of magnitude than those reported for the III-V alloy materials with similar bandgaps.

The longest τ_{SRH} value (~200 μs) for III-V materials was found for the short-wavelength infrared (SWIR) lattice-matched epitaxial InGaAs ternary alloy on an InP substrate with a cutoff wavelength of 1.7 μm.[54] For the most-investigated MWIR InSb alloy, the best τ_{SRH} value was estimated as ~400 ns for LPE-grown material.[55] Throughout the last 50 years, the τ_{SRH} value has not been improved. Similar values of ~400 ns have been reported for MBE-grown InAsSb bulk alloys and InAs/InAsSb superlattices.[56] As was mentioned above, the τ_{SRH} value for T2SLs containing Ga is typically an order of magnitude lower.

It is expected that the SRH recombination mechanism is presumably associated with some departure of the semiconductors from perfect crystallinity.

Table 3.1 Values of τ_{SRH} in $Hg_{1-x}Cd_xTe$ deduced from reported I–V and FPA characteristics (data from Ref. 53).

	x composition	τ_{SRH} (μs)
LWIR	0.225	>100 at 60 K
MWIR	0.30	>1000 at 110 K
MWIR	0.30	~50,000 at 80 K
SWIR	0.455	>3000 at 180 K

Since the ionic bond in II-VI alloys is stronger than in the corresponding III-V materials, the electron wave function around the lattice sites is much more strongly confined, rendering the II-VI lattice more immune to the formation of bandgap states due to deviations from crystalline perfection.[57]

InAs/GaInSb SLSs and QWs are also employed as the active regions of MWIR lasers operated in the 2.5- to 6-μm spectral region. Meyer et al.[57,58] have experimentally determined Auger coefficients for InAs/GaInSb quantum well W interband cascade laser structures with energy gaps corresponding to 2.5–6.5 μm and have compared their values with those for typical III-V and II-VI type-I superlattices. The Auger coefficient is defined by the expression $\gamma_3 \equiv 1/\tau_A n^2$. Figure 3.14 summarizes the Auger coefficients at $\sim$300 K for different material systems: a wide variety of type-I materials, including bulk and quantum well III-V semiconductors as well as HgCdTe. We can see that the room-temperature Auger coefficients for seven different InAs/GaInSb SLs are found to be nearly an order of magnitude lower than typical type-I results for the same wavelength, indicating a significant supression of Auger losses in the antimonide T2SLs. Note that all of the more recently fabricated interband cascade lasers with $\lambda > 3$ μm[58] exhibit a significant further reduction in γ_3 in comparison with those described earlier.[57] The data imply that at this temperature the Auger rate is relatively insensitive to details of the band structure. In contrast to MWIR devices, the promise of Auger suppression has not yet been observed in practical LWIR type-II device materials.

3.5. InAs/GaSb versus InAs/InAsSb Superlattice Systems

Infrared detectors made of InAs/InAsSb SLs on GaSb substrates are in the early stage of development and are less studied than their InAs/GaSb SL

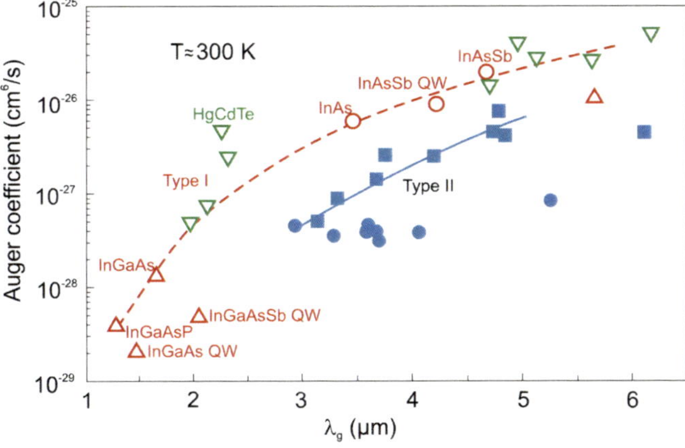

Figure 3.14 Experimental Auger coefficient versus gap wavelength for type-II W interband cascade laser structures (solid circles—Ref. 58) and various T2SL QWs taken from Ref. 57 (solid squares), along with typical data for a variety of conventional III-V and HgCdTe type-I materials (adapted from Ref. 57). The solid and dashed lines are guides to the eye.

counterparts. Due to only two common elements (In and As) being used in superlattice layers and a relatively simple interface structure with Sb-changing elements, InAs/InAsSb superlattice growth follows with better controllability and simpler manufacturability.

The interest in InAs/InAsSb SLs stemmed mainly from a need to overcome the carrier lifetime limitations imposed by the GaSb layer in InAs/GaSb SLs. A significant longer minority-carrier lifetime has been obtained in an InAs/InAsSb SL system compared to an InAs/GaSb SL system operating at the same wavelength range (for MWIR material at 77 K, $\sim$1 μs, and $\sim$100 ns, respectively). An increase in minority-carrier lifetime suggests lower dark currents for InAs/InAsSb SL photodiodes in comparison with InAs/GaSb SL detectors. In practice, however, the dark currents are not as low as expected and are higher than for InAs/GaSb SL photodiodes.

As was mentioned previously, the InAs/GaSb T2SL band alignment results in an overlap energy (estimated as 140–170 meV) between the conduction-band (CB) minimum and the valence-band (VB) maximum of the two materials. For InAs/InAsSb T2SLs, the band offset is defined in terms of the VB offset between InAs and InSb (considered to be about 0.620 meV). The primary difference in the profiles of the CB and VB in the InAs/GaSb and InAs/InAsSb T2SLs is illustrated in Fig. 3.15.

The band offsets in conduction ΔE_c and valence ΔE_v bands in the InAsSb SL ($\Delta E_c \sim 142$ meV, $\Delta E_v \sim 226$ meV) are much smaller compared to those of the InAs/GaSb SL ($\Delta E_c \sim 930$ meV, $\Delta E_v \sim 510$ meV).[26] This situation suggests a higher contribution of tunneling currents in the dark current of InAs/InAsSb SL photodiodes operated at a higher temperature. Also, experimental data and a theoretical estimation of absorption coefficients in both superlattices given by Klipstein et al. indicate a lower absorption coefficient of InAs/InAsSb SL in comparison with the InAs/GaSb SL.[32] Table 3.2 compares the essential properties of both superlattice systems.

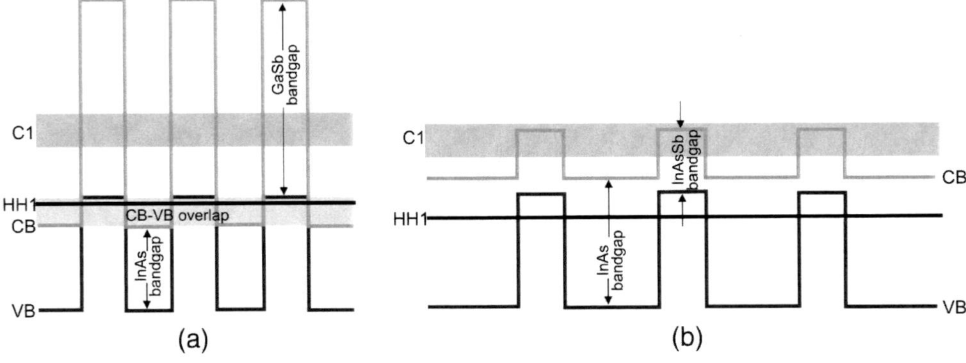

Figure 3.15 Bandgap diagram for (a) InAs/GaSb and (b) InAs/InAsSb T2SLs (reprinted from Ref. 59).

Table 3.2 Essential properties of InAs/GaSb and InAs/InAsSb superlattice systems at 77 K.

Parameter	InAs/GaSb SL	InAs/InAsSb SL
ΔE_c; ΔE_v	$\sim$930 meV; $\sim$510 meV	$\sim$142 meV; $\sim$226 meV
Background doping	$<10^{15}$ cm^{-3}	$>10^{15}$ cm^{-3}
Quantum efficiency	$\sim$50–60%	$\sim$30%
Thermal generation lifetime	$\sim$0.1 μs	$\sim$1
R_0A product ($\lambda_c = 10$ μm)	300 Ωcm^2	?
R_0A product ($\lambda_c = 5$ μm)	10^7 Ωcm^2	?
Detectivity ($\lambda_c = 10$ μm, FOV $=0$)	1×10^{12} cm Hz$^{1/2}$W^{-1}	1×10^{11} cm Hz$^{1/2}$W^{-1}

References

1. L. Esaki and R. Tsu, "Superlattice and negative conductivity in semiconductors," *IBM J. Res. Develop.* **14**, 61–65 (1970).
2. J. Tersoff, "Theory of semiconductor heterojunctions: The role of quantum dipoles," *Phys. Rev. B* **30**, 4874 (1984).
3. W. Pollard, "Valence-band discontinuities in semiconductor heterojunctions," *J. Appl. Phys.* **69**, 3154–3158 (1991).
4. H. Krömer, "Band offsets and chemical bonding: The basis for heterostucture applications," *Phys. Scr.* **68**, 10–16 (1996).
5. G. Duggan, "A critical review of semiconductor heterojunction band offsets," *J. Vac. Sci. Technol. B* **3**, 1224 (1985).
6. T. W. Hickmott, "Electrical measurements of band discontinuits at heterostructure interfaces," in *Two Dimensional Systems: Physics and New Devices*, edited by G. Bauer, F. Kuchar, and H. Heinrich, Springer Series in Solid State Sciences, Vol. **67**, Springer, Berlin, p. 72 (1986).
7. J. Menendez and A. Pinczuk, "Light scattering determinations of band offsets in semiconductor heterojunctions," *IEEE J. Quant. Electron.* **24**, 1698–1711 (1988).
8. A. Rogalski, *Infrared Detectors*, 2nd edition, CRC Press, Boca Raton, Florida (2011).
9. M. A. Kinch, *Fundamentals of Infrared Detectors*, SPIE Press, Bellingham, Washington (2007) [doi: 10.1117/3.741688].
10. G. C. Osbourn, "InAsSb strained-layer superlattices for long wavelength detector applications," *J. Vac. Sci. Technol. B* **2**, 176–178 (1984).
11. D. L. Smith and C. Mailhiot, "Proposal for strained type II superlattice infrared detectors," *J. Appl. Phys.* **62**, 2545–2548 (1987).
12. C. Mailhiot and D. L. Smith, "Long-wavelength infrared detectors based on strained InAs-GaInSb type-II superlattices," *J. Vac. Sci. Technol.* **A7**, 445–449 (1989).
13. C. H. Grein, P. M. Young, and H. Ehrenreich, "Minority carrier lifetimes in ideal InGaSb/InAs superlattice," *Appl. Phys. Lett.* **61**, 2905–2907 (1992).

14. G. C. Osbourn, "Design of III-V quantum well structures for long-wavelength detector applications," *Semicond. Sci. Technol.* **5**, S5–S11 (1990).

15. G. A. Sai-Halasz, R. Tsu, and L. Esaki, "A new semiconductor superlattice," *Appl. Phys. Lett.* **30**, 651–653 (1977).

16. G. A. Sai-Halasz, L. L. Chang, J. M. Welter, and L. Esaki, "Optical absorption of $In_{1-x}Ga_xAs$-$GaSb_{1-y}As_y$ superlattices," *Solid State Commun.* **27**, 935–937 (1978).

17. L. L. Chang, N. J. Kawai, G. A. Sai-Halasz, P. Ludeke, and L. Esaki, "Observation of semiconductor-semimetal transition in InAs/GaSb superlattices," *Appl. Phys. Lett.* **35**, 939–942 (1979).

18. A. Rogalski, *New Ternary Alloy Systems for Infrared Detectors*, SPIE Press, Bellingham, Washington (1994).

19. H. R. Jen, K. Y. Ma, and G. B. Stringfellow, "Long-range order in InAsSb," *Appl. Phys. Lett.* **54**, 1154–1156 (1989).

20. S. R. Kurtz, L. R. Dawson, R. M. Biefeld, D. M. Follstaedt, and B. L. Doyle, "Ordering-induced band-gap reduction in $InAs_{1-x}Sb_x$ $(x \approx 0.4)$ alloys and superlattices," *Phys. Rev. B* **46**, 1909–1912 (1992).

21. R. A. Stradling, S. J. Chung, C. M. Ciesla, C.J.M. Langerak, Y. B. Li, T. A. Malik, B. N. Murdin, A. G. Norman, C. C. Philips, C. R. Pidgeon, M. J. Pullin, P. J. P. Tang, and W. T. Yuen, "The evaluation and control of quantum wells and superlattices of III-V narrow gap semiconductors," *Mater. Sci. Eng. B* **44**, 260–265 (1997).

22. Y.-H. Zhang, A. Lew, E. Yu, and Y. Chen, "Microstructural properties of $InAs/InAs_xSb_{1-x}$ ordered alloys grown by modulated molecular beam epitaxy," *J. Crystal Growth* **175/176**, 833–837 (1997).

23. Y.-H. Zhang, "$InAs/InAs_xSb_{1-x}$ type-II superlattice midwave infrared lasers," in *Optoelectronic Properties of Semiconductors and Superlattices*, edited by M. O. Manasreh, Gordon and Breach Science Publishers, Amsterdam, pp. 461–500, (1997).

24. G. J. Brown, S. Elhamri, W. C. Mitchel, H. J. Haugan, K. Mahalingam, M. J. Kim, and F. Szmulowicz, "Electrical, optical, and structural studies of InAs/InGaSb VLWIR superlattices," in *The Wonder of Nanotechnology: Quantum Optoelectronic Devices and Applications*, edited by M. Razeghi, L. Esaki, and ad K. von Klitzing, SPIE Press, Bellingham, Washington, pp. 41–58 (2013) [doi: 10.1117/3.1002245.ch2].

25. E. H. Steenbergen, O. O. Cellek, H. Li, S. Liu, X. Shen, D. J. Smith, and Y.-H. Zhang, "$InAs/InAs_xSb_{1-x}$ superlattices on GaSb substrates: A promising material system for mid- and long-wavelength infrared detectors," in *The Wonder of Nanotechnology: Quantum Optoelectronic Devices and Applications*, edited by M. Razeghi, L. Esaki, and K. von Klitzing, SPIE Press, Bellingham, Washington, pp. 59–83 (2013) [doi: 10.1117/3.1002245.ch3].

26. H. Kröemer, "The 6.1 Å family (InAs, GaSb, AlSb) and its hetero-structures: a selective review," *Physica E* **20**, 196–203 (2004).

27. Y. Wei and M. Razeghi, "Modelling of type-II InAs/GaSb superlattices using an empirical tight-binding method and interface engineering," *Phys. Rev. B* **69**, 085316 (2004).

28. F. Rutz, R. Rehm, J. Schmitz, M. Wauro, J. Niemasz, J.-M. Masur, A. Wörl, M. Walther, R. Scheibner, J. Wendler, and J. Ziegler, "InAs/GaSb superlattices for high-performance infrared detection," *Sensor + Test Conferences 2011, IRS² Proceedings*, pp. 16–20 (2011).

29. E. Machowska-Podsiadlo and M. Bugajski, "Superlattices: Design of InAs/GaSb superlattices for optoelectronic applications—Basic theory and numerical methods," in *CRC Concise Encyclopedia of Nanotechnology*, edited by B. I. Kharisov, O. V. Kharissova, and U. Ortiz-Mendez, CRC Press, Boca Raton, Florida, pp. 1008–1024 (2015).

30. F. Szmulowicz, H. Haugan, and G. J. Brown, "Effect of interfaces and the spin-orbit band on the band gaps of InAs/GaSb superlattices beyond the standard envelope-function approximation," *Physical Review B* **69**(15), 155321 (2004).

31. Y. Livneh, P. C. Klipstein, O. Klin, N. Snapi, S. Grossman, A. Glozman, and E. Weiss, "**k·p** model for the energy dispersions and absorption spectra of InAs/GaSb type-II superlattices," *Physical Review B* **86**, 235311 (2012).

32. P. C. Klipstein, Y. Livneh, A. Glozman, S. Grossman, O. Klin, N. Snapi, and E. Weiss, "Modeling InAs/GaSb and InAs/InAsSb superlattice infrared detectors," *J. Electron. Materials* **43**(8), 2984–2990 (2014).

33. P. C. Klipstein, E. Avnon, Y. Benny, R. Fraenkel, A. Glozman, S. Grossman, O. Klin, L. Langoff, Y. Livneh, I. Lukomsky, M. Nitzani, L. Shkedy, I. Shtrichman, N. Snapi, A. Tuito, and E. Weiss, "InAs/GaSb type II superlattice barrier devices with a low dark current and a high-quantum efficiency," *Proc. SPIE* **9070**, 90700U (2014) [doi: 10.1117/12.2049825].

34. X. Cartoixà, D. Z. Y. Ting, and T. C. McGill, "Description of bulk inversion asymmetry in the effective bond-orbital model," *Physical Review B* **68**(23), 235319 (2003).

35. L. W. Wang, S. H. Wei, T. Mattila, A. Zunger, I. Vurgaftman, and J. R. Meyer, "Multiband coupling and electronic structure of $(InAs)_n/(GaSb)_n$ superlattices," *Physical Review B* **60**(8), 5590 (1999).

36. A. J. Williamson and A. Zunger, "InAs quantum dots: Predicted electronic structure of free-standing versus GaAs-embedded structures," *Physical Review B* **59**(24), 15819 (1999).

37. D. C. Dente and M. L. Tilton, "Comparing pseudopotential predictions for InAs/GaSb superlattices," *Physical Review B* **66**(16), 165307 (2002).

38. P. Piquini, A. Zunger, and R. Magri, "Pseudopotential calculations of band gaps and band edges of shortperiod $(InAs)_n/(GaSb)_n$ superlattices

with different substrates, layer orientations, and interfacial bonds," *Physical Review B* **77**(11), 115314 (2008).

39. P. Harrison, *Quantum Wells, Wires and Dots*, Wiley, Chichester, United Kingdom (2009).

40. G. Ariyawansa, J. M. Duran, M. Grupen, J. E. Scheihing, T. R. Nelson, and M. T. Eismann, "Multispectral imaging with type II superlattice detectors," *Proc. SPIE* **8353**, 83530E (2012) [doi: 10.1117/12.917300].

41. D. Z.-Y. Ting, A. Soibel, L. Höglund, J. Nguyen, C. J. Hill, A. Khoshakhlagh, and S. D. Gunapala, "Type-II superlattice infrared detectors," in *Semiconductors and Semimetals*, Vol. **84**, edited by S. D. Gunapala, D. R. Rhiger, and C. Jagadish, Elsevier, Amsterdam, pp. 1–57 (2011).

42. G. A. Umana-Membreno, B. Klein, H. Kala, J. Antoszewski, N. Gautam, M. N. Kutty, E. Plis, S. Krishna, and L. Faraone, "Vertical minority carrier electron transport in p-type InAs/GaSb type-II super-lattices," *Appl Phys. Lett.* **101**, 253515 (2012).

43. E. H. Steenbergen, O. O. Cellek, D. Labushev, Y. Qiu, J. M. Fastenau, A. W. K Liu, and Y.-H. Zhang, "Study of the valence band offsets between InAs and $InAs_{1-x}Sb_x$ alloys," *Proc. SPIE* **8268**, 82680K (2012) [doi: 10.1117/12.907101].

44. I. Vurgaftman, G. Belenky, Y. Lin, D. Donetsky, L. Shterengas, G. Kipshidze, W. L. Sarney, and S. P. Svensson, "Interband absorption strength in long-wave infrared type-II superlattices with small and large superlattice periods compared to bulk materials," *Appl. Phys. Lett.* **108**, 222101 (2016).

45. K. Jóźwikowski, M. Kopytko, and A. Rogalski, "Numerical estimations of carrier generation-recombination processes and the photon recycling effect in HgCdTe heterostructure photodiodes," *J. Electron. Mater.* **41**, 2766–2774 (2012).

46. E. R. Youngdale, J. R. Meyer, C. A. Hoffman, F. J. Bartoli, C. H. Grein, P. M. Young, H. Ehrenreich, R. H. Miles, and D. H. Chow, "Auger lifetime enhancement in $InAs-Ga_{1-x}In_xSb$ superlattices," *Appl. Phys. Lett.* **64**, 3160–3162 (1994).

47. C. H. Grein, P. M. Young, M. E. Flatté, and H. Ehrenreich, "Long wavelength InAs/InGaSb infrared detectors: Optimization of carrier lifetimes," *J. Appl. Phys.* **78**, 7143–7152 (1995).

48. A. Rogalski, *Infrared Detectors*, 2nd edition, CRC Press, Boca Raton, Florida (2010).

49. M. A. Kinch, F. Aqariden, D. Chandra, P.-K. Liao, H. F. Schaake, and H. D. Shih, "Minority carrier lifetime in p-HgCdTe," *J. Electron. Mater.* **34**, 880–884 (2005).

50. D. Zuo, P. Qiao, D. Wasserman, and S. L. Chuang, "Direct observation of minority carrier lifetime improvement in InAs/GaSb type-II

superlattice photodiodes via interfacial layer control," *Appl. Phys. Lett.* **102**, 141107 (2013).

51. M. Z. Tidrow, L. Zheng, and H. Barcikowski, "Recent success on SLS FPAs and MDA's new direction for development," *Proc. SPIE* **7298**, 72981O (2009) [doi: 10.1117/12.822879].
52. Y. Aytac, B. V. Olson, J. K. Kim, E. A. Shaner, S. D. Hawkins, J. F. Klem, M. E. Flatte, and T. F. Boggess, "Effects of layer thickness and alloy composition on carrier lifetimes in mid-wave infrared InAs/InAsSb superlattices," *Appl. Phys. Lett.* **105**, 022107 (2014).
53. M. A. Kinch, *State-of-the-Art Infrared Detector Technology*, SPIE Press, Bellingham, Washington (2014) [doi: 10.1117/3.1002766].
54. J. G. Pellegrino, R. DeWames, P. Perconti, C. Billman, and P. Maloney, "HOT MWIR HgCdTe performance on CZT and alternative substrates," *Proc. SPIE* **8353**, 83532X (2012) [doi: 10.1117/12.923836].
55. S. R. Jost, V. F. Meikleham, and T. H. Myers, "InSb: A key material for IR detector applications," *Mat. Res. Symp. Proc.* **90**, 429–436 (1987).
56. E. H. Steenbergen, B. C. Connelly, G. D. Metcalfe, H. Shen, M. Wraback, D. Lubyshev, Y. Qiu, J. M. Fastenau, A.W.K. Liu, S. Elhamri, O. O. Cellek, and Y.-H. Zhang, "Significantly improved minority carrier lifetime observed in a long-wavelength infrared III-V type-II superlattice comprised of InAs/InAsSb," *Appl. Phys. Lett.* **99**, 251110 (2011).
57. J. R. Meyer, C. L. Felix, W. W. Bewley, I. Vurgaftman, E. H. Aifer, L. J. Olafsen, J. R. Lindle, C. A. Hoffman, M. J. Yang, B. R. Bennett, B. V. Shanabrook, H. Lee, C. H. Lin, S. S. Pei, and R. H. Miles, "Auger coefficients in type-II InAs/Ga$_{1-x}$In$_x$Sb quantum wells," *Appl. Phys. Lett.* **73**, 2857–2859 (1998).
58. W. W. Bewley, J. R. Lindle, C. S. Kim, M. Kim, C. L. Canedy, I. Vurgaftman, and J. R. Meyer, "Lifetimes and Auger coefficients in type-II W interband cascade lasers," *Appl. Phys. Lett.* **93**, 041118 (2008).
59. G. Ariyawansa, E. Steenbergen, L. J. Bissell, J. M. Duran, J. E. Scheihing, and M. T. Eismann, "Absorption characteristics of mid-wave infrared type-II superlattices," *Proc. SPIE* **9070**, 90701J (2014) [doi: 10.1117/12.2057506].

Chapter 4
Antimonide-based Infrared Photodiodes

In the middle and late 1950s, it was discovered that InSb had the smallest energy gap of any semiconductor known at that time, and its applications as a mid-wavelength infrared detector became obvious.[1–3] The energy gap of InSb is less well matched to the 3- to 5-μm band at higher operating temperatures, and better performance can be obtained from $Hg_{1-x}Cd_xTe$. InAs is a similar compound to InSb but has a larger energy gap such that the threshold wavelength is 3–4 μm.

Indium antimonide detectors have been extensively used in high-quality detection systems and have found numerous applications in the defense and space industry for more than 50 years. One of the best known (and most successful) of these systems is the Sidewinder™ air-to-air anti-aircraft missile. The most significant recent advance in infrared technology has been the development of large 2D FPAs for use in staring arrays. Array formats are available with readouts suitable for both high-background $F/2$ operation and low-background astronomy applications. L3 Cincinnati Electronics fabricates InSb sensors of 16 Mp (4096 × 4096 pixels) currently in use by U.S. assets in overseas combat zones.

GaSb-related ternary and quaternary alloys are also established as materials for developing MWIR photodiodes for near-room-temperature operation and for the next generation of very low-loss fiber communication systems. Their current status is presented in Krier's monograph.[4]

The photoelectrical properties of narrow-gap photodiodes have been studied extensively; more details can be found in Refs. 5 and 6, especially Rogalski's monograph,[6] published in 2011, which covers the comprehensive range of subjects necessary to understand infrared detector theory and technology. Here we emphasize our consideration of achievements made in the last two decades.

4.1 Recent Progress in Binary III-V Photodiodes

4.1.1 InSb photodiodes

III-V photodiodes are generally fabricated by impurity diffusion, ion implantation, LPE, MBE, and MOCVD. Initially, p–n junctions in InSb were made by diffusing Zn or Cd into n-type substrates with a net donor concentration in the range of 10^{14}–10^{15} cm^{-3} at 77 K.[7]

At present in the InSb photodiode fabrication process, the standard manufacturing technique begins with bulk n-type single-crystal wafers with a donor concentration of about 10^{15} cm^{-3}. Relatively large, bulk-grown crystals with 6-inch diameters are available on the market. Fabrication of large array hybrid size pixels is possible because the InSb detector material is thinned (after surface passivation and hybridization to a readout chip), which allows it to accommodate the InSb/silicon thermal mismatch. As shown in Fig. 4.1(a), the backside-illuminated InSb p-on-n detector is a planar structure with an ion-implanted junction. After hybridization, epoxy is wicked between the detector and the Si ROIC, and the detector is thinned to 10 µm or less by diamond point turning. One important advantage of a thinned InSb detector is that no substrate is needed; these detectors also respond in the visible portion of the spectrum.

The best-quality InSb photodiodes are generation–recombination (GR) limited. In this limit, SRH traps created by imperfections in the

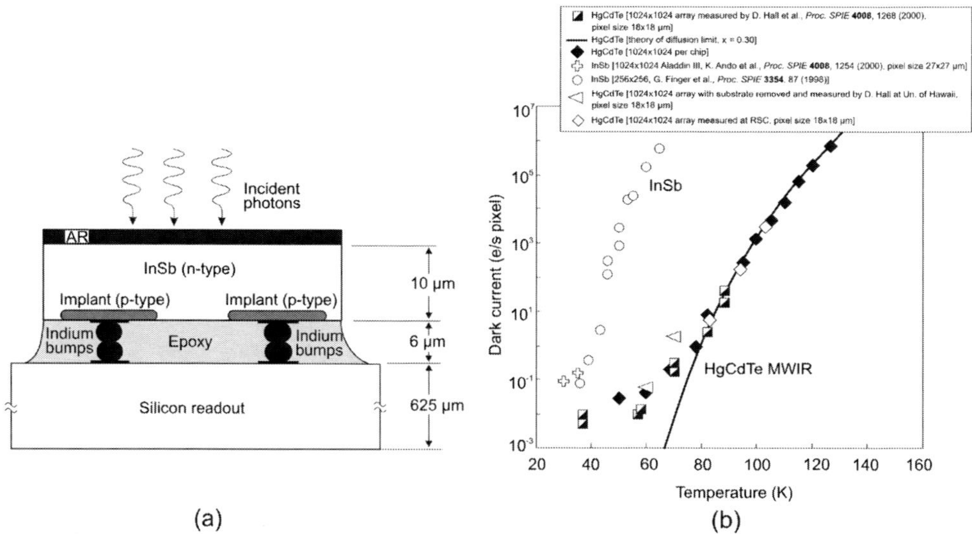

Figure 4.1 InSb photodiode: (a) architecture of an InSb sensor chip assembly (diagram adapted from Ref. 8) and (b) the comparison of dependence of dark current on temperature between the highest reported value for InSb arrays and MBE-grown HgCdTe MWIR FPAs (with 18 × 18 µm pixels) assembly (reprinted from Ref. 9).

semiconductor crystal lattice provide energy states located in the semiconductor bandgap. In standard planar technology, p-on-n junctions are created by ion implantation into n-type substrates. A new approach involving MBE growth has been adopted for reducing the dark current. With *in situ* MBE epilayer growth of p–n structures, ion implantation damage is avoided, so the diodes have a much lower concentration of GR centers than in standard planar p–n junctions.[10] The dark current is thus reduced according to the ratio of concentrations of GR centers in the standard and MBE-grown structures. After growth of high-quality epilayer homojunctions on an InSb substrate, the diodes are isolated by etching mesa structures through the p–n junctions.

An example of the improvement in dark current characteristics is shown in Fig. 4.2, where the temperature dependence of the dark current in planar ion-implanted InSb and epi-InSb FPAs with a 15-μm pitch is compared. The dark current is normalized to that at the epi-InSb photodiode operating at 95 K. The solid lines are fitted assuming GR-limited behavior with activation energy of 0.12 eV, which corresponds to approximately half the bandgap of InSb at low temperatures. Note that the same dark current is achieved at 80 K in planar InSb and 95 K in epi-InSb. The dark current has been decreased about 17× by using the MBE-based technology.

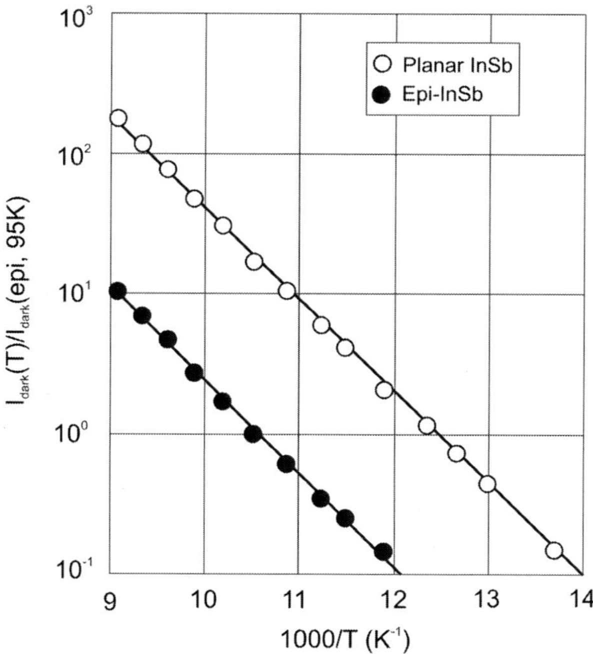

Figure 4.2 Temperature dependence of the dark currents of 15-μm-pitch InSb diodes fabricated using planar (circles) and MBE technology (solid circles). Fitting lines are calculated assuming the GR formula (adapted from Ref. 10).

4.1.2 InAs photodiodes

InAs have been mainly fabricated by ion implantation and the diffusion method.[6,7,11] Recently, high-performance InAs photodiodes have been fabricated using epitaxial techniques: LPE, MBE, and MOCVD.[12–14] Significant breakthroughs in growth and investigations of narrow-bandgap $A^{III}B^V$ hetcrostructures in the InAs-InSb-GaSb system were made at the Ioffe Physical-Technical Institute, St. Petersburg, Russia.[12,15] The performance of these photodiodes can be compared with commercially available Hamamatsu detectors (see Fig. 4.3).

The research group at the Ioffe Physical-Technical Institute has developed InAs immersion-lens photodiodes operating at near room temperatures.[12] InAs heterostructure photodiodes (see Fig. 4.4) were LPE grown onto n^+-InAs transparent substrates (due to the Burstein–Moss effect) and consisted of $\sim$3-μm-thick n-InAs layers and $\sim$3-μm-thick p-InAs$_{1-x-y}$Sb$_x$P$_y$ cladding layers that were lattice matched with InAs substrate ($y \sim 2.2x$). Due to an energy step at the n^+-InAs/n-InAs interface, a beneficial hole confinement for the photodiode operation was expected. Flip-chip mesa devices with a diameter of 280 μm were processed by a multistage wet photolithography process. Cathode and anode contacts were formed by sputtering of Cr, Ni, Au(Te), and Cr, Ni, and Au(Zn) metals followed by an

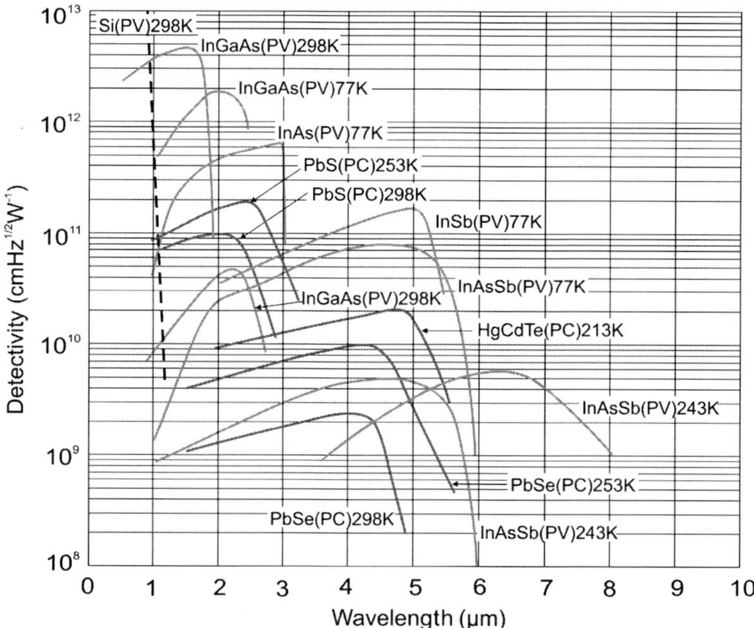

Figure 4.3 Spectral detectivity curves of commercially available infrared detectors operating at different temperatures (PC – photoconductive detectors, PV – photovoltaic detectors).

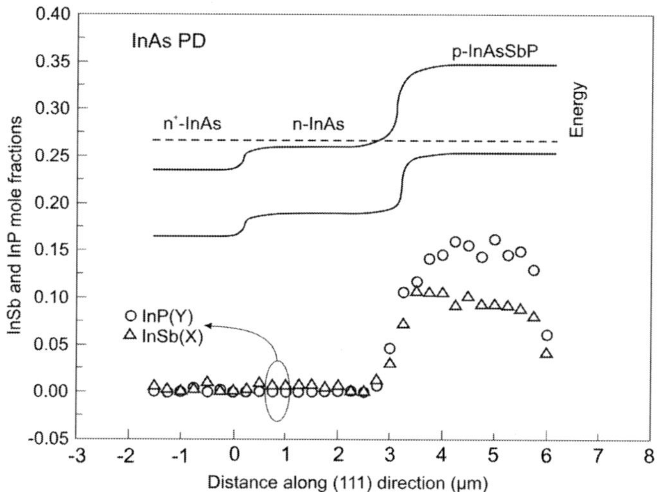

Figure 4.4 Alloy composition in an InAs photodiode structure (composition profile) together with the band diagram (reprinted from Ref. 12).

electrochemical deposition of a 1- to 2-µm thick gold layer. Next, the substrate was thinned down to 150-µm, and chips were soldered onto silicon submounts with Pb-Sn contact pads. Finally, the 3.5-mm-wide silicon lens was attached to the substrate side of a chip by a chalcogenide glass with a high refractive index ($n = 2.4$); see Fig. 4.5(a). It is obvious that the FOV of immersion photodiodes is considerably lower (decreased to 15 deg) than for an uncoated device.

Figure 4.5(b) shows the detectivity spectra of an InAs heterostructure immersion photodiode. Superior detectivity of this photodiode in comparison to commercial Hamamatsu and Judson devices reflects improvements

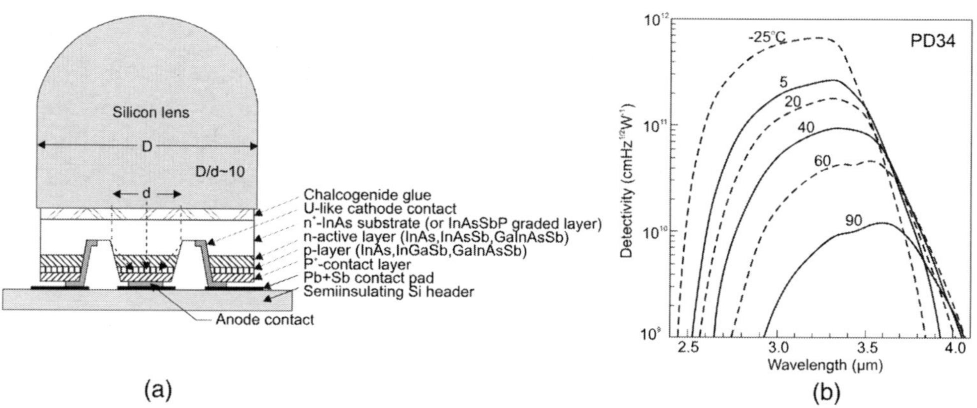

Figure 4.5 InAs heterostructure immersion photodiode: (a) construction of the immersion photodiode and (b) detectivity spectra at near room temperatures (adapted from Ref. 12).

associated with board mirror contact, asymmetric doping, immersion effect, and radiation collection by inclined mesa walls. The narrow spectral responses are a result of filtering in the substrate and intermediate layers. Peak wavelengths shift to long wavelengths as temperature is increased due to bandgap narrowing at higher temperatures. However, the short-wavelength spectra are more sensitive to temperature than the long-wavelength spectra, probably because of the progressively poor transparency of n^+-InAs near the absorption edge at elevated temperatures due to elimination of the conduction band electron degeneration.[12]

4.1.3 InAs avalanche photodiodes

For high-sensitivity applications, semiconductor avalanche photodiode detectors (APD) are most useful because they provide internal gain on the detectors. APDs are most commonly used for communication and active-sensing applications. According to the local-field model,[16] the noise power spectral density for mean gain $\langle M \rangle$ and mean photocurrent $\langle I_{ph} \rangle$ is given by the expression

$$\langle I_n^2 \rangle = 2qI_{ph}\langle M \rangle^2 F(M), \tag{4.1}$$

where $F(M)$ is the excess noise factor, which arises from the random nature of impact ionization. Under the conditions of uniform electric fields and pure electron injection, the excess noise factor can be expressed in terms of $\langle M \rangle$ and the ratio of the hole ionization coefficients α_h to the electron ionization coefficients α_e ($k = \alpha_h/\alpha_e$) as

$$F_e(M_e) = k\langle M_e \rangle + (1 - k)\left(2 - \frac{1}{\langle M_e \rangle}\right). \tag{4.2}$$

The electron and hole ionization rates are defined as the number of ionizing collisions per unit distance. The ionization rates depend strongly (exponentially) on the threshold electric fields required to overcome different carrier-scattering effects. Examples of the experimental ionization rates at 300 K for different material systems versus electric field are given in Fig. 4.6. The ionization rates can be equal for electrons and holes, as in the case of GaP, or they can significantly differ from each other, as in the cases of Si, Ge, and most compound semiconductors.

Currently, the following materials have proved to be appropriate for the fabrication of high-performance APDs:

- silicon (for wavelengths of 0.4 to 1.1 μm). The electron ionization rate is much higher than the hole ionization rate ($\alpha_e \gg \alpha_h$);

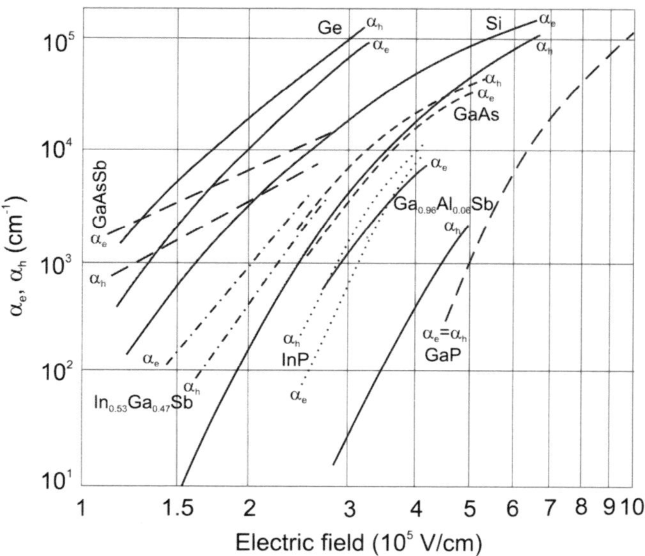

Figure 4.6 Ionization coefficients of electrons α_e and holes α_h as a function of the electric field for some semiconductors used in avalanche photodiodes.

- germanium (for the wavelengths of up to 1.65 μm). Since the bandgap in Ge is lower than in Si, and the ionization rates for electrons and holes are approximately equal ($\alpha_e \approx \alpha_h$), the noise is considerably higher, which limits the applications of Ge-based APDs;
- GaAs-based devices. Most compound materials have $\alpha_e \approx \alpha_h$, so designers usually use heterostructures such as GaAs/Al$_{0.45}$Ga$_{0.55}$As, for which α_e(GaAs) $\gg \alpha_e$(AlGaAs). The large increase in gain occurs due to the avalanche effect that occurs in GaAs layers. GaAs/Al$_{0.45}$Ga$_{0.55}$As heterostructures are in the spectral range below 0.9 μm. Applying InGaAs layers allows the sensitivity to extend to $\approx$1.4 μm;
- InP-based devices used in the wavelength range of 1.2–1.6 μm. In the lattice-matched double heterostructure n$^+$-InP/n-GaInAsP/p-GaInAsP/ p$^+$-InP, either of the carriers are injected into the high-field region—this structure is essential for low-noise operation. The second structure, p$^+$-InP/ n-InP/n-InGaAsP/n$^+$-InP, is similar to a Si reach-through device. The absorption occurs in the relatively wide InGaAsP layers, and avalanche multiplication of the minority carriers proceeds in the n-InP layer;
- Hg$_{1-x}$Cd$_x$Te APDs. These devices are electron-initiated, and their operation has been demonstrated for a board range of compositions from $x = 0.7$ to 0.21, corresponding to cutoff wavelengths from 1.3 μm to 11 μm. Thus, HgCdTe APD at gain $= 100$ provides 10 to 20 times less noise than InGaAs or InAlAs APDs and 4 times less noise than Si APDs.

Beck et al. were the first to report APD characteristics consistent with $k = 0$ in 2001,[17] when they published results from Hg$_{0.7}$Cd$_{0.3}$Te APDs. They have since shown that for a number of compositions, the impact ionization for holes remains essentially zero in $Hg_{1-x}Cd_xTe$ APDs detecting in the short-, mid-, and longwave infrared.[18] They coined the phrase electron-APD (e-APD) to describe such APDs where only electrons undergo impact ionization. In this case, the excess noise factor is <2 and independent of gain. However, low-bandgap HgCdTe also results in relatively high dark current at room temperature, which necessitates operation at low temperatures.

New breakthrough in the development of InAs APDs has been obtained recently.[19,20] Similarly to HgCdTe devices, InAs APDs have also demonstrated $k \approx 0$ with moderately low dark current at room temperature. It has been shown that in InAs p–i–n diodes, significant electron-initiated multiplication can be achieved, while hole-initiated multiplication in InAs n-i-p diodes remains negligible across the same electric field range.[21]

Figure 4.7(a) shows an InAs APD MBE-grown mesa structure with an i-region as thick as 6 μm. Beryllium and silicon were used as acceptors and donors, respectively. The n-type background doping concentration in the i-region was below 1×10^{15} cm^{-3}. To suppress surface leakage current, devices with diameters up to 500 μm were wet etched using 1:1:1 (phosphoric acid, hydrogen peroxide, deionized water), followed by 30 s of etching in 1:8:80 (sulphuric acid, hydrogen peroxide, deionized water). The etched mesa sidewalls were additionally covered with SU-8 passivation. Good ohmic contact is easily formed by depositing Ti/Au (20/150 nm) without annealing.

Figure 4.7(b) shows that the measured gain in InAs APDs increases exponentially with reverse bias, showing no sign of breakdown, which is a signature of $k \approx 0$.[22,23] Devices yielded room-temperature multiplication gains >300.[20] The bandwidth is transit-time limited in the range 2–3 GHz independent of gain.

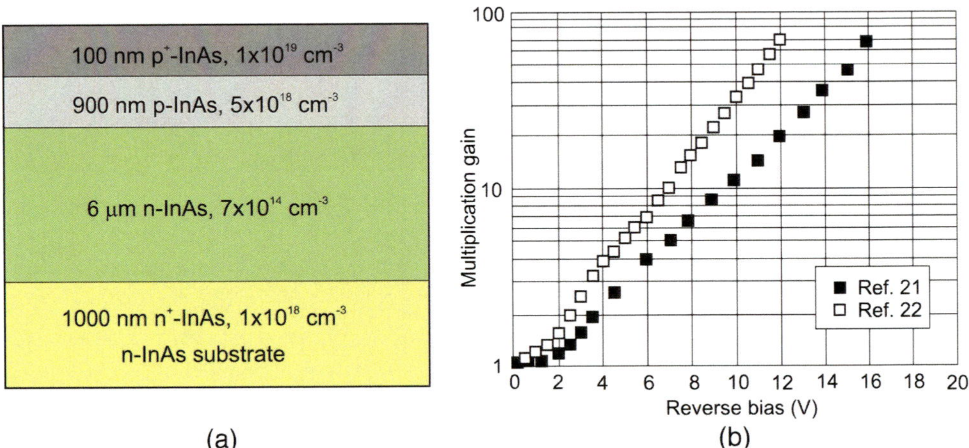

Figure 4.7 InAs avalanche photodiode: (a) mesa structure with unintentionally doped i-region and (b) measured gain.

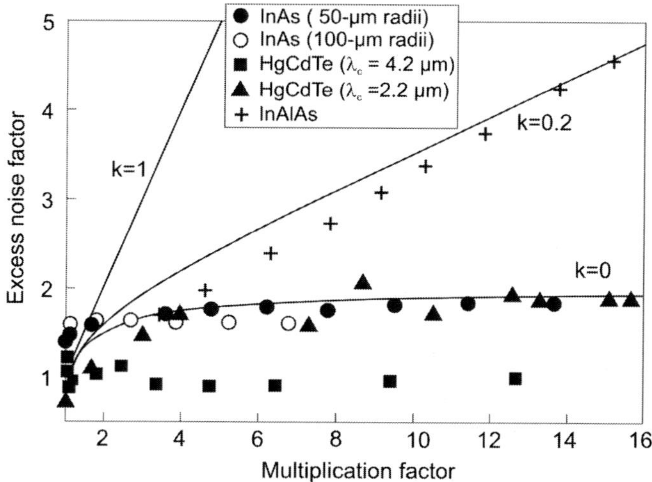

Figure 4.8 A comparison among the F_e values reported on APDs of different materials, including InAs diodes with a 3.5-µm intrinsic width and radii of 50 µm and 100 µm, HgCdTe photodiodes with cutoff wavelengths of 4.2 µm and 2.2 µm, and an InAlAs diode (reprinted from Ref. 19).

Figure 4.8 shows the excess noise factor versus multiplication factor reported in APDs of different materials. The F_e measured on InAs p-i-n diodes falls slightly below the local-field model prediction for $k = 0$ as given by Eq. (4.2). This is comparable to the F_e value reported for SWIR HgCdTe e-ADPs, although it is somewhat higher than that reported for MWIR HgCdTe e-APDs. To explain such excess noise below the lower limit case of the local-field model, it is necessary to consider the influence of "dead space"— the distance in which no impact ionization occurs for electrons—which is neglected from the local-field model. If the multiplication region is thick, the dead space can be neglected and the local field model provides an accurate description of the APD characteristics. The excess noise in the InAlAs APD rises with increasing multiplication, as occurs for all conventional APDs in which both carriers undergo impact ionization.

The realization of InAs e-APDs has brought the ideal avalanche multiplication and excess noise characteristics into the readily available III-V material system for a more widespread application previously only achievable in the less-readily available HgCdTe system. These properties make InAs APDs attractive for a number of near- and mid-infrared sensing applications, including remote gas sensing, light detection and ranging (LIDAR), and both active and passive imaging.

4.2 InAsSb Bulk Photodiodes

Ternary and quaternary III-V compound materials are suitable for fabricating optoelectronic devices in the near- and mid-infrared wavelength range. The availability of binary substrates such as InAs and GaSb allows the growth of multilayer homo- and heterostructures, where lattice-matched ternary and quaternary layers can be tailored to detect wavelengths in the range of 0.8 to 5 μm. The bandgap of $Ga_xIn_{1-x}As_ySb_{1-y}$ can be continuously tuned from about 475 to 730 meV while remaining lattice matched to a GaSb substrate[24,25] (as shown in Fig. 4.9), in contrast to leading ternary materials in this range such as InGaAs on InP. Both ternary (InGaSb and InAsSb) and quaternary (InGaAsSb and AlGaAsSb) materials indicate good performance for a wavelength range ≥2 μm; however, the research is still underway, and they are yet commercially available. The availability of ternary InGaSb virtual substrates has a promising potential for developing high-performance detectors[26–28] without the influence of the binary substrates generally used for processing the ternary materials.

4.2.1 Technology and properties

A variety of InAsSb photodiode configurations have been proposed, including mesa and planar, n-p, n-p$^+$, p$^+$-n, and p-i-n structures. The techniques used to form p–n junctions include diffusion of Zn, Be ion implantation, and the creation of p-type layers on n-type material by LPE, MBE, and MOCVD. The photodiode technology essentially relies on n-type material with concentrations generally about 10^{16} cm^{-3}. A summary of the research on the fabrication of InAsSb photodiodes in the period before 2010 is given in Rogalski's monograph.[6]

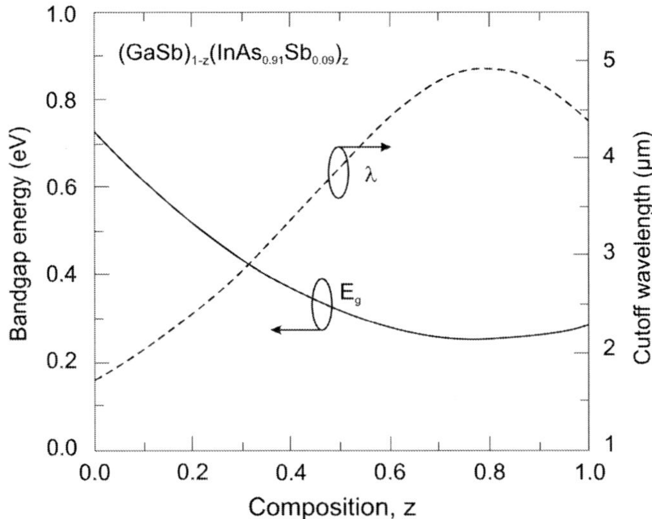

Figure 4.9   The bandgap of $Ga_xIn_{1-x}As_ySb_{1-y}$ with *x* and *y* concentrations chosen in the ratio $(GaSb)_{1-z}(InAs_{0.91}Sb_{0.09})_z$ can be continuously tuned from about 475 to 730 meV while remaining lattice matched to a GaSb substrate.

A significant development in InAsSb photodiode research was made in 1980 when lattice-matched $InAs_{1-x}Sb_x$/GaSb $(0.09 \leq x \leq 0.15)$ device structures were used.[11] Lattice mismatch up to 0.25% for the $InAs_{0.86}Sb_{0.14}$ epitaxial layer was accommodated in terms of low etch-pit density $(\approx 10^4 \text{ cm}^{-2})$. The structure of a backside-illuminated $InAs_{1-x}Sb_x$/GaSb photodiode is shown in Fig. 4.10(a). The photons enter through the GaSb transparent substrate and reach the $InAs_{1-x}Sb_x$ active layer, where they are absorbed. The GaSb substrate determines the short-wavelength cut-on value, which is 1.7 µm at 77 K, whereas the active region establishes the long-wavelength cutoff value [see Fig. 4.10(b)]. The p–n junctions were obtained as homojunctions using the LPE technique. The carrier concentrations, both in the undoped n-type layer and in the Zn-doped p-type layer, were approximately 10^{16} cm^{-3}. High-quality $InAs_{0.86}Sb_{0.14}$ photodiodes were demonstrated by a high R_0A product in excess of 10^9 Ωcm^2 at 77 K.

In order to improve device performance (lower dark current and higher detectivity), several groups have developed P-i-N heterostructure devices comprising an unintentionally doped InAsSb active layer sandwiched between P and N layers of larger-bandgap materials. The lower minority-carrier concentration in the high-bandgap layers results in a lower diffusion dark current and higher R_0A product and detectivity. Figure 4.11 shows a schematic band diagram of the N-i-P double-heterostructure antimonide-based III-V photodiodes together with the different combinations of active and cladding layers in the device structure. Depending on contact configurations and transparency of substrates, both backside and frontside illumination

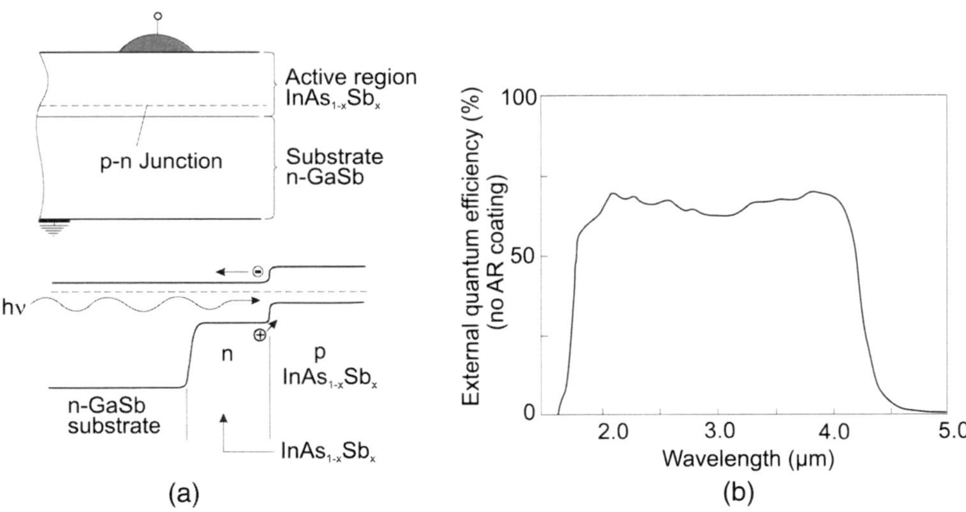

Figure 4.10 Backside-illuminated $InAs_{0.86}Sb_{0.14}$/GaSb photodiode: (a) device structure and energy-band diagram of the structure, and (b) spectral response at 77 K (reprinted from Ref. 11 with permission from AIP Publishing).

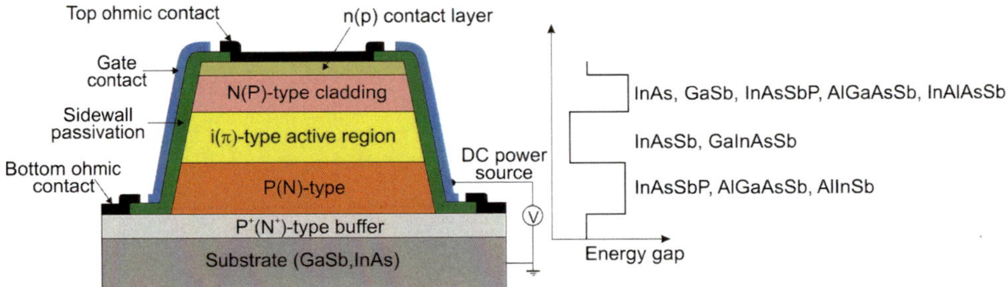

Figure 4.11   Schematic band diagram of the N-i-P double-heterostructure-antimonide-based III-V photodiodes. Different combinations of active and cladding layers are also shown on the right.

can be used. Usually p-type GaSb and n-type InAs are used. Despite the relatively low absorption coefficients, substrates require thinning to small thicknesses, even less than 10 μm. InAs is fragile, and many fabrication processes are not possible. This obstacle can be overcome by using heavily doped n^+-InAs substrates where strong degeneracy of the electrons in the conduction band occurs at relatively low electron concentration ($>10^{17}$ cm^{-3}). For example, the Burstein–Moss shift in heavily doped n^+-InAs ($n = 6 \times 10^{18}$ cm^{-3}) makes the corresponding substrates transparent to the 3.3 μm wavelength.[12]

Antimonide-based ternary and quaternary alloys are well established as materials for development of MWIR photodiodes for near-room-temperature operation. Many articles have discussed the properties of mid-IR photodiodes, and many of the investigations were made in the Ioffe Institute; see, for example, Refs. 12, 15, and 29. These LPE-grown heterostructure devices consist of n-type InAs(100) substrates with $n = 2 \times 10^{16}$ cm^{-3} (for undoped) or $n^+ = 2 \times 10^{18}$ cm^{-3} (for Sn-doped), ≈10-μm-thick undoped n-InAs$_{1-x}$Sb$_x$ active layers; and, finally, p-InAs$_{1-x-y}$Sb$_x$P$_y$(Zn) claddings (contact layer) (see Fig. 4.12). The narrowgap InAsSb active layer is surrounded by semiconductors with wider energy gaps.

In device processing, standard optical photolithography and wet chemical etching processes were implemented to obtain 26-μm-high circular mesas ($\varnothing_m = 190$ μm) and 55-μm-deep grooves for separation of the 580×430 μm rectangular chips. Next, circular Au- or Ag-based reflective anode ($\varnothing_a = 170$ μm) and cathode contacts were formed on the same chip side by sputtering and thermal evaporation in vacuum followed by 3-μm-thick gold-plating deposition. Finally, a flip-chip bonding/packaging procedure was implemented using the 1800×900 μm submount made from a semi-insulating Si wafer with Pb-Sn bonding pads. Photodiode chips were mounted upside down, with the n^+-InAs side being an "entrance window" for the incoming radiation, as shown in the insert in Fig. 4.12. Some chips were equipped with aplanatic hyperhemispherical Si immersion lenses ($\varnothing = 3.5$ mm)

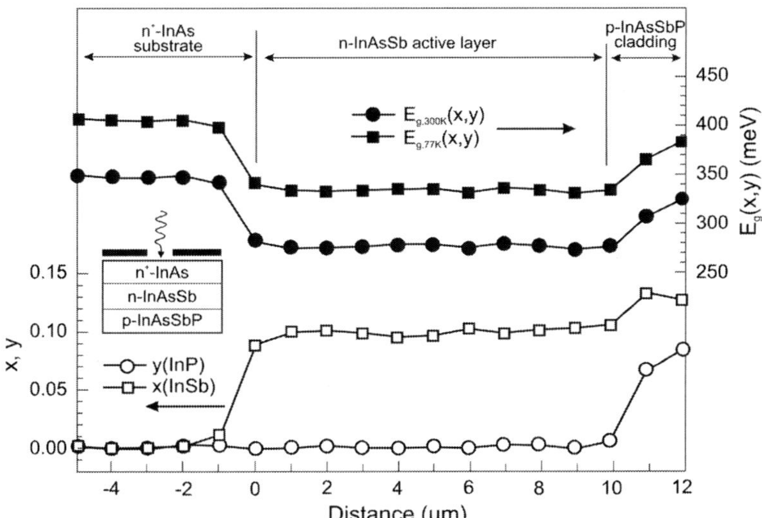

Figure 4.12 Alloy composition and energy bandgap structure versus distance for an n$^+$-InAs/n-InAsSb/p-InAsSbP double-heterostructure photodiode (reprinted from Ref. 15).

with antireflection coating using a chalcogenide glass as an optical glue between the Si and the n$^+$-InAs (or n-InAs). The final construction of the immersion photodiode is similar to that shown in Fig. 4.5(a).

Figure 4.13 summarizes experimental data of zero-bias resistivity and R_0A product versus photon energy for n$^+$-InAs/n-InAsSb/p-InAsSbP double-heterostructure photodiodes. An exponential dependence of R_0A product, approximated by exp(E_g/kT), indicates that the diffusion current determines the transport properties of the heterojunctions with negligible leakage current

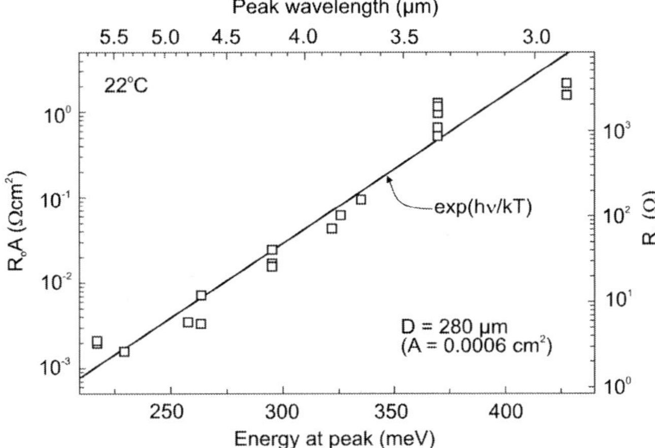

Figure 4.13 Plot of R_0A product versus photon energy for a series of InAsSb double-heterostructure photodiodes at room temperature (adapted from Ref. 12).

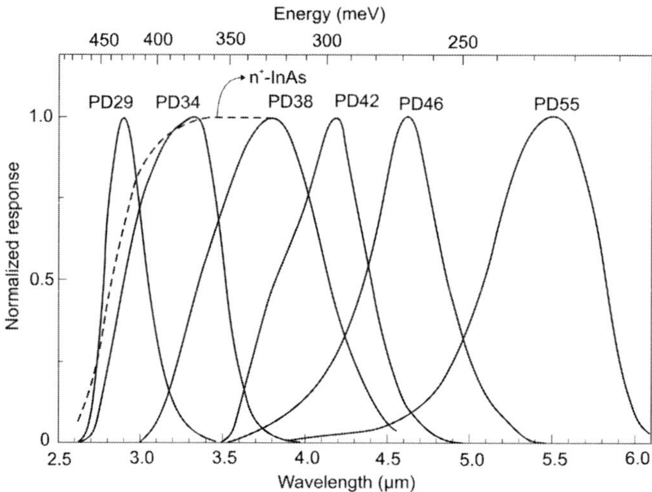

Figure 4.14 Room-temperature spectral response of InAsSb double-heterostucture photodiodes and normalized transmission of 175-μm-thick n^+-InAs [$n^+ = (3–6) \times 10^{18}$ cm^{-3}] (adapted from Ref. 12).

flow mechanisms at $T > 190$ K. At lower temperatures, GR at high bias and tunnelling at low bias prevail.

Figure 4.14 shows normalized spectral responsivity curves for InAsSb double-heterostructure photodiodes (PDs) with different cutoff wavelengths at room temperature. The photodiodes are characterized by a narrow spectral response (with FWHM of about 0.3–0.8 μm) resulting from spectral filtering in the substrate and intermediate layers. The narrowest response of the PD29 device is due to poor transmission of n^+-InAs at short wavelengths. On the other hand, the spectral response of the longwave PDs exhibits a broad shoulder at short wavelengths that originates from the diffusion of carriers created in high-energy InAsSbP regions towards the narrow-gap p–n junction.

Figure 4.15 shows the Johnson-limited detectivity spectra of InAsSb double-heterostucture photodiodes at different temperatures. As seen from this figure, the responsivity spectra bears four distinct regions: (1) the cutoff region (4.7 < λ < 5.5 μm), (2) the sharp longwave response decline region, (3) the smooth response decline region, and (4) the fast shortwave response decline region. Region 4 is due to transmission degradation in heavily doped n^+-InAs substrate. In this case the Moss–Burstein effect associated absorption edge shift in n^+-InAs is as large as 1 μm (note the difference in short-wavelength spectral responsivity between the heavily doped sample #878 and the undoped sample #877).

Figure 4.16(a) summarizes the peak detectivity of photodiodes depending on temperature and wavelength. Figure 4.16(b) presents detectivities depending peak wavelength for the backside illuminated (BSI) and the coated photodiodes [with immersion lens (IL)]. For the immersion lens photodiodes,

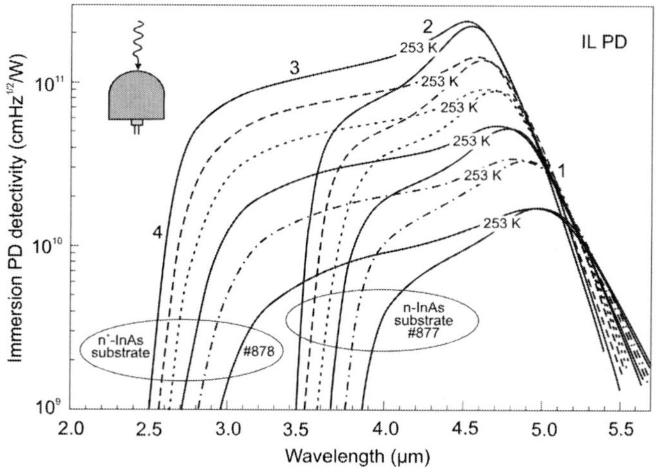

Figure 4.15 Detectivity spectra at different temperatures for InAsSb double-heterostucture photodiodes with n-InAs and n^+-InAs substrates (reprinted from Ref. 15).

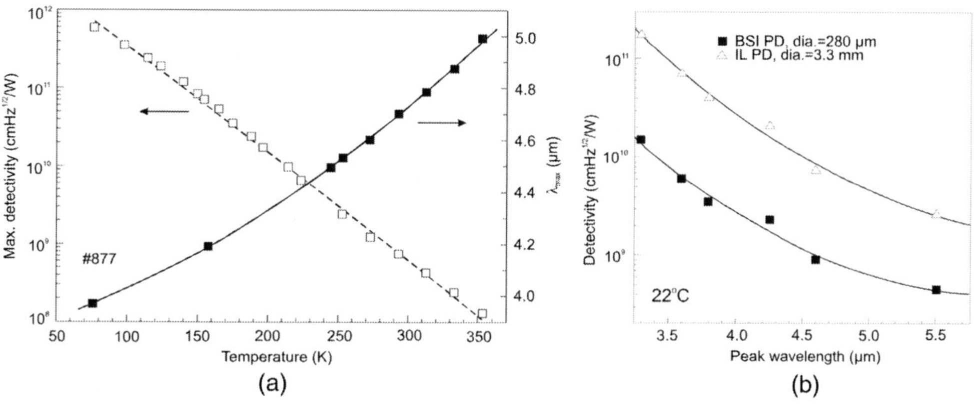

Figure 4.16 Detectivity of InAsSb double-heterostucture photodiodes: (a) maximum detectivity and peak wavelength versus temperature (adapted from Ref. 15); (b) spectral peak detectivity of photodiodes without (back-side illuminated – SI) and with Si lenses (immersion illuminated – IL) (adapted from Ref. 12).

the peak detectivities are generally about one decade higher than those for bare chip PDs; see Fig. 4.16(b). The photodiodes developed at the Ioffe Physical-Technical Institute are superior to the majority of those published in the literature. At the same time, the achievable R_0A products are lower than that given in Ref. 11.

4.2.2 Performance limits

Despite the promise of the III-V-based detectors, HgCdTe remains the highest-performing IR material technology for a number of applications. The main obstacles in the rapid development of InAsSb photodiodes are difficulties in the preparation of single crystals and epitaxial layers. During the last 20 years high-quality InAsSb photodiodes for the 3- to 5-μm spectral region have been developed. However, a long-standing hope that InAsSb might become a useful material in the 8- to 12-μm spectral band has not been realized to date. In comparison with HgCdTe, the main obstacles to achieving this goal are: poor-quality crystal structure (there is no ideal III-V substrate/epitaxial combination that is appropriate for LWIR spectral band), poor SRH lifetimes, and relatively high background carrier concentration (above 10^{15} cm^{-3}). Moreover, the observed energy gaps of InAs$_{1-x}$Sb$_x$ alloys with middle Sb compositions [close to minimum energy gap (see Fig. 2.11)] are not inherent and well controlled due to CuP-type ordering and residual strain effects.

In 1996 Rogalski et al. reported a theoretical analysis of MWIR InAs$_{1-x}$Sb$_x$ ($0 \leq x \leq 0.4$) photodiodes with operation extending to the temperature range of 200–300 K.[30] It has been shown that the theoretical performance of high-temperature InAsSb photodiodes is comparable to that of HgCdTe photodiodes. Figure 4.17 presents a theoretical limit to the R_0A product for p$^+$-n InAsSb photodiodes operated at 200 and 300 K in a spectral

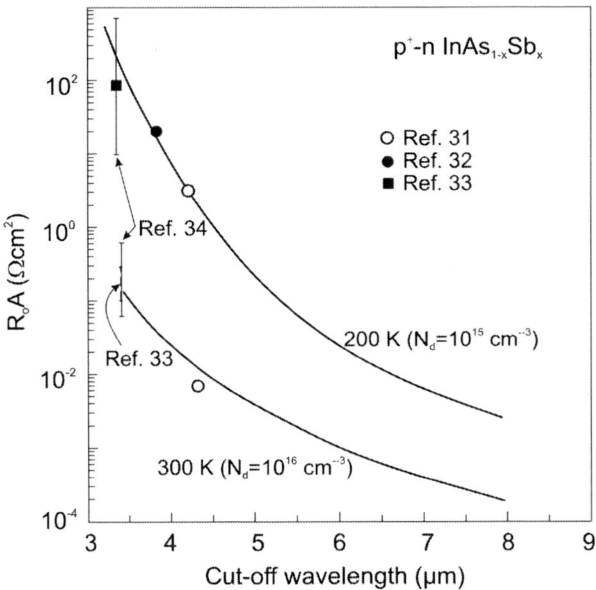

Figure 4.17 The dependence of the R_0A product on the long-wavelength cutoff for p$^+$–n InAs$_{1-x}$Sb$_x$ photodiodes at temperatures of 200 and 300 K. The curves are calculated for two doping concentrations (10^{15} cm^{-3} and 10^{16} cm^{-3}) in the base region of the photodiode with thickness of 15 μm; the carrier concentration of 10^{18} cm^{-3} in the p$^+$-cap layer with a thickness of 1 μm is assumed. The experimental data are taken from Refs. 31 (circles), 32 (solid circles), 33 (solid squares), and 34 (reprinted from Ref. 30).

range between 3 and 8 μm when the doping level in the base n-type layer (10^{15} cm^{-3} and 10^{16} cm^{-3}) is close to the level available in practice. For comparison with theoretical predictions, only limited experimental data is marked. The agreement is satisfactory.

More recently, Wróbel et al. considered the effects of doping profiles on room-temperature MWIR InAsSb photodiode parameters (R_0A product and detectivity).[35] In theoretical estimations, a new insight into composition dependence of spin–orbit splitting bandgap energy is taken into account (Fig. 2.13).[40]

Figure 4.18 shows the dependence of the R_0A product on the long-wavelength cutoff for InAs$_{1-x}$Sb$_x$ photodiodes at 300 K. The theoretical lines are calculated for doping concentration of 10^{16} cm^{-3} in both p-on-n and n-on-p 5-μm-thick active regions. The figure clearly shows that the influence of Auger S mechanisms in n-on-p photodiodes with an active-region composition close to that of InAs ($0 \leq x \leq 0.15$; $\lambda_c < 4.5$ μm) considerably decreases the R_0A product in comparison with p-on-n devices, where the influence of Auger S mechanisms is eliminated. However, if the composition of the active region is $x \geq 0.15$ ($\lambda_c > 4.5$ μm), the structure based on p-type material is more optimal than p-on-n structure.

The theoretically predicted performance is comparable with the experimental values of p-on-n photodiodes with an n-type active region. The agreement between both types of data is good. Some of experimental data are located above the theoretical line. It is observed if the thickness of the photodiode active region is smaller than the minority-carrier diffusion length.

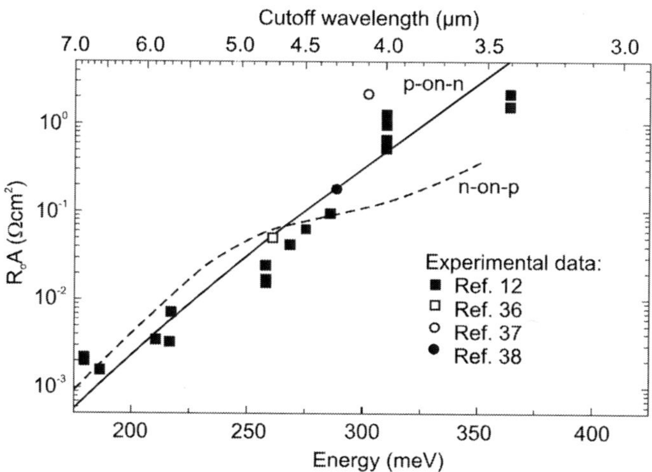

Figure 4.18 The dependence of the R_0A product on the long-wavelength cutoff for InAs$_{1-x}$Sb$_x$ photodiodes at room temperature. The theoretical lines are calculated for a doping concentration of 10^{16} cm^{-3} in both p-on-n and n-on-p 5-μm-thick active regions. The gathered experimental data concerns p-on-n photodiodes with an n-type active region (reprinted from Ref. 35).

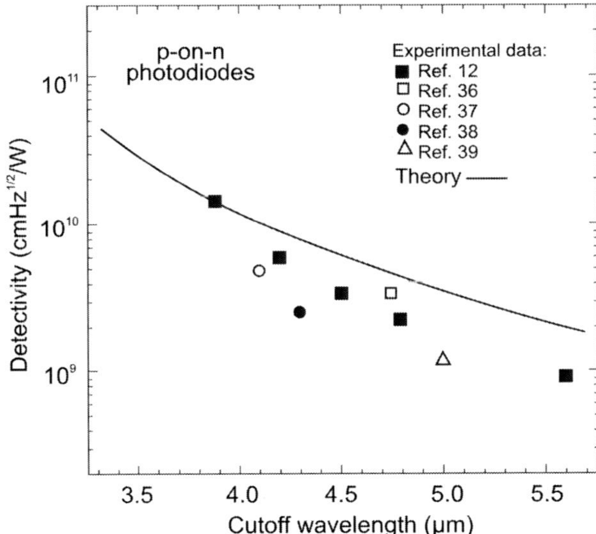

Figure 4.19   Dependence of detectivity on the long-wavelength cutoff for p-on-n InAs$_{1-x}$Sb$_x$ photodiodes at room temperature. The theoretical line is calculated for 5-µm-thick active region with a doping concentration of 10^{16} cm^{-3}. The gathered experimental data are taken from the indicated literature (reprinted from Ref. 35).

Reducing the volume in which the diffusion current is generated causes the corresponding dark current to decrease and the R_0A product to increase.

Figure 4.19 presents the spectral dependence of thermal-noise-limited detectivity of p-on-n InAsSb photodiodes operated at room temperature. The theoretical line is calculated using the formula

$$D^* = \frac{\eta \lambda q}{2hc}\left(\frac{R_0 A}{kT}\right)^{1/2},$$

(4.3)

assuming quantum efficiency $\eta = 0.7$ and a doping concentration in the active region of 10^{16} cm^{-3}.

This figure shows that the upper detectivity experimental data coincides well with the theoretical prediction. The discrepancy between both types of results increases with cutoff wavelength increase, which is mainly caused by decreasing the experimentally measured quantum efficiency.

InAs$_{1-x}$Sb$_x$ material is potentially capable of operating at the longest cutoff wavelength (≈ 12.0 µm at 77 K) of the entire III-V alloy family. To realize IR detectors in all of the potential operating regions, lattice-matched substrates are necessary. This problem seems to be resolved by using Ga$_{1-x}$In$_x$Sb substrates. In this case, the lattice parameter can be tuned to between 6.095 Å (GaSb) and 6.479 Å (InSb). Several research groups have succeeded in growing GaInSb single crystals. One composition worth noting is

$Ga_{0.38}In_{0.62}Sb$, which is lattice matched to $InAs_{0.35}Sb_{0.65}$, which has a bandgap minimum.

Recently, bulk unrelaxed InAsSb alloys with Sb compositions up to 65% were MBE metamorphically grown on compositionally graded GaInSb and AlInSb buffers on GaSb substrates.[41] The graded buffer layer had a total thickness of up to 3.5 μm. The lattice constant of InAsSb layers equaled the lateral lattice constant at the top of the buffer layer, resulting in a low residual strain (<0.1%). As a result, unrelaxed InAsSb epilayers grown on graded buffer layers were found to be ordering-free with random a distribution of group-V atoms.

References

1. C. Hilsum and A. C. Rose-Innes, *Semiconducting III-V Compounds*, Pergamon Press, Oxford (1961).
2. O. Madelung, *Physics of III-V Compounds*, Wiley, New York (1964).
3. T. S. Moss, G. J. Burrel, and B. Ellis, *Semiconductor Optoelectronics*, Butterworths, London (1973).
4. *Mid-Infrared Semiconductor Optoelectronics*, edited by A. Krier, Springer, London (2006).
5. W. F. M. Micklethwaite and A. J. Johnson, "InSb: Materials and devices," in *Infrared Detectors and Emitters: Materials and Devices*, edited by P. Capper and C. T. Elliott, Kluwer Academic Publishers, Boston, pp. 178–204 (2001).
6. A. Rogalski, *Infrared Detectors*, 2nd edition, CRC Press, Boca Raton, Florida (2010).
7. T. S. Moss, G. J. Burrel, and B. Ellis, *Semiconductor Optoelectronics*, Butterworths, London (1973).
8. P. J. Love, K. J. Ando, R. E. Bornfreund, E. Corrales, R. E. Mills, J. R. Cripe, N. A. Lum, J. P. Rosbeck, and M. S. Smith, "Large-format infrared arrays for future space and ground-based astronomy applications," *Proc. SPIE* **4486**, 373 (2002) [doi: 10.1117/12.455119].
9. M. Zandian, J. D. Garnett, R. E. DeWames, M. Carmody, J. G. Pasko, M. Farris, C. A. Cabelli, D. E. Cooper, G. Hildebrandt, J. Chow, J. M. Arias, K. Vural, and D. N. B. Hall, "Mid-wavelength infrared p-on-on $Hg_{1-x}Cd_xTe$ heterostructure detectors: 30–120 Kelvin state-of-the-art performance," *J. Electron. Mater.* **32**, 803–809 (2003).
10. I. Shtrichman, D. Aronov, M. Ben Ezra, I. Barkai, E. Berkowicz, M. Brumer, R. Fraenkel, A. Glozman, S. Grossman, E. Jacobsohn, O. Klin, P. Klipstein, I. Lukomsky, L. Shkedy, N. Snapi, M. Yassen, and E. Weiss, "High operating temperature epi-InSb and XBn-InAsSb photodetectors," *Proc. SPIE* **8353**, 83532Y (2012) [doi: 10.1117/12. 918324].

11. L. O. Bubulac, A. M. Andrews, E. R. Gertner, and D. T. Cheung, "Backside-illuminated InAsSb/GaSb broadband detectors," *Appl. Phys. Lett.* **36**, 734–736 (1980).

12. M. A. Remennyy, B. A. Matveev, N. V. Zotova, S. A. Karandashev, N. M. Stus, and N. D. Ilinskaya, "InAs and InAs(Sb)(P) (3–5 μm) immersion lens photodiodes for potable optic sensors," *Proc. SPIE* **6585**, 658504 (2007) [doi: 10.1117/12.722847].

13. C. H. Kuan, R. M. Lin, S. F. Tang, and T. P. Sun, "Analysis of the dark current in the bulk of InAs diode detectors," *J. Appl. Phys.* **80**, 5454–5458 (1996).

14. X. Zhou, J. S. Ng, and C. H. Tan, "InAs photodiode for low temperature sensing," *Proc. SPIE* **9639**, 96390V (2015) [doi: 10.1117/12.2197343].

15. P. N. Brunkov, N. D. Il'inskaya, S. A. Karandashev, A. A. Lavrov, B. A. Matveev, M. A. Remennyi, N. M. Stus, and A. A. Usikov, "InAsSbP/InAs$_{0.9}$Sb$_{0.1}$/InAs DH photodiodes ($\lambda_{0.1} = 5.2$ μm, 300 K) operating in the 77–353 K temperature range," *Infrared Phys. Technol.* **73**, 232–237 (2015).

16. R. J. McIntryre, "Multiplication noise in uniform avalanche diodes," *IEEE Trans. Electron Devices* **ED-13**, 164–168 (1966).

17. J. D. Beck, C.-F. Wan, M. A. Kinch, and J. E. Robinson, "MWIR HgCdTe avalanche photodiodes," *Proc. SPIE* **4454**, 188 (2001) [doi: 10.1117/12.448174].

18. J. Beck, C. Wan, M. Kinch, J. Robinson, P. Mitra, R. Scritchfield, F. Ma, and J. Campbell, "The HgCdTe electron avalanche photodiode," *J. Electronic Materials* **35**, 1166–1173 (2006).

19. A. R. J. Marshall, "The InAs Electron Avalanche Photodiode," in *Advances in Photodiodes*, edited by G. F. D. Betta, InTech, 2011, http://www.intechopen.com/books/advances-in-photodiodes/the-inas-electron-avalanche-photodiode.

20. S. Bank, S. J. Maddox, W. Sun, H. P. Nair, and J. C. Campbell, "Recent progress in high gain InAs avalanche photodiodes," *Proc. SPIE* **9555**, 955509 (2015) [doi: 10.1117/12.2189149].

21. A. R. J. Marshall, C. H. Tan, and J. P. R. David, "Impact ionization in InAs electron avalanche photodiodes," *IEEE Trans. Electron. Dev.* **57** (10), 2631–2638 (2010).

22. A. R. J. Marshall, P. J. Ker, A. Krysa, J. P. R. David, and C. H. Tan, "High speed InAs electron avalanche photodiodes overcome the conventional gain-bandwidth product limit," *Opt. Exp.* **19**(23), 23341–23349 (2011).

23. W. Sun, Z. Lu, X. Zheng, J. C. Campbell, S. J. Maddox, H. P. Nair, and S. R. Bank, "High-gain InAs avalanche photodiodes," *IEEE J. Quant. Electron.* **49**(2), 154–161 (2013).

24. S. Adachi, *Physical Properties of III-V Semiconducting Compounds: InP, InAs, GaAs, GaP, InGaAs, and InGaAsP*, Wiley-Interscience, New York, 1992; *Properties of Group-IV, III-V and II-VI Semiconductors*, John Wiley & Sons, Ltd., Chichester, UK (2005).

25. I. Vurgaftman, J. R. Meyer, and L. R. Ram-Mohan, "Band parameters for III–V compound semiconductors and their alloys," *J. Appl. Phys.* **89**, 5815–5875 (2001).

26. W. F. Micklethwaite, R. G. Fines, and D. J. Freschi, "Advances in infrared antimonide technology," *Proc. SPIE* **2554**, 167 (1995) [doi: 10.1117/12.218185].

27. A. Tanaka, J. Shintani, M. Kimura, and T. Sukegawa, "Multi-step pulling of GaInSb bulk crystal from ternary solution," *J. Crystal Growth* **209**, 625–629 (2000).

28. P. S. Dutta, "III-V ternary bulk substrate growth technology: a review," *J. Crystal Growth* **275**, 106–112 (2005).

29. M. P. Mikhailova and I. A. Andreev, "High-speed avalanche photodiodes for the 2–5 μm spectral range," in *Mid-infrared Semiconductor Optoelectronics*, edited by A. Krier, Springer-Verlag, London, pp. 547–592 (2006).

30. A. Rogalski, R. Ciupa, and W. Larkowski, "Near room-temperature InAsSb photodiodes: Theoretical predictions and experimental data," *Solid-State Electron.* **39**, 1593–1600 (1996).

31. L. O. Bubulac, E. E. Barrowcliff, W. E. Tennant, J. P. Pasko, G. Williams, A. M. Andrews, D. T. Cheung, and E. R. Gertner, "Be ion implantation in InAsSb and GaInSb," *Inst. Phys. Conf.* Ser. No. 45, 519–529 (1979).

32. M. P. Mikhailova, N. M. Stus, S. V. Slobodchikov, N. V. Zotova, B. A. Matveev, and G. N. Talalakin, "InAs$_{1-x}$Sb photodiodes for 3–5 μm spectral range," *Fiz. Tekh. Poluprovodn.* **30**, 1613–1619 (1996).

33. EG&G Optoelectronics, Inc., Data Sheet (1995).

34. A. I. Andrushko, A. V. Pencov, Ch. M. Salichov, and S. V. Slobodchikov, "R$_0$A product of p–n InAs junctions," *Fiz. Tekh. Poluprovodn.* **25**, 1696–1690 (1991).

35. J. Wróbel, R. Ciupa, and A. Rogalski, "Performance limits of room-temperature InAsSb photodiodes," *Proc. SPIE* **7660**, 766033 (2010) [doi: 10.1117/12.855196].

36. A. Krier and W. Suleiman, "Uncooled photodetectors for the 3–5 μm spectral range based on III-V heterojunctions," *Appl. Phys. Lett.* **89**, 083512 (2006).

37. Y. Sharabani, Y. Paltiel, A. Sher, A. Raizman, and A. Zussman, "InAsSb/GaSb heterostructure based mid-wavelength-infrared detector for high temperature operation," *Appl. Phys. Lett.* **90**, 232106 (2007).

38. H. Shao, W. Li, A. Torfi, D. Moscicka, and W. I. Wang, "Room-temperature InAsSb photovoltaic detectors for mid-infrared applications," *IEEE Photnics Technol. Lett.* **18**, 1756–1758 (2006).
39. H. H. Gao, A. Krier, and V. V. Sherstnev, "Room-temperature $InAs_{0.89}Sb_{0.11}$ photodetectors for CO detection at 4.6 µm," *Appl. Phys. Lett.* **77**, 872–874 (2000).
40. S. A. Cripps, T. J. C. Hosea, A. Krier, V. Smirnov, P. J. Batty, Q. D. Zhuang, H. H. Lin, P. W. Liu, and G. Tsai, "Determination of the fundamental and spin-orbit-splitting band gap energies of InAsSb-based ternary and pentenary alloys using mid-infrared photoreflectance," *Thin Solid Films* **516**, 8049–8058 (2008).
41. Y. Lin, D. Donetsky, D. Wang, D. Westerfeld, G. Kipshidze, L. Shterengas, W. L. Sarney, S. P. Svensson, and G. Belenky, "Development of bulk InAsSb alloys and barrier heterostructures for long-wave infrared detectors," *J. Electronic Mater.* **44** (10), 3360–3366 (2015).

Chapter 5
Type-II Superlattice Infrared Photodiodes

Many types of optoelectronic devices can be significantly enhanced through the introduction of quantum confinement in reduced-dimensionality hetero-structures. This was the main motivation for the study of superlattices as alternative infrared detector materials. The HgTe/CdTe SL system proposed in 1979[1,2] was the first from a new class of quantum-sized structures for IR photoelectronics, which was proposed as a promising new alternative structure for the construction of LWIR detectors to replace those of HgCdTe alloys. It was anticipated that superlattice infrared materials would have several advantages over bulk HgCdTe for these reasons:

- a higher degree of uniformity, which is importance for detector arrays;
- smaller leakage current due to the suppression of tunneling (larger effective masses) available in superlattices; and
- lower Auger-recombination rates due to substantial splitting of the light- and heavy-hole bands and increased electron effective masses.

More recently, significant interest has been shown in multiple quantum well AlGaAs/GaAs photoconductors.[3] However, these detectors are extrinsic in nature and have been predicted to be limited to performances inferior to those of intrinsic HgCdTe detectors.[4] On account of this, in addition to the use of intersubband absorption, two additional physical principles are utilized to directly shift bandgaps into the infrared spectral range:

- superlattice strain-induced bandgap reduction: InSb/InAsSb, InAs/InAsSb; and
- superlattice-induced broken bandgap eduction: InAs/GaInSb.

These types of superlattice rely on an intrinsic valence-to-conduction band absorption process. Early attempts to realize superlattices with properties suitable for infrared detection were unsuccessful, largely because of the difficulties associated with epitaxial deposition of HgTe/CdTe superlattices.

At present, performance of HgTe/HgCdTe SL photodiodes is inferior in comparison with high-quality HgCdTe photodiodes with comparable cutoff wavelengths. As a result, a lack of research funding has led to an industry-wide suspension of further efforts to develop HgTe/HgCdTe SL infrared detectors.

Apart from using intersubband transitions, III-V compounds can be effectively used for infrared applications by favorable realization of valence-to-conduction band optical transitions between the states in alternative layers due to the overlap of the envelope wave functions. In this way appreciable absorption coefficients at normal incidence can be provided in infrared devices.[5–8] In 1990, Miles and co-workers reported the first $Ga_{1-x}In_xSb$/InAs T2SL material with high structural quality.[8] Soon thereafter, high-performance InAs/GaInSb SL photovoltaic detectors were predicted by the theoretical promise of longer intrinsic lifetimes due to the suppression of the Auger-recombination mechanism.[9,10]

5.1 InAs/GaSb Superlattice Photodiodes

The first InAs/InGaSb SLS photodiodes with photoresponse out to 10.6 μm were presented by Johnson et al.[11] The detectors consist of double heterojunctions (DH) of the SL with n-type and p-type GaSb grown on GaSb substrates. The use of heterojunctions in photodiodes offers several advantages over homojunctions. In 1997, researchers from Fraunhofer Institute demonstrated good detectivity (approaching HgCdTe, 8-μm cutoff, 77 K) on individual devices, initiating renewed interest in LWIR detection with type-II SLs.[12] While theoretical predictions of detector performance seem to favor the InAs/InGaSb system due to the additional strain provided by the InGaSb layer, the majority of the research in the past ten years has been focused on the binary InAs/GaSb system. This is attributed to the complexity of structures grown with the large mole fraction of In.

As shown in Fig. 3.6, the InAs/GaSb T2SL consists of alternating layers of nanoscale materials. Typically, their thicknesses vary from 6 to 20 monolayers (MLs). The overlap of electron (hole) wave functions between adjacent InAs (GaSb) layers results in the formation of electron (hole) minibands in the conduction (valence) band. Optical transitions in two mid-bandgap semiconductors—between holes localized in GaSb layers and electrons confined in InAs layers—are employed in the wide spectral IR detection process between 3 and 30 μm.

Taalat et al. have shown[13] strong influence of the superlattice composition on both the material properties and MWIR photodetector performance, such as the background doping concentration, the shape of the spectral responsivity, and the dark-current value. As shown in Fig. 5.1, a bandgap energy around 248 meV ($\lambda_c = 5$ μm at 77 K) is achieved at three different SL

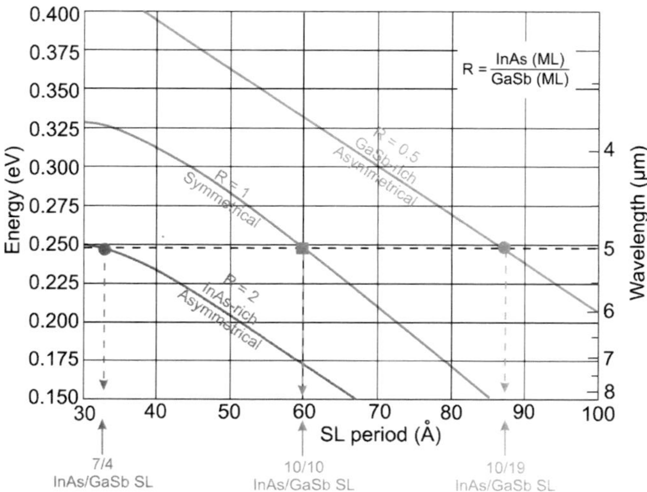

Figure 5.1 Calculated superlattice bandgap at 77 K as a function of the period thickness for different ratios R = InAs/GaSb (reprinted from Ref. 13).

periods: with a GaSb-rich composition (10 MLs InAs/19 MLs GaSb per period), symmetric with the same InAs and GaSb thicknesses (10 MLs), and InAs-rich composition (7 MLs InAs/4 MLs GaSb).

5.1.1 MWIR photodiodes

Superlattice photodiodes are typically based on p–i–n double heterostructures with an unintentionally doped, intrinsic region between the heavily doped contact portions of the device. The sample presented in this work is the InAs/GaSb T2SL p–i–n detector with the SU-8 passivation fabricated at the Center for High Technology Materials, University of New Mexico, Albuquerque.[14] The device structure was grown on Te-doped epiready (100) GaSb substrates. It consists of 100 periods of ten monolayers (10 MLs) of InAs:Si ($n = 4 \times 10^{18}$ cm^{-3})/10 MLs of GaSb as the bottom contact layer. This is followed by 50 periods of graded n-doped 10 MLs of InAs:Si/10 MLs of GaSb, 350 periods of absorber, 25 periods of 10 MLs of InAs:Be ($p = 1 \times 10^{18}$ cm^{-3})/10 MLs of GaSb, and finally, 17 periods of 10 MLs InAs:Be ($p = 4 \times 10^{18}$ cm^{-3})/10 MLs GaSb, which formed a p-type contact layer. Twenty-five periods of the SL structure with graded doping layers are added between the absorber and the contact layer in order to improve transport of minority carriers in the detector structure. By varying the beryllium concentration in the InAs layer of the active region, the residually n-type superlattice is compensated to become slightly p-type. Consequently, increasing the RA product and quantum efficiency was observed in similar structures.[15]

Figure 5.2 shows the schematic photodiode structure and its design. A normal-incidence single-pixel mesa photodiode with a 450×450 μm^2

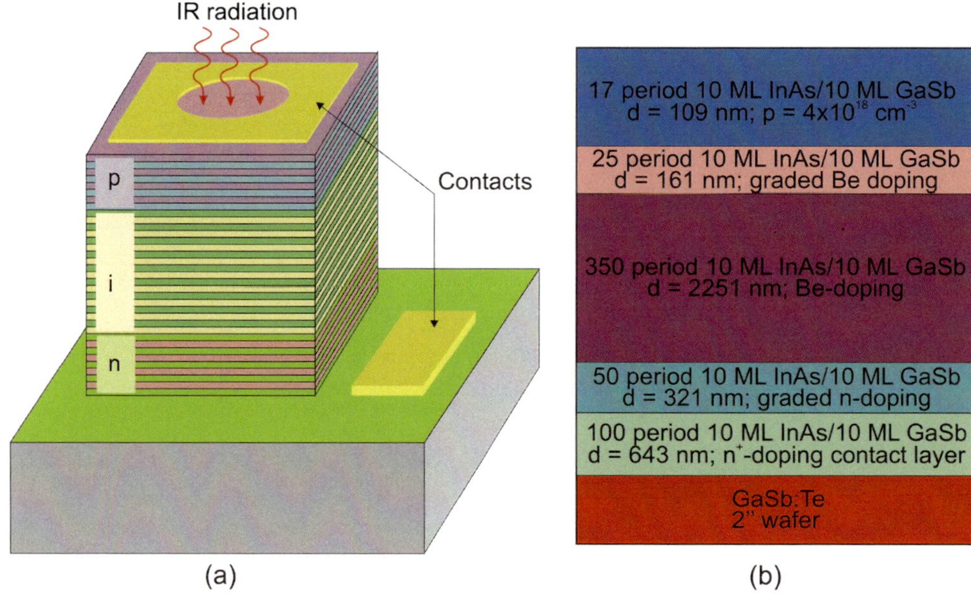

Figure 5.2 MWIR InAs/GaSb type-II superlattice photodiode: (a) schematic device structure, and (b) photodiode design (reprinted from Ref. 19).

electrical area was fabricated by photolithography and inductively coupled plasma etching. The rest of the fabricated devices were dipped in a phosphoric-acid-based solution to remove the native oxide film on the etched mesa sidewalls, then covered with SU-8 ($\sim$1.5-μm thickness) to act as the passivation layer. Ohmic contacts were made by depositing Ti/Pt/Au on the contact layers.[16,17]

The cutoff wavelength increases with temperature increase, assuming 5.6 μm at 120 K and 6.2 μm at 230 K. This shift can be attributed to the dependence of a bandgap on temperature according to the Varshni formula. In order to explain the current–voltage characteristics of the MWIR type-II SL photodiodes, a bulk-based model with an effective bandgap of SL material is used.

It is well recognized that the photodiode dark current can be found as a superposition of several mechanisms (see Fig. 5.3):

$$I_{dark} = I_{diff} + I_{gr} + I_{btb} + I_{tat} + I_{Rshunt}, \tag{5.1}$$

including four main mechanisms: diffusion I_{diff}, generation–recombination I_{gr}, band-to-band tunnelling I_{btb}, and trap-assisted tunnelling I_{tat}. The remaining mechanism is current due to the shunt resistance I_{Rshunt}, which originates from the surface and bulk leakage current and shows the presence in the reverse-bias region. The avalanche current, which occurs in diodes with large depletion widths and a high reverse bias voltage, is omitted in our

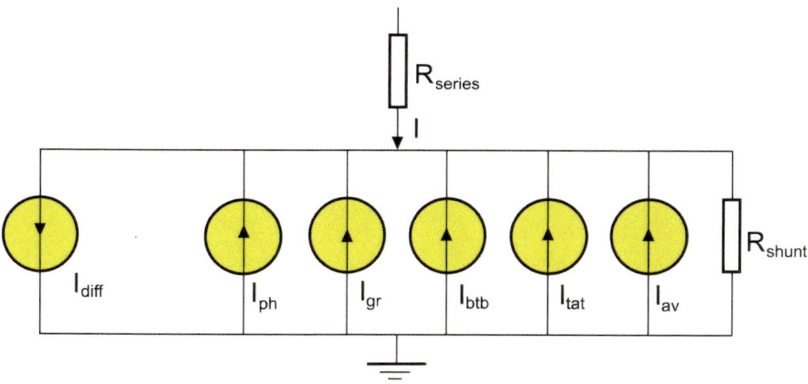

Figure 5.3 Possible currents operating in the photodiode. I_{diff} is the ideal diffusion current, I_{ph} is the photocurrent, I_{gr} is due to the generation–recombination mechanism, I_{btb} is due to band-to-band tunnelling, I_{tat} is due to trap-to-band tunnelling, I_{av} is the avalanche current, and R_{shunt} is due to surface and bulk leakage shunt resistance. Limiting currents act in opposition to the diffusion current.

considerations. A summary of these well-known contributions to dark current that have been taken into account in our modelling is given in Refs. 18–20.

Several examples of the measured and fitting characteristics are presented below. Figures 5.4 and 5.5 present comparisons of experimental and theoretically predicted characteristics of the dark current density J–V and the resistance area product versus bias voltage $RA(V)$ for MWIR InAs/GaSb superlattice photodiodes at temperatures 160 K and 230 K, respectively. As can be seen, in a wide region of bias voltages (between +0.1 V and –1.6 V) and temperatures (also below 160 K, not shown), excellent agreement between both types of results has been obtained.

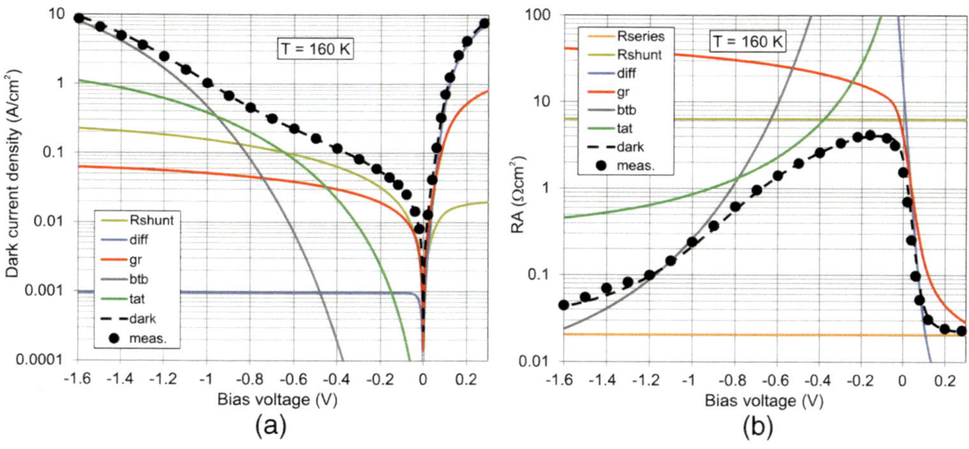

Figure 5.4 Measured and modeled characteristics of a MWIR p–i–n InAs/GaSb type-II superlattice photodiode at 160 K: (a) dark current density versus bias voltage and (b) resistance area product versus bias voltage (reprinted from Ref. 20).

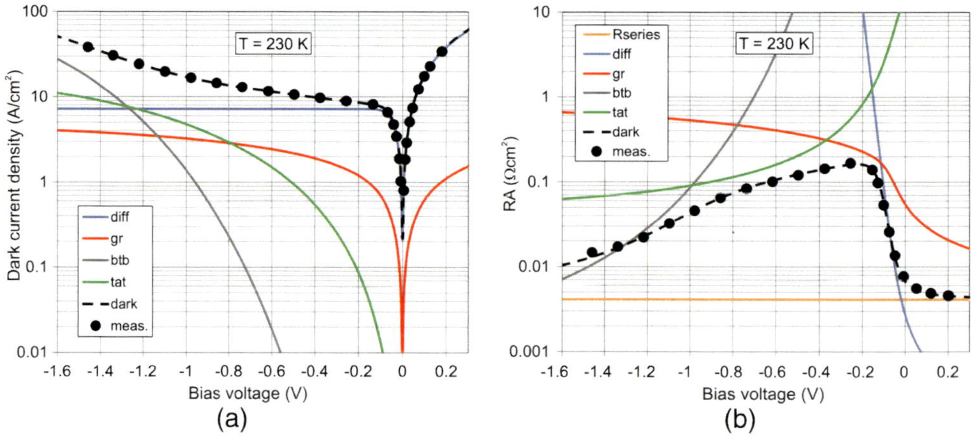

Figure 5.5 Measured and modeled characteristics of a MWIR p–i–n InAs/GaSb type-II superlattice photodiode at 230 K: (a) dark current density versus bias voltage and (b) resistance area product versus bias voltage (reprinted from Ref. 20).

At a low temperature ($\leq$120 K) in reverse-bias voltage the current conducted through the shunt resistance dominates the reverse characteristic of the diode. It can be explained here that the dislocations that intersect the junction and/or surface leakage currents are generally responsible for shunt currents in a diode. In the region of higher reverse voltages (above 1 V), the influence of band-to-band tunnelling is decisive. The influence of shunt resistance is negligible in thermoelectrically cooled (TE-cooled) photodiodes (in a temperature range above 170 K).

With temperature increase, the contribution of diffusion and generation–recombination currents increases in the zero-bias and the low-bias regions and dominates at 230 K. At medium values of reverse bias (between 0.6 and 1.0 V at 160 K), the dark current is mostly due to a trap-assisted tunnelling. At high values of reverse bias (above 1 V at 160 K), bulk band-to-band tunnelling dominates. In the region of forward-bias voltage above 0.1 V, the influence of series resistance is decisive.

Wróbel et al.[21] examined a near-mid-gap-trap energy level in InAs$_{10ML}$/GaSb$_{10ML}$ type-II superlattices using thermal analysis of the dark current, Fourier transform photoluminescence, and low-frequency noise spectroscopy. Several wafers and diodes with similar period designs and the same macroscopic construction were investigated. All characterization techniques gave nearly the same value of about 140 meV independent of substrate type. Additionally, photoluminescence spectra show that the transition related to the trap center is temperature independent.

Figure 5.6 presents a comparison of the R_0A product versus the cutoff wavelength for the HgCdTe photodiodes fabricated in our laboratory (at the Institute of Applied Physics, Military University of Technology, Warsaw, Poland) and type-II InAs/GaSb superlattice photodiodes operating at

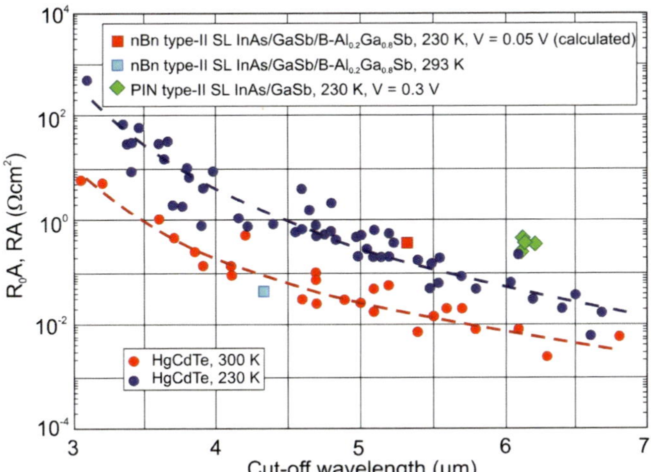

Figure 5.6 Dependence of the RA and R_0A products on cutoff wavelength for MWIR InAs/GaSb/B-Al$_{0.2}$Ga$_{0.8}$Sb type-II SL nBn detectors, HgCdTe bulk diodes, and InAs/GaSb type-II SL p–i–n diodes operated at near room temperature (adapted from Ref. 19).

230 K.[19,22] It can be clearly seen that the performance of superlattice devices has reached a level that is comparable to the performance of state-of-the-art HgCdTe detectors. In fact, the RA product of 6.2-μm superlattice devices is even higher, but it was measured at 0.3 V reverse bias. The dashed lines are trend lines of HgCdTe device experimental data. HgCdTe photodiodes still generally offer better performance, especially at lower temperatures, e.g., 77 K,[23] mainly due to a larger quantum efficiency. Typically, the quantum efficiency of our type-II photodiodes is about 30% (at 0.3 reverse bias) in comparison with 70% for the HgCdTe devices. The active region of the superlattice devices is thinner than that of HgCdTe photodiodes. As a result, the dark current and the R_0A product of devices with high-bandgap contacts are comparable to those of HgCdTe photodiodes at 230 K.

A research group at Montpellier University[13,24,25] has elaborated on MWIR T2SL p–i–n device structures similar to those shown in Fig. 4.11. As was noted, several InAs/GaSb SL material properties depend strongly on the chosen SL period, such as the temperature bandgap energy, absorption coefficient, residual doping level, and carrier lifetime. The superlattice period influences carrier localization, which results in different mini-band widths and thus different joint density of states and line shape of absorption coefficients. For example, Fig. 5.7 reports photoresponse spectra of symmetrical InAs/GaSb SL structures having emission in the MWIR wavelength range.

The important observation is the independence of minority-carrier lifetime from interface density. The recombination centers limiting the SL lifetime are located in the binary materials rather than at the interfaces.[13,26] The worst material properties are obtained for the largest GaSb content. In SL samples

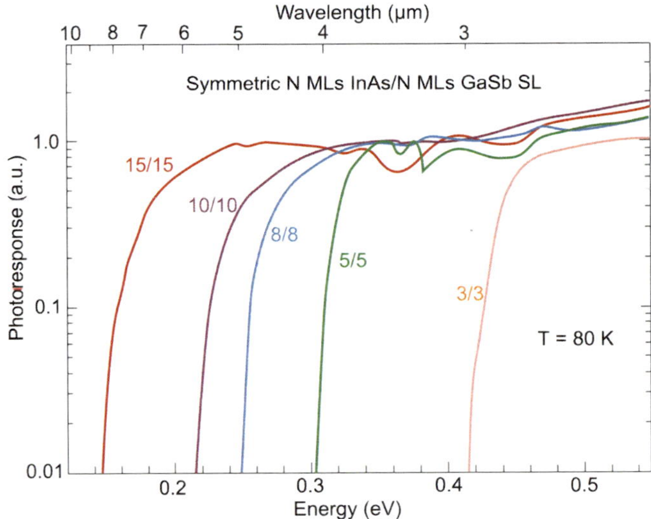

Figure 5.7 Normalized photoresponse spectra of InAs(N)/GaSb(N) symmetrical MWIR SL detector structures with $N = 3, 5, 8, 10$, and 15 MLs. The spectra are recorded at 80 K (adapted from Ref. 24).

at 77 K with a cutoff wavelength close to 5 μm, the residual doping level was found to increase from 6×10^{14} cm^{-3} to 5.5×10^{15} cm^{-3} when the GaSb content in each SL period increased from 36% to 65%. Since the GR-limited current that dominated at 77 K is proportional to $n_i/\tau n^{1/2}$ (n is the majority-carrier density), an increase of τ correspondingly contributes to a decrease by the same amount of I_{gr}. Figure 5.8 gathers the experimental data of dark

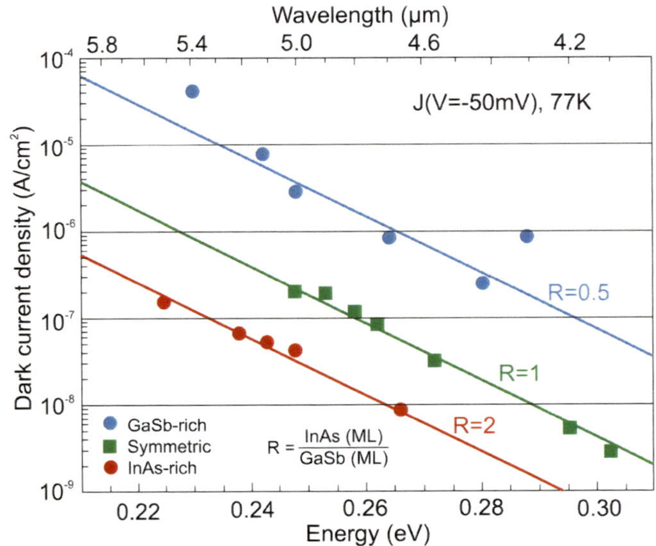

Figure 5.8 Dark current density at 77 K as a function of the period thickness of p–i–n InAs/GaSb T2SL photodiodes for different ratios of $R = $ InAs/GaSb (adapted from Ref. 25).

current density for GaSb-rich, symmetric, and InAs-rich SL compositions. In the case of an asymmetric (7/4) InAs-rich InAs/GaSb T2SL structure ($R = 1.75$) with a cutoff wavelength of 5.5 μm at 77 K, an R_0A product as high as 7×10^6 Ωcm^2 at 77K was reported.[24]

5.1.2 LWIR photodiodes

Figure 5.9 shows a cross-sectional scheme of a completely processed mesa detector and design of 10.5-μm InAs/GaSb SL photodiode. The layers are usually grown by MBE at substrate temperatures around 400 °C on undoped (001)–oriented GaSb substrates. With the addition of cracker cells for the group-V sources, the superlattice quality is significantly improved. Despite the relatively low absorption coefficients, GaSb substrates require reduction of the thickness to below 25 μm in order to transmit appreciable IR radiation.[27] Since the GaSb substrates and buffer layers are intrinsically p-type, the p-type contact layer—intentionally doped with beryllium at an acceptor concentration of 1×10^{18} atoms/cm^3—is grown first.

Sensors for the LWIR spectral ranges are based on binary InAs/GaSb short-period superlattices.[28,29] The layers needed are already so thin that there is no benefit to using GaInSb alloys. The oscillator strength of the InAs/GaSb SL is weaker than that of InAs/GaInSb; however, the InAs/GaSb SL, which uses unstrained and minimally strained binary semiconductor layers, may also have material quality advantages over the SL, which uses a strained ternary semiconductor (GaInSb). For the formation of p–i–n photodiodes, the lower periods of the InAs/GaSb SL are p-doped with 1×10^{17} cm^{-3} Be in the GaSb layers. These acceptor-doped SL layers are followed by a 1- to 2-μm-thick, nominally undoped superlattice region. The width of the intrinsic region does vary in the designs. The width used should be correlated to the carrier diffusion lengths for improved performance. The upper layer of the SL stack is doped with silicon (1×10^{17} to 1×10^{18} cm^{-3}) in the InAs layers and is typically 0.5 μm thick. The top of the SL stack is then capped with an InAs:Si ($n \approx 10^{18}$ cm^{-3}) layer to provide good ohmic contact.

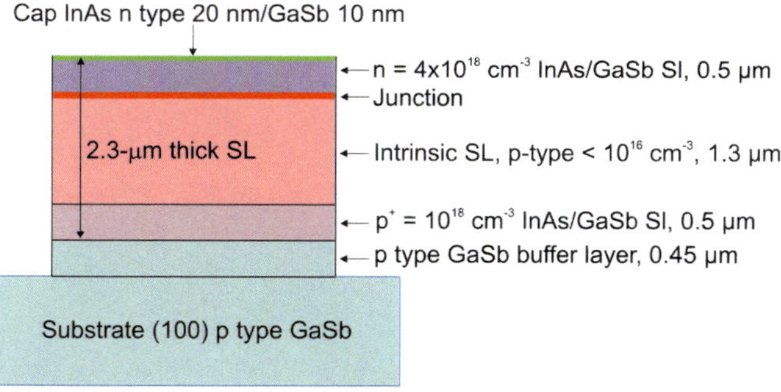

Figure 5.9 Schematic design of p–i–n double-heterojunction InAs/GaSb photodiode with a 10.5-μm cutoff wavelength.

The main technological challenge for the fabrication of photodiodes is the growth of thick SL structures without degrading the material quality. High-quality SLS materials that are thick enough to achieve acceptable quantum efficiency are crucial to the success of the technology.

Figure 5.10 shows the experimental data and theoretical prediction of the R_0A product as a function of temperature for an InAs/GaSb photodiode with a 10.5-μm cutoff wavelength at 78 K. The photodiode is depletion-region (GR)–limited in the temperature range below 100 K. Space-charge recombination currents dominate reverse bias at 78 K and with the dominant recombination centers located at the intrinsic Fermi level, as shown in Fig. 5.10(a). The trap-assisted tunneling is dominant at $T \leq 40$ K. The performance of LWIR photodiodes in the high temperature range is limited by the diffusion process. At a low temperature and near-zero-bias voltage, the currents are diffusion limited. At larger biases, trap-assisted-tunneling currents dominate.

An additional insight in dark current mechanisms in LWIR InAs/GaSb SL photodiodes has been given by Rehm et al.[31,32] By assuming that the device is affected by sidewall leakage, bulk and sidewall contributions can be described by the well-known relation for the total dark current density:

$$I_{dark} = I_{dark,bulk} + \sigma \times P/A, \qquad (5.2)$$

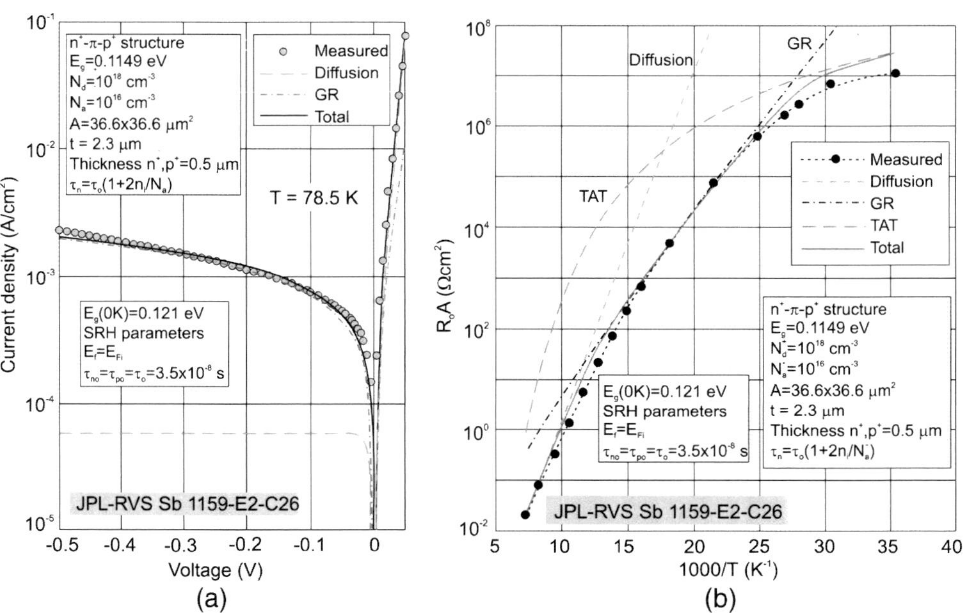

Figure 5.10 Experimental data and theoretical characteristics for an InAs/GaSb photodiode with $\lambda_c = 10.5$ μm at 78.5 K: (a) J–V characteristic at 78 K and (b) R_0A product as a function of temperature (adapted from Ref. 30).

where σ is the sidewall current per unit length of the mesa sidewall, and P/A is the perimeter-to-area ratio of the device. In general, $\sigma(V,T)$ is a function of the applied bias voltage V and temperature T. The bulk dark current, $I_{dark,bulk}$ contains the components described by Eq. (5.1).

Figure 5.11(a) shows that the sidewall current does not dominate the behavior of the $I(V)$ characteristic at low reverse bias at 77 K. The temperature dependence of the $I(V)$ characteristic at 77 K at low reverse bias is diffusion limited [see Fig. 5.11(b)].

Heterojunction device concepts help to considerably reduce dark current,[32] as was shown by the design of a p^+-InAs/GaSb SL absorber combined with an N^--high-bandgap part, which was realized with a second, conduction-band-matched InAs/GaSb SL with a higher bandgap.

Figure 5.12 compares the R_0A values of InAs/GaSb SLS and HgCdTe photodiodes in the long-wavelength spectral range. The solid line denotes the theoretical diffusion-limited performance of the p-type HgCdTe material. As seen in the figure, the most recent photodiode results for SL devices rival those of practical HgCdTe devices, indicating that substantial improvement has been achieved in SL detector development.

The quantum efficiency of the p–i–n photodiode structure shown in Fig. 5.13 critically depends on the thickness of the i (π)–region. By fitting the quantum efficiencies of a series of photodiodes with i-region thicknesses varying from 1 to 4 μm, Aifer et al.[34] determined that the minority-carrier electron diffusion length in LWIR is 3.5 μm. This value is considerably lower in comparison with a typical value for high-quality HgCdTe photodiodes. More recently, an external quantum efficiency of 54% has been obtained for 12-μm-cutoff-wavelength photodiodes by extending the thickness of the π-region to 6 μm. Figure 5.13(a) shows the dependence of quantum efficiency on the thickness of the π-region, and Fig. 5.13(b) presents the spectral current

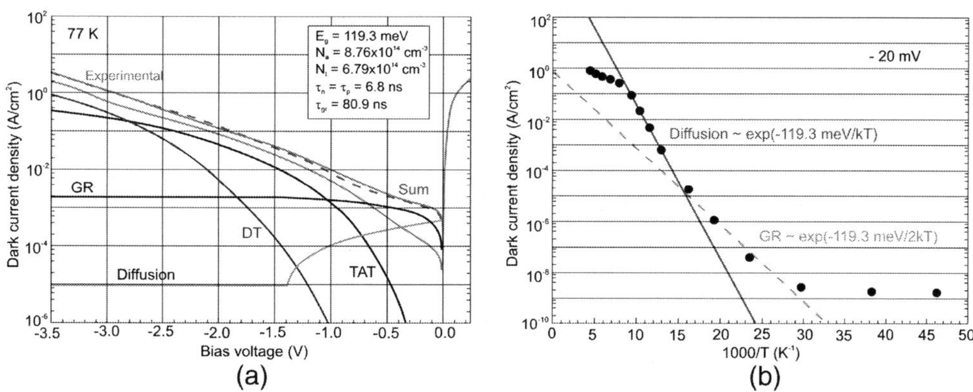

Figure 5.11 Dark current density analysis of a homojunction LWIR InAs/GaSb SL photodiode: (a) versus bias voltage at 77 K, and (b) versus temperature (reprinted from Ref. 32).

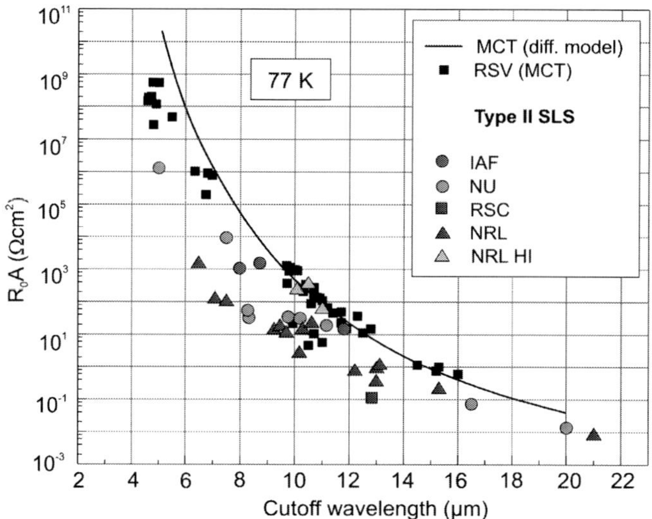

Figure 5.12 Dependence of the R_0A product of InAs/GaSb SLS photodiodes on cutoff wavelength compared to theoretical and experimental trendlines for comparable HgCdTe photodiodes at 77 K (adapted from Ref. 33).

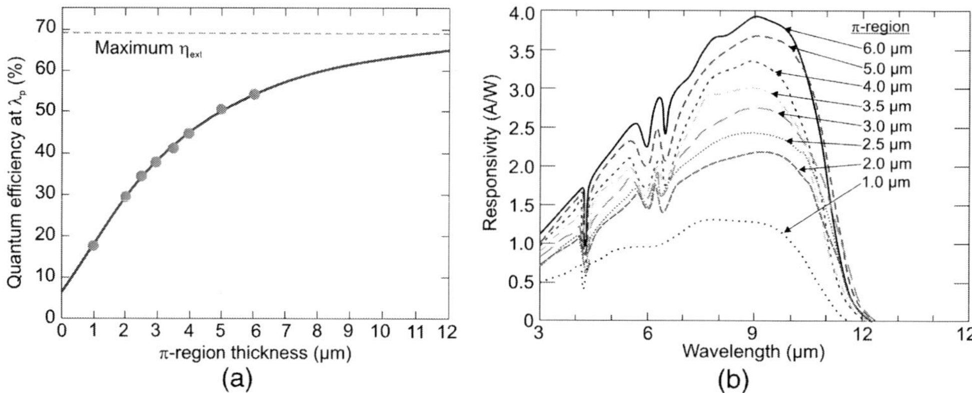

Figure 5.13 Spectral characteristics of InAs/GaInSb SL photodiodes at 77 K: (a) quantum efficiency as a function of π-region thickness, where the dashed line represents the maximum possible quantum efficiency without antireflective coating; (b) measured current responsivity of photodiodes with π-region thickness ranging from 1 to 6 µm. The CO_2 absorption at 4.2 µm is visible as well as the water vapor absorption between 5 and 8 µm (reprinted from Ref. 35 with permission from AIP Publishing).

responsivity of eight of these structures with different thicknesses of the π-region.[35]

Figure 5.14 compares the calculated detectivity of type-II and p-on-n HgCdTe photodiodes as a function of wavelength and temperature of operation with the experimental data of type-II detectors operated at 78 K.[36] The solid lines are theoretical thermally limited detectivities for HgCdTe

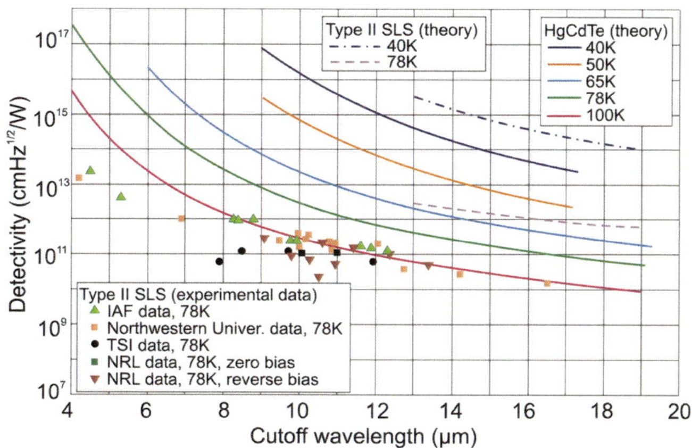

Figure 5.14 The predicted detectivity of type-II and p-on-n HgCdTe photodiodes as a function of wavelength and temperature. Experimental data are taken from several sources (adapted from Ref. 36).

photodiodes, calculated using a one-dimensional (1D) model that assumes the diffusion current from the narrower-bandgap n-side to be dominant and a minority-carrier recombination via Auger and radiative processes. These calculations used typical values for the n-side donor concentration ($N_d = 1 \times 10^{15}$ cm^{-3}), a narrow-bandgap active layer thickness (10 µm), and a quantum efficiency of 60%. The predicted thermally limited detectivities of the T2SL are larger than those for HgCdTe.[10,37,38]

From the Fig. 5.14 results, it can be seen that the measured thermally limited detectivities of T2SL photodiodes are as yet inferior to the current HgCdTe photodiode performance, and the T2SL photodiode performance has not achieved theoretical values. This limitation appears to be due to two main factors: relatively high background concentrations (about 5×10^{15} cm^{-3}, although values below 10^{15} cm^{-3} have been reported)[39] and a short minority-carrier lifetime (typically tens of nanoseconds in lightly doped p-type material). Until now, non-optimized carrier lifetimes have been observed and at desirably low carrier concentrations are limited by the SRH recombination mechanism. The minority-carrier diffusion length is in the range of several micrometers. Improving these fundamental parameters is essential to realize the predicted performance of type-II photodiodes.

5.2 InAs/InAsSb Superlattice Photodiodes

Interest in InAs/InAsSb superlattice materials for IR detection is motivated by carrier lifetime limitations imposed by the GaSb layer in InAs/GaSb SLs. As was mentioned in section 3.2, a significant longer minority-carrier lifetime has been obtained in an InAs/InAsSb SL system as compared to an InAs/GaSb SL operating in the same wavelength range and temperature. It is expected that such an increase in minority-carrier lifetime results in lower dark current for InAs/InAsSb SL detectors in comparison with their InAs/GaSb SL counterparts.[40] In addition, with two common elements (indium and arsenic) in the SL layers, the InAs/InAsSb SL has a relatively simple interface structure with only one changing element (antimony), which promises better controllability in epitaxial growth and simpler manufacturability.

Both MWIR[41] and LWIR[42] InAs/InAsSb SL photodiodes have been demonstrated. The experimentally measured dark current density of MWIR photodiodes with a cutoff wavelength of 5.4 μm at 77 K was larger than that of conventional InAs/GaSb SL detectors. This was attributed to the increased probability of carrier tunneling due to reduced valence- and conduction-band offsets in the InAs/InAsSb SL system.

Higher-quality LWIR InAs/InAs$_{1-x}$Sb$_x$ SL photodiodes have been demonstrated by Hoang et al.[42] Despite the introduction of large amounts of Sb ($x = 0.43$), the material quality is still good and leads to high-performance photodetectors. The cutoff wavelength of the active region is mainly determined by the valence band level in the InAs$_{1-x}$Sb$_x$ layer, which is directly related to the amount of Sb. The samples were MBE grown on Te-doped (001) GaSb substrate. The device structure consists of 0.5-μm-thick InAsSb buffer layer, followed by a 0.5-μm-thick bottom n-contact ($n \sim 10^{18}$ cm^{-3}), a 0.5-μm-thick slightly n-doped barrier, a 2.3-μm-thick slightly p-doped active region ($\sim 10^{15}$ cm^{-3}), and a 0.5-μm-thick top p-contact ($p \sim 10^{18}$ cm^{-3}). Finally, the structure is capped with a 200-nm-thick p-doped GaSb layer. The n-type and p-type dopants are silicon and beryllium, respectively.

The effective passivation of InAs/InAsSb T2SL photodiodes is in a very early stage of development. Usually, photodiodes are not passivated. The simplest passivation process involves common dielectric insulators (such as an oxide or nitride of silicon) being deposited onto the exposed surface of the device, as is utilized in the silicon industry.

Figure 5.15 shows the electrical characteristics of the LWIR InAs/InAsSb SL photodiode in the temperature range of 25 to 77 K. At 77 K the R_0A product is 0.84 Ωcm^2, indicating a lower value in comparison with the InAs/GaSb counterpart (see Fig. 5.12). Above 50 K, the diode exhibits Arrhenius-type behavior with an associated activation energy of 39 meV, which is approximately one-half of the active region's bandgap ($\sim$80 meV for 15 μm), indicating that the GR current from the active region is the dark-current-limiting mechanism. Below 50 K, the R_0A deviates from the trend and becomes less sensitive to temperature variation. This behavior suggests that the dark current is limited by other mechanisms—either the tunneling current or the surface leakage at this temperature range.

The spectral characteristics of this photodiode are shown in Fig. 5.16. At 77 K, the sample exhibited a 100% cutoff wavelength at 17 μm and a 50% cutoff wavelength at 14.6 μm. High quantum efficiency is achieved with reverse

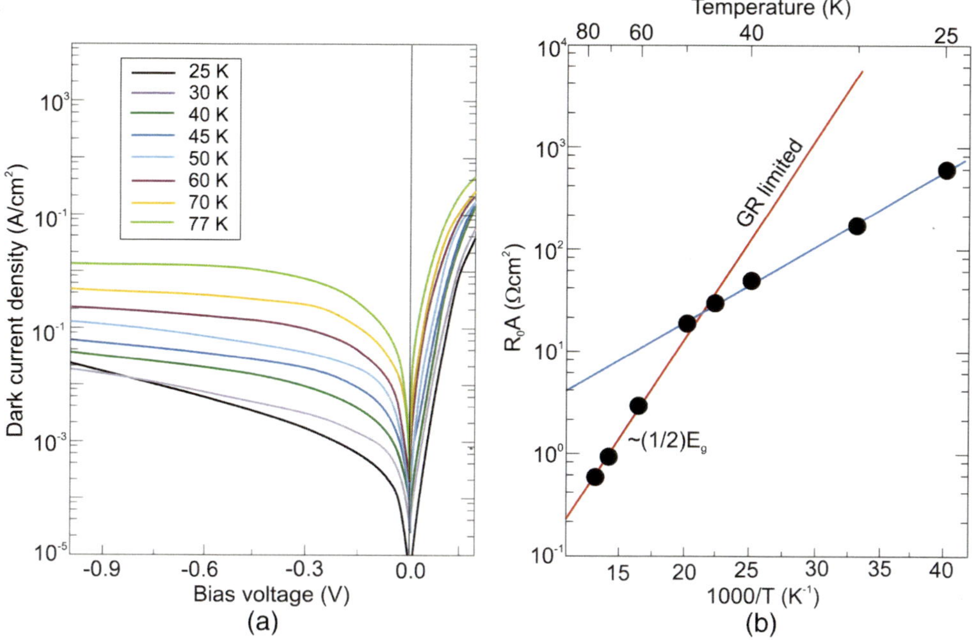

Figure 5.15 Electrical characteristics of LWIR InAs/InAsSb SL photodiodes: (a) current–voltage characteristics as a function of temperatures and (b) R_0A versus temperature (reprinted from Ref. 42 with permission from AIP Publishing).

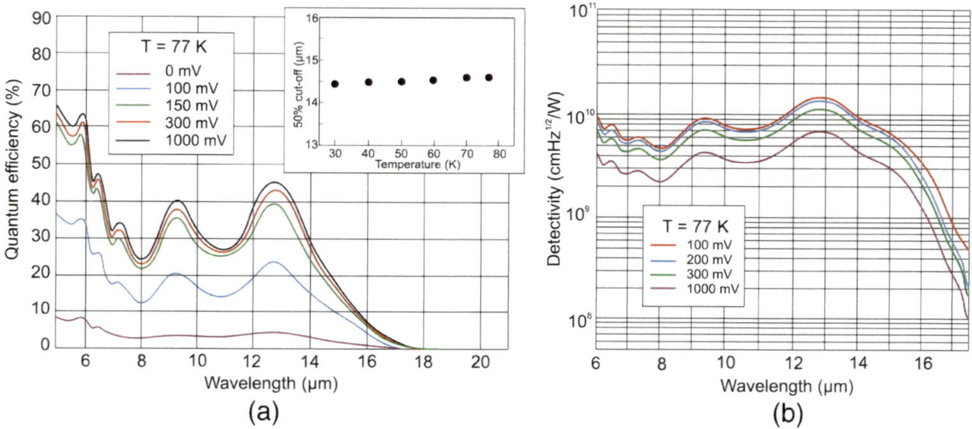

Figure 5.16 Spectral characteristics of the LWIR InAs/InAsSb SL photodiode: (a) quantum efficiency spectrum with different applied biases at 77 K. Inset: 50% cutoff wavelength as a function of temperature; (b) the calculated shot-noise- and Johnson-noise-limited detectivity of the device at 77 K with different applied biases (reprinted from Ref. 42 with permission from AIP Publishing).

bias above 150 mV and becomes saturated at 300 mV. Upon saturation, the current responsivity reaches a peak of 4.8 A/W, corresponding to a quantum efficiency of 46% for a 2.3-μm-thick active region. Figure 5.16(b) shows the calculated shot-noise- and Johnson-noise-limited detectivity of the device at 77 K based on the measured quantum efficiency, the dark current, and the R_0A product.

5.3 Device Passivation

Despite numerous efforts by various research groups to fabricate T2SL devices, the development of an effective device passivation scheme has not been well established. The mesa sidewalls are a source of excess currents. Besides efficient suppression of surface leakage currents, a passivation layer suitable for production purposes must withstand various treatments that occur during the subsequent processing of the device. Considerable surface leakage is attributed to the discontinuity in the periodic crystal structure due to mesa delineation. Focal plane arrays (FPAs) of 1 cm^2 still dominate the IR market, while pixel pitch has decreased to 15 μm during the last few years, now reaching 12 μm, 10 μm, and even 5 μm in test devices.[43] This trend is expected to continue. Pixel reduction is mandatory also to cost reduction of a system (reduction of the optics diameter, reduction of the dewar size and weight, together with a reduction in power, and an increase in reliability). Scaling of the pixel dimensions makes FPA performance strongly dependent on surface effects due to a large pixel surface/volume ratio. Thus, methods for elimination of surface currents need to be developed.

The native oxide of most III-V compounds is not beneficial as a natural passivation material. The oxidation of GaSb at relatively low temperatures occurs in accordance with the reaction

$$2\text{GaSb} + 3\text{O}_2 \rightarrow \text{Ga}_2\text{O}_3 + \text{Sb}_2\text{O}_3. \tag{5.3}$$

In the first step of chemical passivation, the solution removes the native oxides at the surface, and then new atoms occupy the dangling bonds to prevent reoxidation of the material, prevent surface contamination, and minimize band bending.

A review of passivation techniques for T2SLS detectors may be found in the work of Plis et al.,[44] where the techniques are categorized into two types:

- encapsulation of etched detector sidewalls with thick layers of dielectrics, organic materials (polyimide and various photoresists), or a wider-bandgap III-V material, or
- use of chalcogenide passivation, which involves saturation of unsatisfied bonds on semiconductor surfaces by sulphur atoms.

Here we briefly describe the available passivation techniques and their limitations.

The effectiveness of passivation is commonly evaluated using a variable-area diode array method. The dark current density can be expressed as the summation of the bulk component of dark current and the surface leakage current. The inverse of the dynamic R_0A product of the photodiode at zero bias can be approximated as

$$\frac{1}{R_0A} = \frac{1}{(R_0A)_{Bulk}} + \frac{1}{\rho_{Surface}}\frac{P}{A},\qquad(5.4)$$

where $(R_0A)_{Bulk}$ is the bulk R_0A contribution (Ωcm^2), $\rho_{Surface}$ is the surface resisitivity (Ωcm), P is the diode perimeter, and A is the diode area. The slope of the function $(R_0A)^{-1} = f(P/A)$ is directly proportional to the diode's surface-dependent leakage current $1/\rho_{Surface}$. If the bulk current dominates the detector performance, then the dependence $(R_0A)^{-1} = f(P/A)$ has a slope close to zero. If the surface leakage is significant, then an increase in the dark current density is observed for smaller devices.

Figure 5.17 shows exemplary dependencies of $(R_0A)^{-1}$ as a function of P/A for LWIR T2SL photodiodes passivated in different ways and operated at 77 K.[45,46] Detectors with mesa sizes ranging within 100–400 µm etched by inductively coupled plasma (ICP) and electron cyclotron resonance (ECR) with sidewalls encapsulated by polyimide have demonstrated the highest surface resistivity (6.7×10^4 Ωcm) among the four treatments [see Fig. 5.17(a)]. A comparison of the electrical performance of detectors with the same post-etch

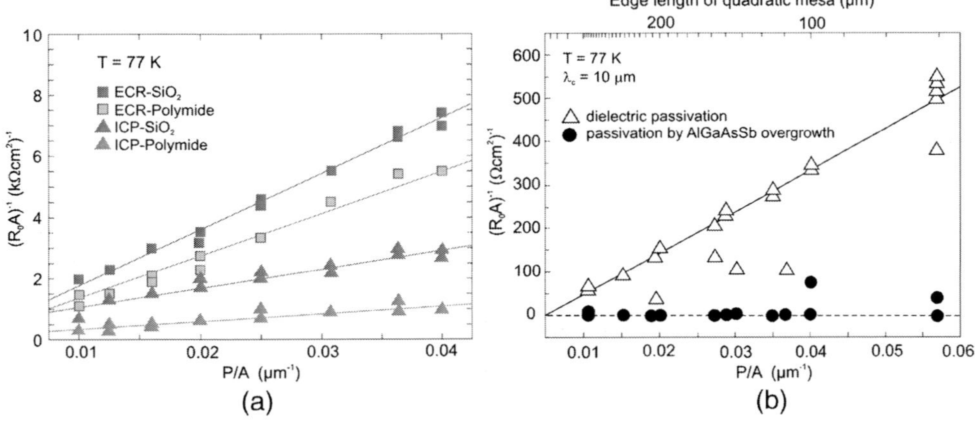

Figure 5.17 Dependence of $(R_0A)^{-1}$ on P/A for LWIR T2SL photodiodes at 77 K: (a) detectors fabricated with ICP and ECR etching techniques and protected by SiO_2 and polymide (reprinted from Ref. 45 with permission from AIP Publishing); (b) detectors passivated by $Al_xGa_{1-x}As_ySb_{1-y}$ overgrowth and by the conventional dielectric layer passivation method (reprinted from Ref. 46 with permission from AIP Publishing).

encapsulation method (polyimide) revealed an order of magnitude lower dark current density for the ICP-polyimide sample. Dielectric passivation with low-fixed and interfacial charge densities at process temperatures substantially lower than the T2LS growth temperature (to prevent T2SL period mixing) presents the challenge of developing high-quality passivation layers. It appears that the band bending at the mesa sidewalls caused by the abrupt termination of the periodic crystal structure induces accumulation or type inversion of charge, which results in surface tunneling currents along the sidewalls. As was demonstrated by Delaunay et al.,[47] native fixed charges present in the dielectric passivation layer (e.g., SiO_2) can either improve or deteriorate the narrow-bandgap device performance. A good method for controling band bending of the SiO_2-T2SL to establish the flat-band condition and suppress the leakage current is to apply a negative-bias voltage along the device sidewalls.[48]

Effective methods of T2SL device passivation involve encapsulation of etched sidewalls with a wide-bandgap material or the "shallow etch" technique, which isolates the neighboring devices but terminates within a wider-bandgap layer. Rehm et al. have used MBE overgrowth of a lattice matched, large-bandgap AlGaAsSb layer over etched mesa sidewalls.[46] In order to prevent the Al-containing passivation layer from oxidizing, a thin silicon nitride layer was deposited after the re-growth process. Figure 5.17(b) shows the dependence of $(R_0A)^{-1}$ on the perimeter-to-area ratio for two similar detector structures—one passivated by the overgrowth and the other by conventional dielectric layer passivation. No surface leakage is observed for overgrowth passivation (the slope is close to zero). In addition, Szmulowicz and Brown[49] proposed mesa sidewall encapsulation with GaSb to eliminate the surface currents. The GaSb encapsulant acts as a barrier to electrons at both the n- and p-type sides of the SL.

It appears that the reproducivity and long-term stability achieved by the SiO_2 passivation layer is more critical for photodiodes in the LWIR range than in the MWIR range. In general, the inversion potentials are larger for higher-bandgap materials; therefore, SiO_2 can passivate high-bandgap materials (MWIR photodiodes) but not low-bandgap materials (LWIR photodiodes). Using this property, a double-heterostructure that prevents the inversion of the high-bandgap p-type and n-type superlattice contact regions has been proposed (see Fig. 4.11).[47] For such structures, the surface leakage channel at the interface between the active region and the p- or n-type contacts is considerably decreased.

Several additional design modifications that dramatically improve the LWIR photodiode dark current and R_0A product have been described. The very shallow slope of the shallow-etch samples demonstrate that it is possible to reduce excess currents that are due to sidewalls.[49]

An alternative method of eliminating excess currents due to sidewalls uses shallow-etch mesa isolation with a band-graded junction.[34,50] The primary effect of the grading is to suppress tunneling and GR currents in the depletion region at low temperatures. Since both processes depend exponentially on bandgap, it is highly advantageous to substitute a wide gap into the depletion region. In this approach, the mesa etch terminates at the point just past the junction and exposes only a very thin (300 nm), wider-bandgap region of the diode. Subsequent passivation is therefore in wider-bandgap material. As a result, this passivation reduces the electrical junction area, increases the optical fill factor, and eliminates deep trenches within the detector array. However, if the lateral diffusion lengths are larger than the distance between neighboring pixels in the FPAs (typically several pixels), crosstalk between the FPA elements can be encountered, degrading the image resolution.

Different organic materials—polyimides and photoresists—are also attractive for passivants due to the simplicity of their integration into the T2SL detector fabrication procedure. Organic passivants are usually spin coated onto a detector at room temperature in thicknesses varying from 0.2 to 100 μm. A more popular passivant is SU-8, a high-contrast, epoxy-based negative photoresist developed by IBM.[51] Photo-polymerized SU-8 is mechanically and chemically stable after a hard bake. In several papers, passivation of MWIR and LWIR T2SL detectors with SU-8 photoresist,[52–54] polyimides,[55] and AZ-1518 photoresist[56] has been reported. Polyimides are polymers of imide monomers, characterized by good thermal stability and chemical resistance, and excellent mechanical properties. For LWIR InAs/GaSb SL photodiodes ($\lambda_c = 11.0$ μm at 77 K with sides ranging in size from 25 to 50 μm) passivated with polymide layers, no surface dependence was observed for diodes with R_0A values within the range of 6–13 Ωcm^2.[55]

Stimulated by the successful application of sulfide passivation of GaAs surfaces, T2SL devices have been also passivated by alkaline sulfides, including Na_2S and $(NH_4)_2S$ in aqueous solutions. It was found that chalcogenide passivation through immersion in a sulfur-containing solution, or deposition of a sulfur-based layer, effectively removes native oxide with minimal surface etching and creates a covalently bonded sulfur layer. However, chalcogen-based passivation does not provide physical protection and encapsulation of the device, and temporal instability of such a passivation layer has been reported.[44]

The most effective technique to reduce the surface leakage current in T2SL devices is currently the gating technique shown schematically in Fig. 4.11. By creating a metal gate electrode on top of a dielectric passivation layer, surface leakage can be tailored by an applied voltage.[48]

Summarizing the technological passivation problems, various passivation techniques have been developed for T2SL detectors. Up until now, however, there is no universal approach that would equally efficiently treat SL detectors with different cutoff wavelengths. In particular, more studies need to be conducted on the long-term stability of proposed passivation schemes to successfully integrate them into the FPA fabrication procedure.

5.4 Noise Mechanisms in Type-II Superlattice Photodetectors

A proper understanding of the noise behavior of T2SL photodetectors is still lacking. The observed noise behavior is very complex and, depending on the circumstances in the diode under test, several mechanisms appear to be present. Detailed data on the noise properties are still sparse in the published literature.[57–65] The last general remark concerns different types of T2SL photodetectors: p–i–n photodiodes, nBn detectors, and interband cascade (IBC) detectors. These last two types of T2SL photoconductors are described in Chapters 6 and 7.

According to classical theory, the fundamental noise current in diffusion and GR-limited photodiodes is given by

$$I_n^2 = 2q(I_{dark} + 2I_s)\Delta f, \qquad (5.5)$$

where q denotes the elementary charge, I_{dark} is the total dark current, I_s is the reverse-bias saturation value of the diffusion current, and Δf is the measurement bandwidth. At zero-bias voltage, Eq. (5.5) reduces to the well-known Johnson noise expression. In the case of a GR-limited photodiode at high bias, $2I_s$ is small compared to I_d, and Eq. (5.5) tends toward the familiar shot-noise expression.

For the high-quality T2SL photodiodes that are limited by the SRH processes, the noise current follows Eq. (5.5). An example of such behavior is shown in Fig. 5.18 for an InAs-rich MWIR InAs/GaSb p–i–n photodiode measured at the 7-kHz bandwidth. At low frequency, $1/f$ noise is the most important. The structure consists of a 200-nm Be-doped (p$^+$-type doping $\sim 1 \times 10^{18}$ cm^{-3}) GaSb buffer layer on a lattice-matched GaSb substrate,

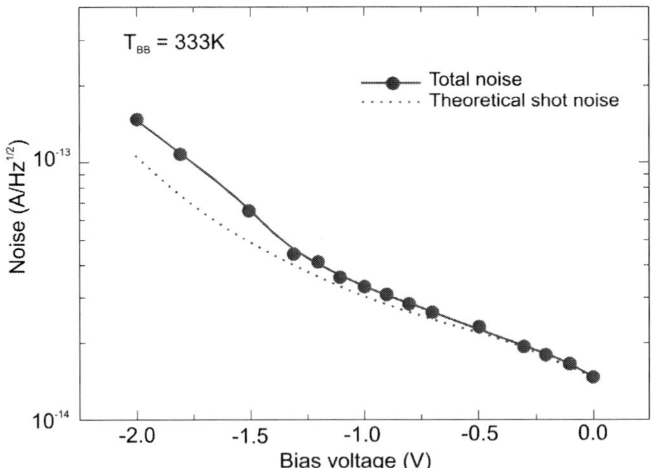

Figure 5.18 Experimental noise versus bias voltage for a 60 °C blackbody at 77 K operating temperature and measured at 7-kHz bandwidth (reprinted from Ref. 64).

several periods of p^+-doped SL, a non-intentionally doped InAs/GaSb SL active region, several periods of n^+-doped SL, and a 20-nm-thick Te-doped (n^+-type doping $\sim 1 \times 10^{18}$ cm^{-3}) InAs cap layer. The nonintentionally doped InAs-rich SL active region is composed of 300 periods of 7.5 InAs MLs and 3.5 GaSb MLs (7.5/3.5 SL structure) for a total thickness of 1 μm. Metallizations are ensured by Cr/Au on the top of the mesa and on the back of the substrate.

In several papers it was shown that 1/f noise is not intrinsically present in T2SL structures. However, large, extraneous, and frequency-dependent noise is generated by sidewall leakage currents.[57,60] As explained, an effective way to eliminate sidewall leakage currents is the development of reliable passivation.

Authors from Fraunhofer Institute in Freiburg[61–63] compared the measured noise from a number of mid- and long-wavelength p–i–n InAs/GaSb photodiodes having a large junction area of 400×400 μm^2 to measurements using smaller reference diodes in a regime of low-frequency white noise. The dark current had varied about four orders of magnitude due to the presence of macroscopic defects in those large-area diodes and strongly increased compared to the GR-limited value of the bulk. The simple shot-noise model completely failed. The shot-noise model only explained the noise of devices with a dark current close to the GR-limited bulk level, and the deviation of the experimentally observed noise from the expected shot noise increased with increasing dark current (see Fig. 5.19). To explain these experimental data, McIntyre's excess-noise model for electron-initiated avalanche multiplication was successfully implemented. It was suggested that the increase in dark current and excess noise is caused by the presence of high-electric-field domains

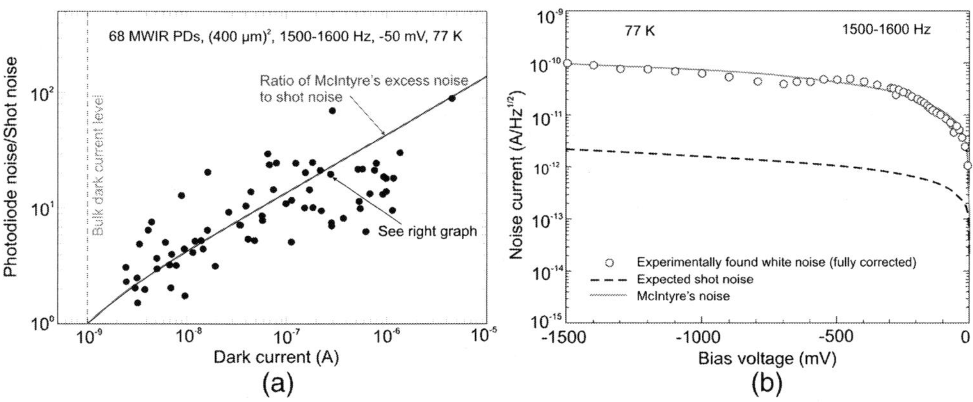

Figure 5.19 Noise data of MWIR homojunction InAs/GaSb T2SL photodiodes: (a) ratio of the experimentally found photodiode white noise to the expected shot noise versus dark current at 77 K and ~50 mV reverse bias for a set of 68 photodiodes with a size of 400×400 μm^2. The dashed vertical line marks the GR-limited bulk dark current level found in small sized diodes; (b) photodiode white noise at 77 K versus bias voltage for the device indicated by the arrow in part (a) (adapted from Ref. 63).

at the sites of crystallographic defects, which give rise to avalanche multiplication processes.

Ciura et al.[65] have investigated the role of GR and diffusion currents in the generation of $1/f$ noise in different MWIR infrared photodetectors with different InAs/GaSb SL absorbers. Measurements of $1/f$ noise at constant, small, reverse-bias voltage versus temperature show that noise intensity follows squared leakage current and that there is no contribution to $1/f$ noise from GR or diffusion currents, or else it is too small to be observed. This general observation should be attributed to InAs/GaSb SL material rather than to any device-specific feature, since the batch of examined devices contained specimens with various architectures (p-i-n photodiode, nBn detector, and IBC detector), passivation methods, and substrates. Figure 5.20 shows samples of noise power spectral density (PSD) for reverse-biased photodetectors. Only PSDs for HgCdTe and T2SL p-i-n devices have pure $1/f$ noise. For nBn and IBC detectors, the general $1/f$ dependence is distributed by processes with a thermally Lorentzian spectrum with different corner frequencies.

New insight into $1/f$ noise was given by a paper published by Kinch et al. in 2013.[66] A simple n^{+}-n^{-}-p-p^{+} diode geometry shown in Fig. 5.21 was considered for which the following assumptions were made: the fixed charge in the passivation is positive and generates an accumulation layer on the n-side and depletion layer at the surface on the p-side; the donor

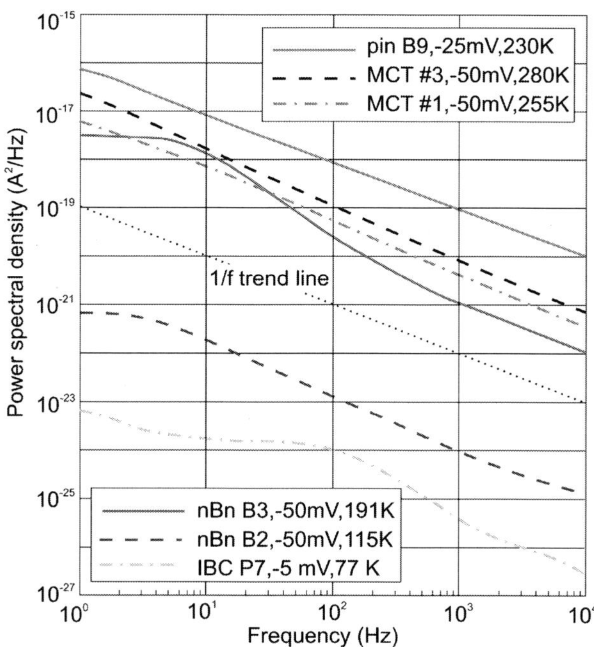

Figure 5.20 Samples of noise PSD measured for reverse-biased specimens (reprinted from Ref. 65 with permission from IEEE).

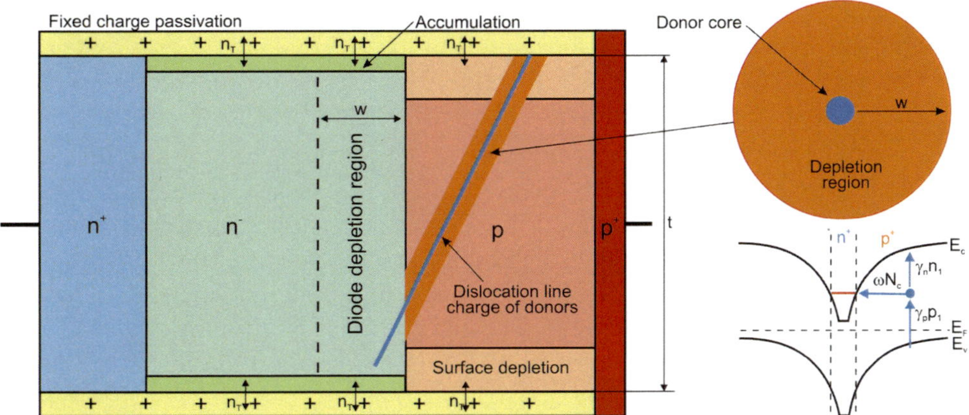

Figure 5.21 n$^+$-n$^-$-p-p$^+$ diode architecture with fixed, positively charged, passivated surface together with the donor-pipe dislocation concept in p-semiconductor volume (repinted from Ref. 66).

concentration n^- is less than the acceptor concentration p, i.e., $n^- \ll p$ (the main depletion region is formed completely on the n-side); the surface on the p-side may or may not support an inversion layer, depending on the magnitude of the fixed positive charge in the passivation and the reverse bias on the diode.

Two models of the 1/f noise components were designed based on charge fluctuations out of McWhorter-like surface traps by tunneling.[67] The first model presents the components as systemic, being associated with passivated external surfaces of the diodes, and the second presents them as isolated defects associated with the internal surfaces of built-in physical defects such as dislocations.

The noise current spectral density can be expressed as

$$S_I = \left[\frac{qH}{2\tau^i_{A1}} - \frac{qn_i^2}{(n+n_i)^2\tau_{SRH}} - \frac{qn_i}{\tau_{SRH}}\frac{w}{2nt}\right]\frac{n_T A_{dif}}{f}$$
$$+ \left[\frac{qn_i}{\tau_{SRH}n} - \frac{q}{2\tau^i_{A1}} - \frac{qn_i^2}{(n+n_i)\tau_{SRH}}\right]\frac{N_T A_{dep}}{f}, \qquad (5.6)$$

where H represents the possible suppression of surface charge modulation by the screening effect of the accumulation layer that is generated by the fixed positive charge; w is the depletion region width; t is the thickness of the n-region; A_{dif} and A_{dep} represent the surface areas of the diffusion and depletion volumes, respectively, on the n-side of the diode contributing trap fluctuations; and N_T is the effective surface trap density. Every unit of charge that tunnels into the accumulated and depleted semiconductor surface region generates a change in the charge of this region, resulting in a possible modulation of some of the minority-carrier dark current components flowing through the diode, i.e., the $1/f$ spectrum in dark current.

From Eq. (5.6) it can be concluded that if the depletion current dominates, then the systemic $1/f$ noise will vary as n_i, whereas for diffusion current it will vary as n_i^2. Typically, at lower temperatures, where depletion currents prevail, the diode $1/f$ performance will be dominated by the side with the lower doping. However, at higher temperatures, where Auger generation dominates, the $1/f$ noise will be independent of the doping concentration on both sides of the junction. These general conclusions are supported by experimental data for both MWIR and LWIR HgCdTe photodiodes.[66] For an accumulated photoconductor, the depletion terms in Eq. (5.6) are absent, and the $1/f$ noise varies as n_i^2 at all temperatures.

Kinch et al.[66] have also modeled isolated defect noise attributed to the effect of the donor-pipe dislocation concept shown in Fig. 5.21. Their model built on the work of Baker and Maxey, who proposed a model for the behavior of threading dislocations in HgCdTe.[68] The dislocation is treated as an n-type pipe of donors along the edge of the dislocation. If the dislocation traverses the n–p junction and protrudes into the p-volume, it will be encased within a surrounding depletion region, as illustrated in Fig. 5.21. However, the effect of the dislocation is minimal on the n-type side of the junction or in a simple n-type photoconductor.

Typically, isolated defect pixels display excess dark current and/or excess noise, and are a leading cause of FPA operability particularly at high temperatures. For example, HgCdTe FPA operability is usually limited not by dark current defects but by noise defects. Pixels with high $1/f$ noise should produce a tail in the rms noise distribution.[69–71] Similar behavior is observed for T2SL photodiodes.[32]

The barrier detectors described in Chapter 6 alleviate the dark current problem associated with the short SRH lifetimes in III-V semiconductors. Absence of depletion regions in the absorption layer may also give the nBn detector some degree of immunity to dislocations and other defects, and may allow growth detector structures on lattice-mismatched substrates. However, the p-type barrier layer inherently causes depletion regions to form in the narrow-gap absorption layer of the nBn detector for all bias voltages, which should be avoided.

The hypothetical systemic $1/f$ noise due to McWhorter-like surface states located on passivated mesa sidewalls without the depletion current term is shown in Fig. 5.22(a) as a function of inverse temperature for a MWIR nBn device with $\tau_{SR} = 400$ ns, $n = 10^{16}$ cm^{-3}, $n_T = 10^{12}$ cm^{-2}, and $H = 1$. Similar calculations for a LWIR T2SL are shown in Fig. 5.22(b). The modeled $1/f$ noise for nBn devices with incorporated depletion regions is considerably higher, which is also shown in Fig. 5.22.

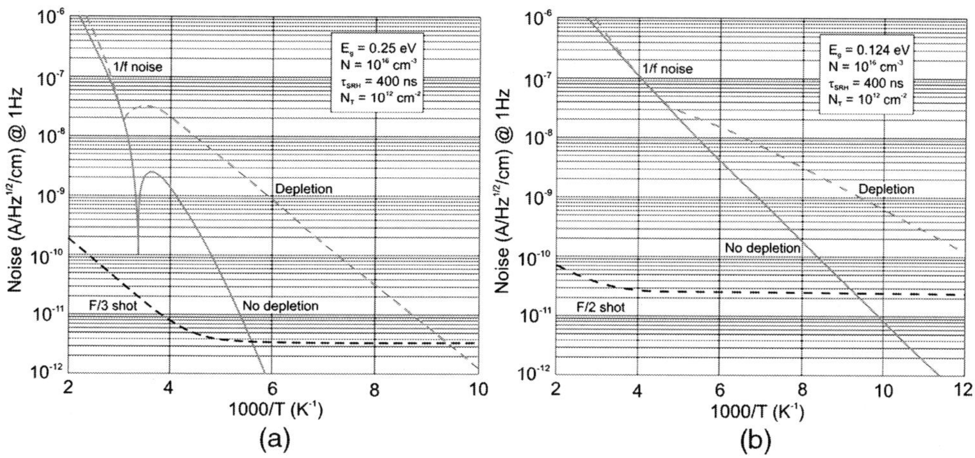

Figure 5.22 $1/f$ noise figure versus inverse temperature for (a) MWIR and (b) LWIR nBn devices with $\tau_{SR} = 400$ ns, $n = 10^{16}$ cm^{-3}, $n_T = 10^{12}$ cm^{-2}, and $H = 1$ (reprinted from Ref. 71).

References

1. J. N. Schulman and T. C. McGill, "The CdTe/HgTe superlattice: Proposal for a new infrared material," *Appl. Phys. Lett.* **34**, 663–665 (1979).
2. D. L. Smith, T. C. McGill, and J. N. Schulman, "Advantages of the HgTe-CdTe superlattice as an infrared detector material," *Appl. Phys. Lett.* **43**, 180–182 (1983).
3. B. F. Levine, "Quantum-well infrared photodetectors," *J. Appl. Phys.* **74**, R1–R81 (1993).
4. A. Rogalski, *Infrared Detectors*, 2nd edition, CRC Press, Boca Raton, Florida (2010).
5. G. C. Osbourn, "InAsSb strained-layer superlattices for long wavelength detector applications," *J. Vac. Sci. Technol. B* **2**, 176–178 (1984).
6. D. L. Smith and C. Mailhiot, "Proposal for strained type II superlattice infrared detectors," *J. Appl. Phys.* **62**, 2545–2548 (1987).
7. C. Mailhiot and D. L. Smith, "Long-wavelength infrared detectors based on strained InAs-GaInSb type-II superlattices," *J. Vac. Sci. Technol. A* **7**, 445–449 (1989).
8. R. H. Miles, D. H. Chow, J. N. Schulman, and T. C. McGill, "Infrared optical characterization of InAs/Ga$_{1-x}$In$_x$Sb superlattices," *Appl. Phys. Lett.* **57**, 801–803 (1990).
9. E. R. Youngdale, J. R. Meyer, C. A. Hoffman, F. J. Bartoli, C. H. Grein, P. M. Young, H. Ehrenreich, R. H. Miles, and D. H. Chow, "Auger lifetime enhancement in InAs-Ga$_{1-x}$In$_x$Sb superlattices," *Appl. Phys. Lett.* **64**, 3160–3162 (1994).

10. C. H. Grein, P. M. Young, M. E. Flatté, and H. Ehrenreich, "Long wavelength InAs/InGaSb infrared detectors: Optimization of carrier lifetimes," *J. Appl. Phys.* **78**, 7143–7152 (1995).

11. J. L. Johnson, L. A. Samoska, A. C. Gossard, J. L. Merz, M. D. Jack, G. H. Chapman, B. A. Baumgratz, K. Kosai, and S. M. Johnson, "Electrical and optical properties of infrared photodiodes using the InAs/$Ga_{1-x}In_xSb$ superlattice in heterojunctions with GaSb," *J. Appl. Phys.* **80**, 1116–1127 (1996).

12. F. Fuchs, U. Wcimcr, W. Plctschcn, J. Schmitz, E. Ahlswcdc, M. Walthcr, J. Wagner, and P. Koidl, "High performance InAs/$Ga_{1-x}In_xSb$ superlattice infrared photodiodes," *Appl. Phys. Lett.* **71**, 3251–3253 (1997).

13. R. Taalat, J.-B. Rodriguez, M. Delmas, and P. Christol, "Influence of the period thickness and composition on the electro-optical properties of type-II InAs/GaSb midwave infrared superlattice photodetectors," *J. Phys. D: Appl. Phys.* **47**, 015101 (2014).

14. H. S. Kim, E. Plis, A. Khoshakhlagh, S. Myers, N. Gautam, Y. D. Sharma, L. R. Dawson, S. Krishna, S. J. Lee, and S. K. Noh, "Performance improvement of InAs/GaSb strained layer superlattice detectors by reducing surface leakage currents with SU-8 passivation," *Appl. Phys. Lett.* **96**, 033502 (2010).

15. D. Hoffman, B. M. Nguyen, P. Y. Delaunay, A. Hood, and M. Razeghi, "Beryllium compensation doping of InAs/GaSb infrared superlattice photodiodes," *Appl. Phys. Lett.* **91**, 143507 (2007).

16. B. Klein, E. Plis, M. N. Kutty, N. Gautam, A. Albrecht, S. Myers, and S. Krishna: "Varshni parameters for InAs/GaSb strained layer superlattice infrared photodetectors," *J. Phys. D: Appl. Phys.* **44**, 075102 (2011).

17. H. S. Kim, E. Plis, N. Gautam, A. Khoshakhlagh, S. Myers, M. N. Kutty, Y. Sharma, L. R. Dawson, and S. Krishna, "SU-8 passivation of type-II InAs/GaSb strained layer superlattice detectors," *Proc. SPIE* **7660**, 76601U (2010) [doi: 10.1117/12.850284].

18. A. Rogalski, K. Adamiec, and J. Rutkowski, *Narrow-Gap Semiconductor Photodiodes*, SPIE Press, Bellingham (2000).

19. J. Wrobel, P. Martyniuk, E. Plis, P. Madejczyk, W. Gawron, S. Krishna, and A. Rogalski, "Dark current modeling of MWIR type-II superlattice detectors," *Proc. SPIE* **8353**, 835316 (2012) [doi: 10.1117/12.925074].

20. J. Wróbel, E. Plis, W. Gawron, M. Motyka, P. Martyniuk, P. Madejczyk, A. Kowalewski, M. Dyksik, J. Misiewicz, S. Krishna, and A. Rogalski, "Analysis of temperature dependence of dark current mechanisms in mid-wavelength infrared pin type-II superlattice photodiodes," *Sensors and Materials* **26**(4), 235–244 (2014).

21. J. Wróbel, Ł. Ciura, M. Motyka, F. Szmulowicz, A. Kolek, A. Kowalewski, P. Moszczyński, M. Dyksik, P. Madejczyk, S. Krishna, and A. Rogalski, "Investigation of a near mid-gap trap energy level in mid-wavelength

infrared InAs/GaSb type-II superlattices," *Semicond. Sci. Technol.* **30**, 115004 (2015).

22. P. Martyniuk, J. Wróbel, E. Plis, P. Madejczyk, A. Kowalewski, W. Gawron, S. Krishna, and A. Rogalski, "Performance modeling of MWIR InAs/GaSb/B-Al$_{0.2}$Ga$_{0.8}$Sb type-II superlattice nBn detector," *Semicond. Sci. Technol.* **27**, 055002 (2012).

23. D. R. Rhiger, "Performance comparison of long-wavelength infrared type II superlattice devices with HgCdTe," *J. Electron. Mater.* **40**, 1815–1822 (2011).

24. P. Christol and J. B. Rodriguez, "Progress on type-II InAs/GaSb superlattice infrared photodetectors: from MWIR to VLWIR spectral domains," *ICSO 2014 International Conference on Space Optics*, Tenerife, Canary Islands, 7–10 October (2014).

25. P. Christol, M. Delmas, R. Rossignol, and J. B. Rodriguez, "Influence of the InAs/GaSb super lattice period composition on the electro-optical performances of T2SL infrared photodiode," *3rd International Conference and Exhibition on Lasers, Optics and Photonics*, Valencia, 1–3 September (2015).

26. S. P. Svensson, D. Donetsky, D. Wang, H. Hier, F. J. Crowne, and G. Belenky, "Growth of type II strained layer superlattice, bulk InAs and GaSb materials for minority lifetime characterization," *J. Crystal Growth* **334**, 103–107 (2011).

27. J. L. Johnson, "The InAs/GaInSb strained layer superlattice as an infrared detector material: An overview," *Proc. SPIE* **3948**, 118 (2000) [doi: 10.1117/12.382111].

28. G. J. Brown, "Type-II InAs/GaInSb superlattices for infrared detection: an overview," *Proc. SPIE* **5783**, 65 (2005) [doi: 10.1117/12.606621].

29. M. Razeghi, Y. Wei, A. Gin, A. Hood, V. Yazdanpanah, M. Z. Tidrow, and V. Nathan, "High performance type II InAs/GaSb superlattices for mid, long, and very long wavelength infrared focal plane arrays," *Proc. SPIE* **5783**, 86 (2005) [doi: 10.1117/12.605291].

30. J. Pellegrino and R. DeWames, "Minority carrier lifetime characteristics in type II InAs/GaSb LWIR superlattice n^{+}–p^{+} photodiodes," *Proc. SPIE* **7298**, 72981U (2009) [doi: 10.1117/12.819641].

31. R. Rehm, F. Lemke, J. Schmitz, M. Wauro, and M. Walther, "Limiting dark current mechanisms in antimony-based superlattice infrared detectors for the long-wavelength infrared regime," *Proc. SPIE* **9451**, 94510N (2015) [doi: 10.1117/12.2177091].

32. R. Rehm, F. Lemke, M. Masur, J. Schmitz, T. Stadelman, M. Wauro, A. Wörl, and M. Walther, "InAs/GaSb superlattice infrared detectors," *Infrared Physics & Technol.* **70**, 87–92 (2015).

33. C. L. Canedy, H. Aifer, I. Vurgaftman, J. G. Tischler, J. R. Meyer, J. H. Warner, and E. M. Jackson, "Antimonide type-II W photodiodes with

long-wave infrared R_0A comparable to HgCdTe," *J. Electron. Mater.* **36**, 852–856 (2007).

34. E. H. Aifer, J. G. Tischler, J. H. Warner, I. Vurgaftman, W. W. Bewley, J. R. Meyer, C. L. Canedy, and E. M. Jackson, "W-structured type-II superlattice long-wave infrared photodiodes with high quantum efficiency," *Appl. Phys. Lett.* **89**, 053519 (2006).

35. B.-M. Nguyen, D. Hoffman, Y. Wei, P.-Y. Delaunay, A. Hood, and M. Razeghi, "Very high quantum efficiency in type-II InAs/GaSb superlattice photodiode with cutoff of 12 μm," *Appl. Phys. Lett.* **90**, 231108 (2007).

36. J. Bajaj, G. Sullivan, D. Lee, E. Aifer, and M. Razeghi, "Comparison of type-II superlattice and HgCdTe infrared detector technologies," *Proc. SPIE* **6542**, 65420B (2007) [doi: 10.1117/12.723849].

37. C. H. Grein, P. M. Young, and H. Ehrenreich, "Minority carrier lifetimes in ideal InGaSb/InAs superlattices," *Appl. Phys. Lett.* **61**, 2905–2907 (1992).

38. C. H. Grein, H. Cruz, M. E. Flatte, and H. Ehrenreich, "Theoretical performance of very long wavelength InAs/In$_x$Ga$_{1-x}$Sb superlattice based infrared detectors," *Appl. Phys. Lett.* **65**, 2530–2532 (1994).

39. A. Hood, D. Hoffman, Y. Wei, F. Fuchs, and M. Razeghi, "Capacitance-voltage investigation of high-purity InAs/GaSb superlattice photodiodes," *Appl. Phys. Lett.* **88**, 052112 (2006).

40. D. Lackner, M. Steger, M. L. W. Thewalt, O. J. Pitts, Y. T. Cherng, S. P. Watkins, E. Plis, and S. Krishna, "InAs/InAsSb strain balanced superlattices for optical detectors: material properties and energy band simulations," *J. Appl. Phys.* **111**, 034507–034510 (2012).

41. T. Schuler-Sandy, S. Myers, B. Klein, N. Gautam, P. Ahirwar, Z.-B. Tian, T. Rotter, G. Balakrishnan, E. Plis, and S. Krishna, "Gallium free type II InAs/InAsSb superlattice photodetectors," *Appl. Phys. Lett.* **101**, 071111 (2012).

42. A. M. Hoang, G. Chen, R. Chevallier, A. Haddadi, and M. Razeghi, "High performance photodiodes based on InAs/InAsSb type-II superlattices for very long wavelength infrared detection," *Appl. Phys. Lett.* **104**, 251105 (2014).

43. A. Rogalski, P. Martyniuk, and M. Kopytko, "Challenges of small-pixel infrared detectors: a review," *Rep. Prog. Phys.* **79**, 046501 (2016).

44. E. Plis, M. N. Kutty, and S. Krishna, "Passivation techniques for InAs/GaSb strained layer superlattice detectors," *Laser Photonics Rev.* **7**(1), 45–59 (2013).

45. E. K. Huang, D. Hoffman, B.-M. Nguyen, P.-Y. Delaunay, and M. Razeghi, "Surface leakage reduction in narrow band gap type-II antimonide-based superlattice photodiodes," *Appl. Phys. Lett.* **94**, 053506 (2009).

46. R. Rehm, M. Walther, F. Fuchs, J. Schmitz, and J. Fleissner, "Passivation of InAs/(GaIn)Sb short-period superlattice photodiodes with 10 μm cutoff

wavelength by epitaxial overgrowth with $Al_xGa_{1-x}As_ySb_{1-y}$," *Appl. Phys. Lett.* **86**, 173501 (2005).

47. P. Y. Delaunay, A. Hood, B. M. Nguyen, D. Hoffman, Y. Wei, and M. Razeghi, "Passivation of type-II InAs/GaSb double heterostructure," *Appl. Phys. Lett.* **91**, 091112 (2007).

48. G. Chen, B.-M. Nguyen, A. M. Hoang, E. K. Huang, S. R. Darvish, and M. Razeghi, "Elimination of surface leakage in gate controlled type-II InAs/GaSb mid-infrared photodetectors," *Appl. Phys. Lett.* **99**, 183503 (2011).

49. F. Szmulowicz and G. J. Brown, "GaSb for passivating type-II InAs/GaSb superlattice mesas," *Infrared Phys. Technol.* **53**, 305–307 (2011).

50. E. H. Aifer, H. Warner, C. L. Canedy, I. Vurgaftman, J. M. Jackson, J. G. Tischler, J. R. Meyer, S. P. Powell, K. Oliver, and W. E. Tennant, "Shallow-etch mesa isolation of graded-bandgap W-structured type II superlattice photodiodes," *J. Electron. Mater.* **39**, 1070–1079 (2010).

51. US Patent No. 4882245, 1989.

52. H. S. Kim, E. Plis, A. Khoshakhlagh, S. Myers, N. Gautam, Y. D. Sharma, L. R. Dawson, S. Krishna, S. J. Lee, and S. K. Noh, "Performance improvement of InAs/GaSb strained layer superlattice detectors by reducing surface leakage currents with SU-8 passivation," *Appl. Phys. Lett.* **96**, 033502–033504 (2010).

53. E. A. DeCuir, Jr., J. W. Little, and N. Baril, "Addressing surface leakage in type-II InAs/GaSb superlattice materials using novel approaches to surface passivation," *Proc. SPIE* **8155**, 815508 (2011) [doi: 10.1117/12.895448].

54. H. S. Kim, E. Plis, N. Gautam, S. Myers, Y. Sharma, L. R. Dawson, and S. Krishna, "Reduction of surface leakage current in InAs/GaSb strained layer long wavelength superlattice detectors using SU-8 passivation," *Appl Phys. Lett.* **97**, 14351 (2010).

55. A. Hood, P.-Y. Delaunay, D. Hoffman, B.-M. Nguyen, Y. Wei, and M. Razeghi, "Near bulk-limited R_0A of long-wavelength infrared type-II InAs/GaSb superlattice photodiodes with polyimide surface passivation," *Appl Phys. Lett.* **90**, 233513 (2007).

56. R. Chaghi, C. Cervera, H. Ait-Kaci, P. Grech, J. B. Rodriguez, and P. Christol, "Wet etching and chemical polishing of InAs/GaSb super-lattice photodiodes," *Semicond. Sci. Technol.* **24**, 065010 (2009).

57. A. Soibel, D.Z.-Y. Ting, C. J. Hill, M. Lee, J. Nguyen, S. A. Keo, J. M. Mumolo, and S. D. Gunapala, "Gain and noise of high-performance long wavelength superlattice infrared detector," *Appl. Phys. Lett.* **96**, 111102 (2010).

58. V. M. Cowan, C. P. Morath, S. Myers, N. Gautam, and S. Krishna, "Low temperature noise measurement of an InAs/GaSb-based nBn MWIR detector," *Proc. SPIE* **8012**, 801210 (2011) [doi: 10.1117/12.884808].

59. C. Cervera, I. Ribet-Mohamed, R. Taalat, J. P. Perez, and P. Christol, and Rodriguez, "Dark current and noise measurements of an InAs/GaSb

superlattice photodiode operating in the midwave infrared domain," *J. Electron. Mater.* **41**(10), 2714–2718 (2012).

60. T. Tansel, K. Kutluer, A. Muti, O. Salihoglu, A. Aydinli, and R. Turan, "Surface recombination noise in InAs/GaSb superlattice photodiodes," *Applied Physics Express* **6**, 032202 (2013).

61. A. Wörl, R. Rehm, and M. Walther, "Excess noise in long-wavelength infrared InAs/GaSb type-II superlattice pin-photodiodes," *22nd International Conference on Noise and Fluctuations (ICNF)*, 24–28 June 2013.

62. R. Rehm, A. Wörl, and M. Walther, "Noise in InAs/GaSb type-II superlattice photodiodes," *Proc. SPIE* **8631**, 86311M (2013) [doi: 10.1117/12.2013854].

63. M. Walther, A. Wörl, V. Daumer, R. Rehm, L. Kirste, F. Rutz, and J. Schmitz, "Defects and noise in type-II superlattice infrared detectors," *Proc. SPIE* **8704**, 87040U (2013) [doi: 10.1117/12.2015926].

64. E. Giard, R. Taalat, M. Delmas, J.-B. Rodriguez, P. Christol, and I. Ribet-Mohamed, "Radiometric and noise characteristics of InAs-rich T2SL MWIR pin photodiodes," *J. Euro. Opt. Soc. Rap. Public.* **9**, 14022 (2014).

65. Ł. Ciura, A. Kołek, J. Wróbel, W. Gawron, and A. Rogalski, "1/f noise in mid-wavelength infrared detectors with InAs/GaSb superlattice absorber," *IEEE Trans. Electron Devices* **62**(6), 2022–2026 (2015).

66. M. A. Kinch, R. L. Strong, and C. A. Schaake, "1/f noise in HgCdTe focal-plane arrays," *J. Electron. Mater.* **42**(11), 3243–3251 (2013).

67. A. L. McWhorter, "1/f Noise and Germanium Surface Properties," in *Semiconductor Surface Physics*, edited by R. H. Kingston, Pennsylvania University Press, Philadelphia, pp. 207–228 (1957).

68. I. M. Baker and C. D. Maxey, "Summary of HgCdTe 2D array technology in the U.K.," *J. Electron. Mater.* **30**(6), 282–289 (2000).

69. L. O. Bubulac, J. D. Benson, R. N. Jacobs, A. J. Stoltz, M. Jaime-Vasquez, L. A. Almeida, A. Wang, L. Wang, R. Hellmer, T. Golding, J. H. Dinan, M. Carmody, P. S. Wijewarnasuriya, M. F. Lee, M. F. Vilela, J. Peterson, S. M. Johnson, D. F. Lofgreen, and D. Rhiger, "The distribution tail of LWIR HgCdTe-on-Si FPAs: a hypothetical physical mechanism," *J. Electron. Mater.* **40**(3), 280–288 (2011).

70. R. L. Strong and M. A. Kinch, "Quantification and modeling of RMS noise distributions in HDVIP® infrared focal plane arrays," *J. Electron. Mater.* **43**(8), 2824–2830 (2014).

71. M. A. Kinch, *State-of-the-Art Infrared Detector Technology*, SPIE Press, Bellingham, Washington (2014) [doi: 10.1117/3.1002766].

Chapter 6
Infrared Barrier Photodetectors

Historically, the first barrier detector was proposed by A. M. White in 1983[1] as a high-impedance photoconductor. This detector postulates an n-type heterostructure with a narrow-bandgap absorber region coupled to a thin, wide-bandgap layer followed by a narrow-bandgap contact region. A. M. White in his prescient patent also proposed a bias-selectable two-color detector realized and exploited currently in HgCdTe and in T2SL material systems.

The barrier detector concept assumes almost zero one-band offset approximation throughout the heterostructure, allowing flow of only minority carriers in a photoconductor. Little or no valence band offset (VBO) was difficult to realize using standard infrared detector materials such as InSb and HgCdTe. The situation changed dramatically in the middle of first decade of the 21st century after the introduction of the 6.1-Å III-V material detector family, and when the first high-performance detectors and FPAs were demonstrated.[2,3] Introduction of unipolar barriers in various designs based on T2SLs drastically changed the architecture of infrared detectors.[4] In general, unipolar barriers are used to implement the barrier detector architecture for increasing the collection efficiency of photogenerated carriers and reducing dark current generation without inhibiting photocurrent flow. The ability to tune the positions of the conduction and valence band edges independently in a broken-gap T2SL is especially helpful in the design of unipolar barriers.

6.1 Principle of Operation

The term "unipolar barrier" was coined to describe a barrier that can block one carrier type (electron or hole) but allows an unimpeded flow of the other (see Fig. 6.1). Among the different types of barrier detectors the most popular is the nBn detector shown in Fig. 6.1. The n-type semiconductor on one side of the barrier constitutes a contact layer for biasing the device, while the n-type narrow-bandgap semiconductor on the other side of the barrier is a photon-absorbing layer whose thickness should be comparable to the absorption length of light in the device, typically several microns. The same doping type

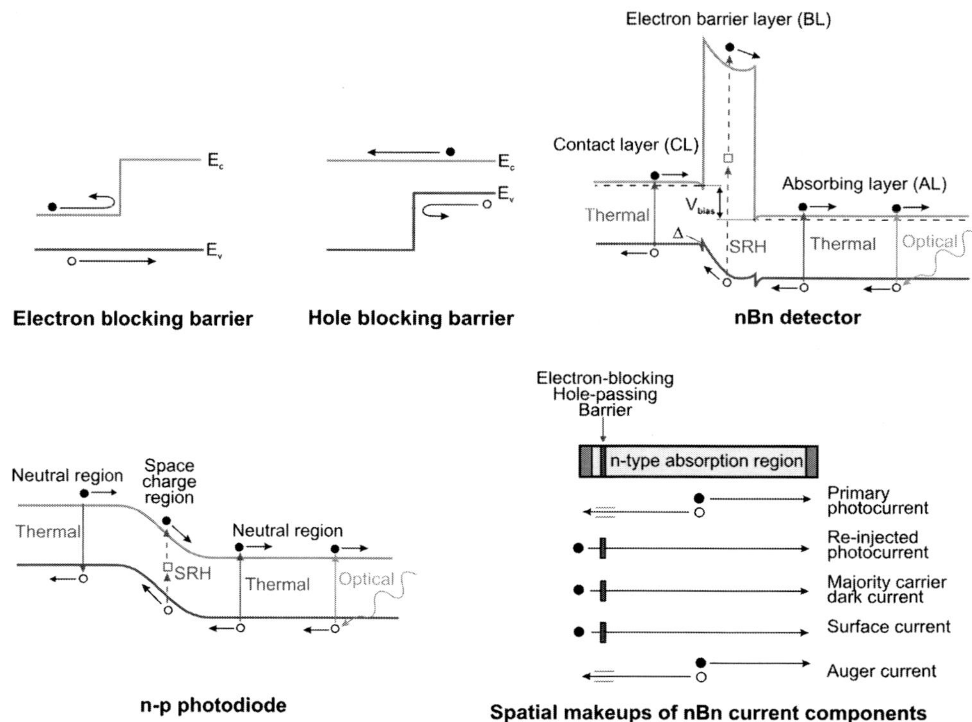

Figure 6.1 Illustrations of (clockwise) electron- and hole-blocking unipolar barriers, bandgap diagram of an nBn barrier detector (the valence band offset Δ is shown explicitly), the p-n photodiode, and the spatial makeups of the various current components and barrier blocking (adapted from Ref. 5).

in the barrier and active layers is key to maintaining a low, diffusion-limited dark current. The barrier needs to be carefully engineered. It must be nearly lattice matched to the surrounding material and have zero offset in one band and a large offset in the other. It should be located near the minority-carrier collector and away from the region of optical absorption. Such a barrier arrangement allows photogenerated holes to flow to the contact (cathode), while majority-carrier dark current, re-injected photocurrent, and surface current are blocked (see bottom right side of Fig. 6.1). Effectively, the nBn detector is designed to reduce the dark current (associated with SRH processes) and noise without impeding the photocurrent (signal). In particular, the barrier serves to reduce the surface leakage current; a benefit of the nBn architecture is self-passivation. Spatial makeups of the various current components and barrier blocking in a nBn detector are shown at the bottom right of Fig. 6.1.[5]

Other key benefits associated with the absence of depletion regions in the narrow-gap absorption layer is immunity of the nBn detector to dislocations and other defects, which may allow growth on lattice-mismatched substrates such as GaAs with reduced penalty of excess dark current generated by misfit dislocations.

Reine and co-workers developed numerical simulations and analytical models to better understand the physics and operation of simple, ideal, defect-free nBn devices with p-type[6] and n-type barriers.[7] For detectors with a p-type barrier, the approximation model is analogous to the well-known depletion approximation for the conventional p–n junction with new boundary conditions for ideal back-to-back photodiodes.

The research presented in Refs. 8 and 9 established a criterion for combining bias voltage and barrier concentration that allows operation with no depletion region in the narrow-gap absorption layer. A valence band barrier is present for an n-type barrier (see Fig. 6.1) that can significantly impede hole current transport between the absorption layer and the contact layer, which can require large bias voltages to overcome. In contrast, a p-type barrier has no barrier, but rather a potential well for holes in the valence band that does not impede hole transport between the absorption layer and the contact layer. However, in the last case, a p-type barrier layer inherently causes depletion regions to form in the narrow-gap absorption layer for all bias voltages, which should be avoided as depletion regions cause excessive GR dark currents.

The nBn detector is essentially a photoconductor with unity gain due to the absence of majority-carrier flow and in this respect is similar to a photodiode—the junction (space charge region) is replaced by an electron-blocking unipolar barrier (B), and the p-contact is replaced by an n-contact. It can be stated that the nBn design is a hybrid photoconductor and photodiode.

Figure 6.2 shows a typical Arrhenius plot of the dark current in a conventional diode and in an nBn detector. The diffusion current typically varies as

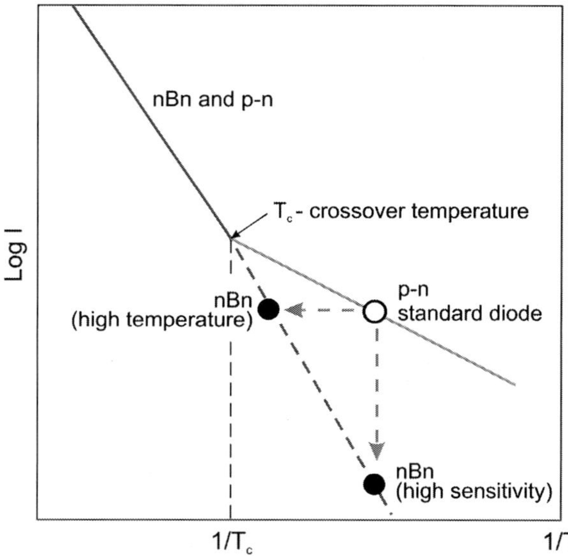

Figure 6.2 A schematic Arrhenius plot of the dark current in a standard diode and in an nBn device (adapted from Ref. 8).

$T^3\exp(-E_{g0}/kT)$, where E_{g0} is the bandgap extrapolated to zero temperature, T is the temperature, and k is Boltzman's constant. The GR current varies as $T^{3/2}\exp(-E_{g0}/2kT)$ and is dominated by the generation of electrons and holes by SRH traps in the depletion region. Because there is no depletion region in an nBn detector, the GR contribution to the dark current from the photon-absorbing layer is totally suppressed. The lower portion of the Arrhenius plot for the standard photodiode has a slope that is roughly half that of the upper portion. The solid line (nBn) is an extension of the high-temperature diffusion-limited region to temperatures below T_c, which is defined as the crossover temperature at which the diffusion and GR currents are equal. In the low-temperature region, the nBn detector offers two important advantages. First, it should exhibit a higher SNR than a conventional diode operating at the same temperature. Second, it will operate at a higher temperature than a conventional diode with the same dark current. The latter is depicted by the horizontal dashed line in Fig. 6.2.

Absence of a depletion region offers a way for materials with relatively poor SRH lifetimes, such as all III-V compounds, to overcome the disadvantage of large depletion dark currents.

The operating principles of the nBn and related detectors have been described in detail in the literature.[3–13] While the idea of an nBn design originated with bulk InAs materials,[3] its demonstration using T2SL-based materials facilitates the experimental realization of the barrier detector concept with better control of band edge alignments.[14]

Klipstein et al.[15] have considered a wide family of barrier detectors, which they divide into two groups: $XB_n\text{n}$ and $XB_p\text{p}$ detectors (see Fig. 6.3). In the

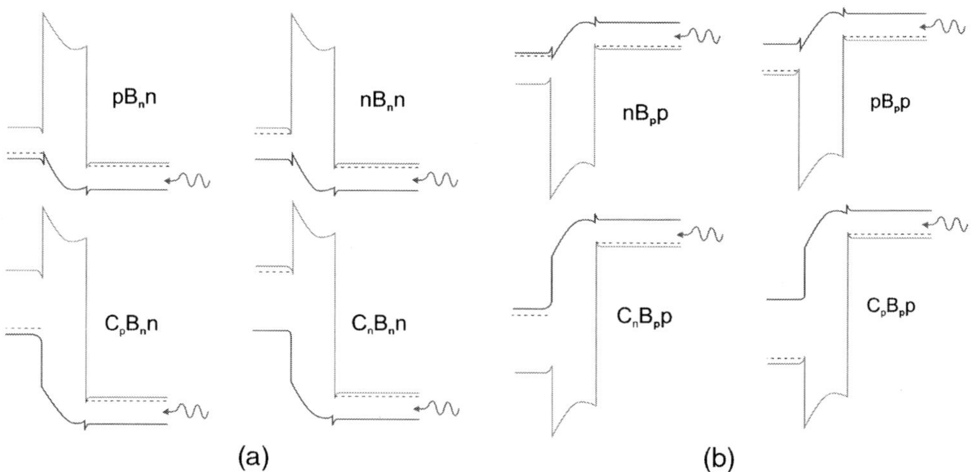

Figure 6.3 Schematic band profile configurations under operating bias for (a) $XB_n\text{n}$ and (b) $XB_p\text{p}$ barrier detector families. In each case the contact layer X is on the left, and infrared radiation is incident on the active layer on the right. When X is composed of the same material as the active layer, both layers have the same symbol (denoting the doping type); otherwise, X is denoted as C (with the doping type as a subscript) (reprinted from Ref. 15).

case of the former group, all designs have the same n-type B_nn structural unit, but use different contact layers X, in which either the doping, material, or both, are varied. If we take, e.g., $C_p B_n$n and nB_nn devices, C_p is the p-type contact made from a material other than the active layer, whereas n is the n-type contact made from the same material as the active layer. In the case of a pB_nn structure, the p–n junction can be located at the interface between the heavily doped p-type material and the lower-doped barrier, or within the lower-doped barrier itself. Our barrier detector family also has p-type members, designated as XB_pp, which are polarity-reversed versions of the n-type detectors. The pBp architecture should be employed when the surface conduction of the materials is p-type and must be used with a p-type absorbing layer. This structure can be realized using, e.g., a p-type InAs/GaSb T2SLs as the absorbing layer.[13,16,17] In addition, the so-called pMp device consists of two p-doped superlattice active regions and a thin M structure with a higher-energy barrier. The bandgap difference between the superlattice M structure falls in the valence band, creating a valence band barrier for the majority holes in the p-type semiconductor.

Unipolar barriers can also be inserted into a conventional p-n photodiode architecture.[5,18] There are two possible locations into which a unipolar barrier can be implemented: (1) outside the depletion layer in the p-type layer or (2) near the junction but at the edge of the n-type absorbing layer (see Fig. 6.4). Depending on the barrier placement, different dark current components are filtered. For example, placing the barrier in the p-type layer blocks surface current, but currents due to diffusion, GR, trap-assisted tunneling (TAT), and band-to-band tunneling (BBT) cannot be blocked. If the barrier is placed in the n-type region, the junction-generated currents and surface currents are effectively filtered out. The photocurrent shares the same spatial makeup as the diffusion current, which is shown in Fig. 6.5.

Unipolar barriers can significantly improve the performance of infrared photodiodes, as shown in Fig. 6.6 for an InAs material system. For InAs, $AlAs_{0.18}Sb_{0.82}$ is an ideal electron-blocking unipolar barrier material.

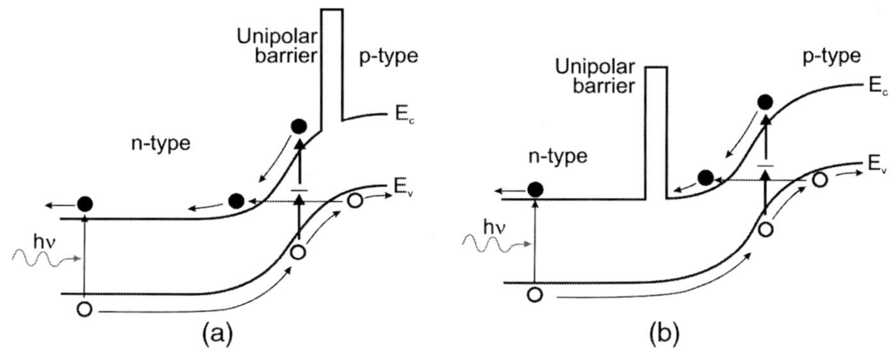

Figure 6.4 Band diagrams of (a) a p-side and (b) an n-side unipolar photodiode under bias (reprinted from Ref. 18 with permission from AIP Publishing).

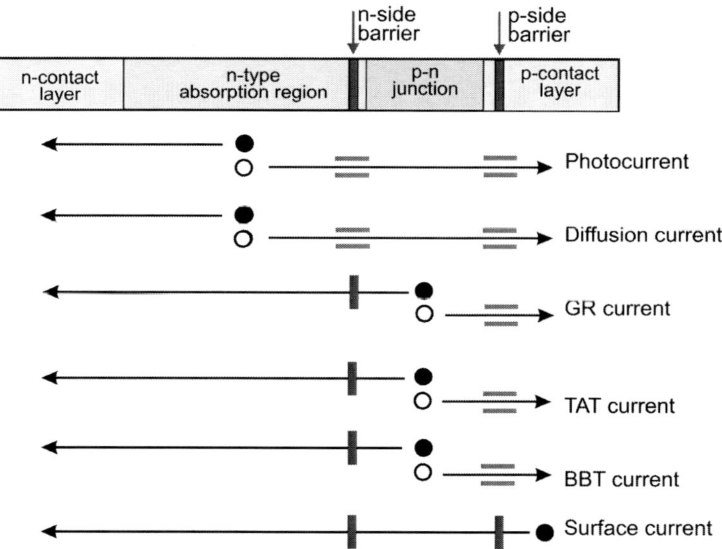

Figure 6.5 Placing the barrier in a unipolar barrier photodiode results in the filtering of surface currents and junction-related currents. The diffusion current is not filtered because it shares the same spatial makeup as the photocurrent (reprinted from Ref. 18 with permission from AIP Publishing).

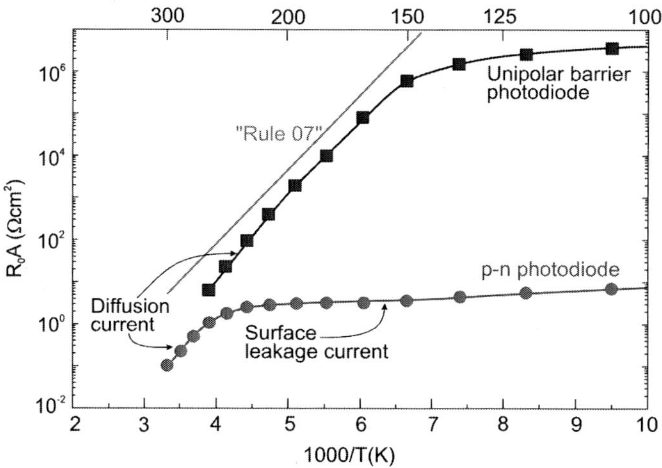

Figure 6.6 R_0A product of a conventional InAs photodiode and a comparable n-side barrier photodiode (reprinted from Ref. 18 with permission from AIP Publishing).

Theoretical predictions suggest that the VBO should be less than kT for the $\mathrm{AlAs}_y\mathrm{Sb}_{1-y}$ barrier composition in the range $0.14 < y < 0.18$. Figure 6.6 compares the temperature-dependent R_0A product data for an n-side unipolar barrier photodiode with that of a conventional p-n photodiode. The unipolar barrier photodiode shows performance near "Rule 07" with an activation energy near the bandgap of InAs, indicating diffusion-limited performance and six orders of magnitude higher R_0A in the low-temperature range compared to that of a conventional p–n junction.

The "Rule 07" criterion manifests the performance of a p-on-n HgCdTe photodiode architecture, which is limited by Auger 1 diffusion current from 10^{15} cm^{-3} n-type material and is a popular mark of reference for comparing the performance of any type of detector with that of a state-of-art HgCdTe detector. Any detector architecture that is limited by Auger 7 p-type diffusion, or by depletion currents, will not behave according to "Rule 07." In fact, the appropriate criterion to use for comparative studies is detector dark current relative to system flux current.

6.2 SWIR Barrier Detectors

An extension of barrier detectors to SWIR region up to 3 μm has been demonstrated using InGaAs and InGaAsSb alloy systems.[19,20] The standard growth method for making SWIR detectors is to utilize the MBE technique.

Savich et al.[20] carried out a comparison of electrical and optical characteristics of conventional photodiodes and nBn architecture detectors with 2.8-μm cutoff wavelengths fabricated with both lattice-mismatched InGaAs and lattice-matched InGaAsSb absorbing layers. In order to minimize the number of defects in the $In_{1-x}Ga_xAs$ absorber on InP substrate, a 2-μm-thick step-graded buffer consisting of AlInAs was growth that the lattice constant was graded from that of InP to that of $In_{0.82}Ga_{0.18}As$. Both the conventional photodiode and nBn detector include this step-graded buffer. The barrier detector includes an additional pseudomorphic AlAsSb unipolar barrier to maintain a larger conduction barrier compared to that of $In_{0.82}Ga_{0.18}As$.

In the case of a lattice-matched solution with GaSb substrate, the quaternary composition of $In_{0.30}Ga_{0.70}As_{0.56}Sb_{0.44}$ at the edge of the miscibility gap has been used to maintain the cutoff wavelength and lattice-matching requirements. In the nBn detector, also a pseudomorphic AlGaSb unipolar barrier with a larger conduction band offset (CBO) and zero VBO compared to the $In_{0.30}Ga_{0.70}As_{0.56}Sb_{0.44}$ absorber was implemented.

Figure 6.7 gathers temperature-dependent dark current characteristics for both lattice-mismatched InGaAs and lattice-matched InGaAsSb detectors at an operating reverse bias of 100 mV.[20] InGaAs-on-InP devices suffer from low material quality, i.e., threading dislocations that occur due to the lattice mismatch, with implications for device dark current performance.

The p-n InGaAs photodiode is limited by the surface leakage current below a temperature of 220 K, while the nBn detector remains diffusion limited down to 150 K. At the room-temperature background photocurrent level the nBn detector shows dark current reduced by a factor greater than 400 compared to the conventional photodiode.

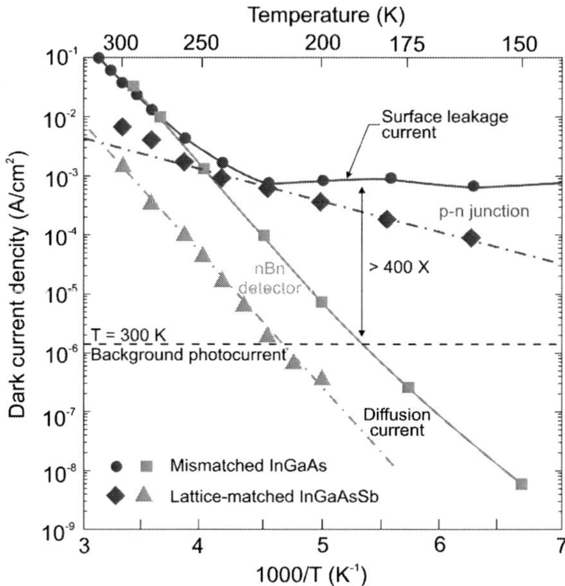

Figure 6.7 Arrhenius plots for both InGaAs and InGaAsSb p–n junctions and nBn detectors with cutoff wavelength at 2.8 μm.

The p–n InGaAsSb junction is limited by the depletion region current below a temperature of 250 K, while the nBn remains diffusion limited as far as measured, down to 250 K. At the 300-K background photocurrent level the nBn detector shows dark current reduced by nearly three orders of magnitude compared to the conventional photodiode.

Figure 6.8 shows that the InGaAsSb nBn detector lattice-matched to a GaSb substrate presents performance near "Rule 07." Its dark current is 10 to 20 times lower in comparison with its lattice-mismatched InGaAs counterpart.

6.3 MWIR InAsSb Barrier Detectors

Detailed growth procedures and device characterization of $InAs_{1-x}Sb_x/$ $AlAs_{1-y}Sb_y$ nBn MWIR detectors were the topic of several papers, e.g., Refs. 8–10, 21, and 22. The n-type doping was usually obtained by either Si or Te elements, and the InAsSb structures were grown on either GaAs(100) or GaSb(100) substrates in a Veeco Gen200 MBE system.[22] The lattice-mismatched structures that used GaAs(100) as the substrate were grown on a 4-μm-thick GaSb buffer layer, whereas the remaining structures were grown directly onto GaSb(100) substrates. The principal layers of the device structures consisted of a thick n-type InAsSb absorption layer (1.5–3 μm), a thin n-type AlSbAs barrier layer (0.2–0.35 μm), and a thin (0.2–0.3 μm) n-type InAsSb contact layer. The bottom contact layer was highly doped.

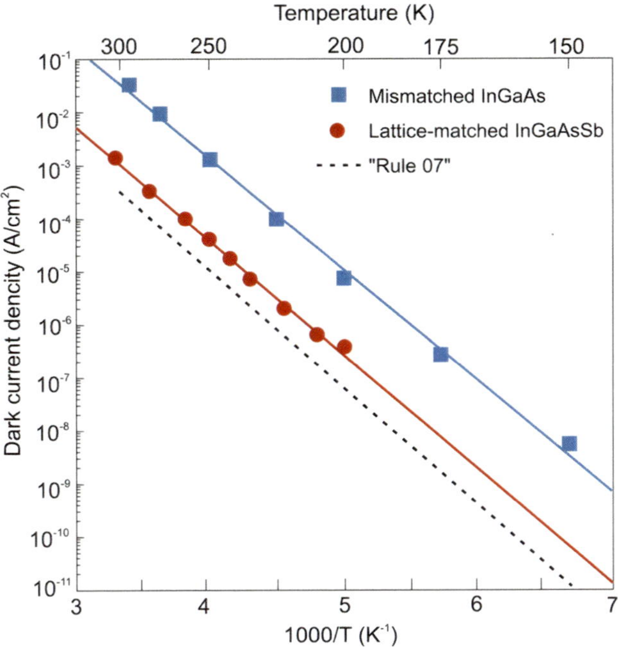

Figure 6.8 Comparison of the dark current characteristics of InGaAs and InGaAsSb nBn detectors. The lattice-matched InGaAsSb barrier detector shows at least an order of magnitude reduction in the dark current compared to the mismatched InGaAs nBn detector (reprinted from Ref. 20).

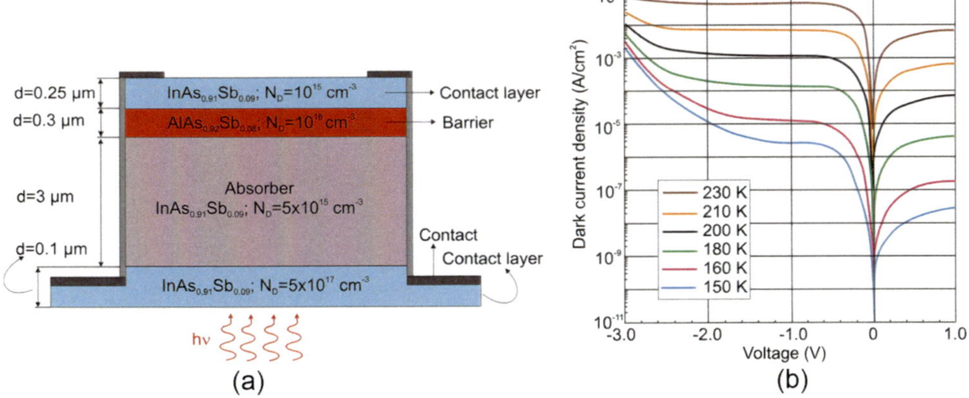

Figure 6.9 InAsSb/AlAsSb nBn MWIR detector: (a) the device structure and (b) dark current density versus bias voltage as a function of temperature for 4096-element (18-μm pitch) detectors ($\lambda_c \approx 4.9$ μm at 150 K) tied together in parallel (reprinted from Ref. 24).

Figure 6.9 shows an example of such an nBn structure that was considered theoretically by Martyniuk and Rogalski,[23] along with the *J–V* characteristics as a function of temperature that were taken from Ref. 24. The alloy composition of $x = 0.09$ or the InAs$_{1-x}$Sb$_x$ absorber layer provided a cutoff wavelength of $\sim$4.9 μm at 150 K. J_{dark} was 1.0×10^{-3} A/cm^2 at 200 K and 3.0×10^{-6} A/cm^2 at 150 K. The detectors are dominated by diffusion currents at -1.0 V bias where the quantum efficiency peaks.

An interesting difference in dark current density versus temperature is shown in Fig. 6.10 for two nominally identical devices with opposite barrier polarities, each operating at a bias of -0.1 V. The nB$_n$n device exhibits a single straight line, characteristic of diffusion-limited behavior, while the nB$_p$n device exhibits two-slope behavior, characteristic of a crossover from diffusion-limited behavior at high temperatures to GR-limited behavior at low temperatures. As shown in the figure, the dark current density at 150 K is more than two orders of magnitude greater for the detector with the p-type barrier because it is already GR limited. It appears, for a typical quantum efficiency of 70% at $F/3$ optics, that the BLIP temperature is about 140 K compared with $\sim$175 K for the detector with the n-type barrier.

Klipstein et al. have presented one of the first commercial nBn array detectors operating in the blue part of the MWIR atmospheric window (3.4–4.2 μm) that were made available on the market by SCD. This detector is known as Kinglet and is a very low-SWaP integrated detector–cooler

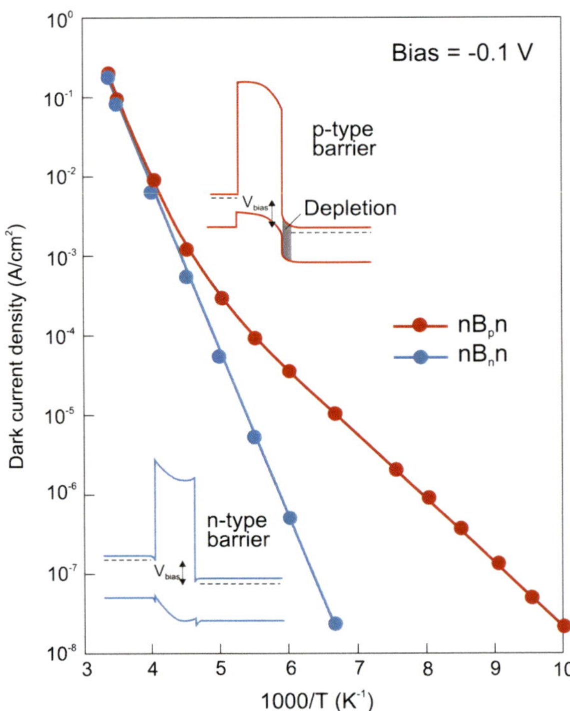

Figure 6.10   Dark current density versus temperature for two identical InAsSb/AlSbAs nBn devices with opposite barrier-doping polarities. The active layer bandgap wavelength is 4.1 μm at 150 K (reprinted from Ref. 13).

assembly (IDCA) with an aperture of *F*/5.5 and an operating temperature of 150 K. The Kinglet digital detector based on SCD's Pelican-D ROIC has an nBn $InAs_{0.91}Sb_{0.09}$/B-AlAsSb 640×512 pixel architecture with a 15-μm pitch. The *NEDT* for *F*/3.2 optics and the pixel operability is shown in Fig. 6.11 as a function of temperature. The *NEDT* is 20 mK at 10-ms integration time, and the operability of nondefective pixels was greater than 99.5% after a standard two-point nonuniformity correction. The *NEDT* and operability begin to change above 170 K, which is consistent with the estimated BLIP temperature of 175 K.

6.4 LWIR InAsSb Barrier Detectors

Recently, Lin et al.[26] have described bulk InAsSb barrier detectors with a cutoff wavelength of about 10 μm at 77 K. Due to a poor-quality crystal structure, the performance of LWIR bulk InAsSb ternary alloy detectors is considerably inferior in comparison with HgCdTe photodiodes.

A schematic cross-sectional device view in shown in Fig. 6.12(a). The device structure consists of a 3-μm-thick compositionally graded GaInSb buffer layer MBE-grown on GaSb substrate, a 1-μm-thick $InAs_{0.60}Sb_{0.40}$ absorber, a 20-nm-thick $Al_{0.6}In_{0.4}Sb_{0.1}Sb_{0.9}$ undoped barrier, and a 20-nm-thick $InAs_{0.60}Sb_{0.40}$ top contact layer Te-doped to a level of 10^{18} cm^{-3}. The undoped AlInAsSb barrier was lattice matched to InAsSb with 40% Sb composition. As is shown, the heterostructures were processed with a window (square with a 250-μm side) for incident radiation on top of the epilayer. The top metal contact layer was a square with a 300-μm size. The InAsSb contact layer outside the metal contact was removed down to the barrier layer by reactive ion etching. For detector passivation, Si_3N_4 was used.

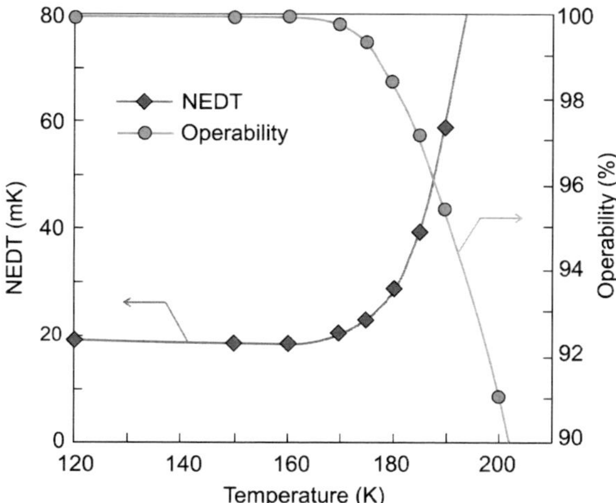

Figure 6.11 Temperature dependence of the *NEDT* (at optics *F*/3.2) and the pixel operability of the Kinglet detector (adapted from Ref. 25).

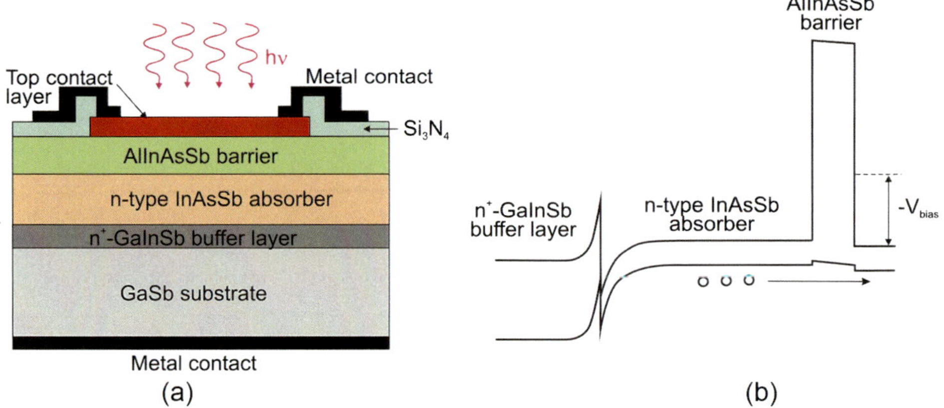

Figure 6.12 LWIR nBn InAsSb detector: (a) a schematic cross-section of the processed detector; (b) a schematic bandgap diagram of the heterostructure with an $InAs_{0.60}Sb_{0.40}$ bulk absorber (reprinted from Ref. 26).

A minority-hole lifetime in the $InAs_{0.60}Sb_{0.40}$ absorber of 185 ns and a diffusion length of 9 μm were determined at 77 K. A hole mobility of 10^3 cm^2/Vs was estimated with frequency response measurements.

The current–voltage characteristics are influenced by the depletion of a part of the absorber adjacent to the barrier, which leads to domination of the GR layer and probably tunneling components. To reach diffusion-limited dark current, the VBO associated with the heterointerfaces must be eliminated.

The spectral detectivity curves of a LWIR nBn InAsSb detector with two absorber doping levels are shown in Fig. 6.13. The demonstrated detectivity of 2×10^{11} cm $Hz^{1/2}$/W at 2π FOV and wavelength of 8 μm was estimated in spite of the significant blue-shift of the absorption edge with doping. For 1-μm-thick $InAs_{0.60}Sb_{0.40}$ absorber at $\lambda = 8$ μm, an absorption coefficient was estimated at 3×10^3 cm^{-1}, which implies a quantum efficiency of 22%. The quantum efficiency increases with bias until it reaches a constant level for a bias of –0.4 V with increasing thickness of the absorption layer (to 40% for 3-μm-thick absorber).

6.5 T2SL Barrier Detectors

Building unipolar barriers for InAs/GaSb superlattices is relatively straight-forward because of the flexibility of the 6.1-Å III-V material family—InAs, GaSb, and AlSb. For SLs with the same GaSb layer widths, their valence band edges tend to line up very closely due to the large heavy-hole mass. As a result, an electron-blocking unipolar barrier for a given InAs/GaSb SL can be formed by using another InAs/GaSb SL with thinner InAs layers or a GaSb/AlSb SL.

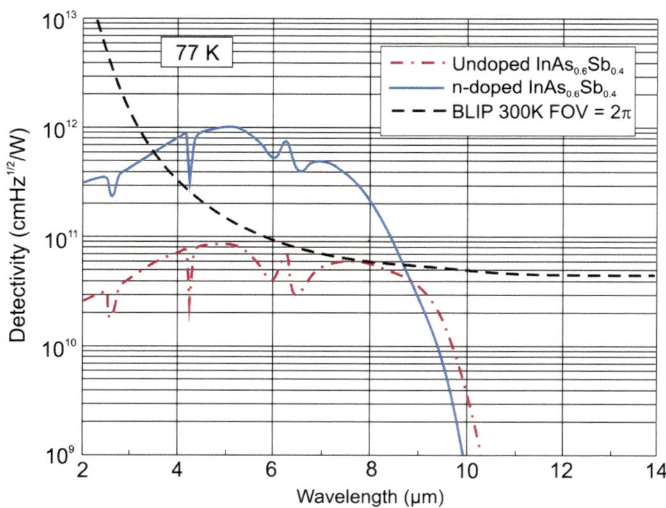

Figure 6.13 Spectral detectivity of barrier detectors with 1-μm-thick $InAs_{0.6}Sb_{0.4}$ absorbers at $T = 77$ K. Solid and dash-dot-dash lines correspond to devices with doped and undoped absorbers, respectively. The dotted line shows the 300-K background-limit for a 2π FOV (reprinted from Ref. 26).

The hole-blocking unipolar barrier can be obtained in different ways using complex supercells, such as the four-layer InAs/GaInSb/InAs/AlGaInSb "W" structure[27] and the four-layer GaSb/InAs/GaSb/AlSb "M" structure.[28] Their designs are shown in Fig. 6.14. The "W" structures, initially developed to

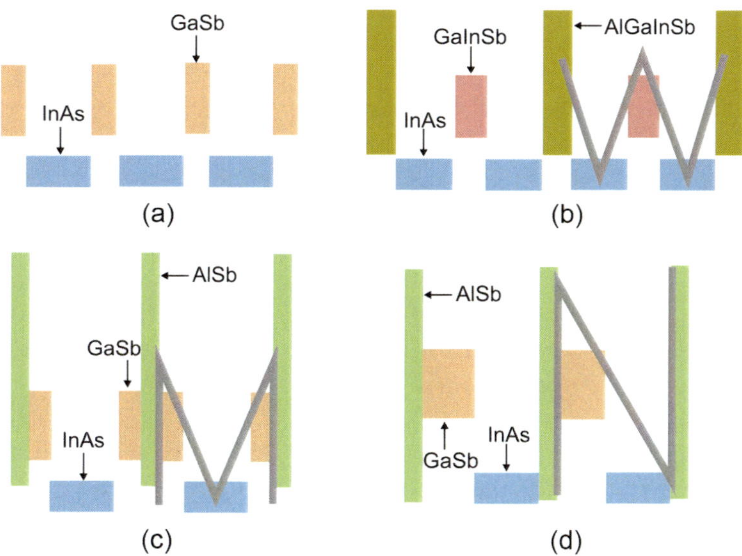

Figure 6.14 Schematic energy band diagrams of (a) InAs/GaSb SL, (b) InAs/GaInSb/InAs/AlGaInSb "W" SL, (c) GaSb/InAs/GaSb/AlSb "M" SL, and (d) GaSb/InAs/AlSb "N" SL (reprinted from Ref. 4).

increase the gain in MWIR lasers, are also promising as LWIR and VLWIR photodiode materials. In these structures two InAs electron wells are located on either side of an InGaSb hole well and are bound on either side by AlGaInSb barrier layers. The barriers confine the electron wavefunctions symmetrically about the hole well, increasing the electron–hole overlap while nearly localizing the wavefunctions. The resulting quasi-dimensional densities of states give strong absorption near the band edge. Due to flexibility in adjusting of the "W" structure, this SL has been used as a hole-blocking unipolar barrier and an absorber, as well as an electron-blocking unipolar barrier.

The newly designed W-structured T2SL photodiodes employ a graded bandgap p-i-n design. The grading of the bandgap in the depletion region suppresses tunneling and GR currents in the depletion region, which has resulted in an order of magnitude improvement in dark current performance, with $R_0A = 216$ Ωcm^2 at 78 K for devices with a 10.5-μm cutoff wavelength. The sidewall resistivity of $\sim$70 kΩcm for untreated mesas apparently indicates self-passivation by the graded bandgap.[29]

In the "M" structure,[30,31] the wider-energy-gap AlSb layer blocks the interaction between electrons in two adjacent InAs wells, thus, reducing the tunneling probability and increasing the electron effective mass. The AlSb layer also acts as a barrier for holes in the valence band and converts the GaSb hole quantum well into a double quantum well. As a result, the effective well width is reduced, and the hole's energy level becomes sensitive to the well dimension. This structure significantly reduces the dark current and does not show any strong degradation of the optical properties of the devices. Moreover, it has a proven control of the positioning of the conduction and valence band energy levels.[31] Consequently, FPAs for imaging within various IR regions, from SWIR to VLWIR, can be fabricated.[32] A device with a cutoff wavelength of 10.5 μm exhibits a R_0A product of 200 Ωcm^2 when a 500-nm-thick "M" structure was used. Using a double M-structure heterojunction at the single device level, a R_0A product of up to 5300 Ωcm^2 has been obtained for a 9.3-μm cutoff wavelength at 77 K.[33]

In the "N" structure,[34] two monolayers of AlSb are inserted asymmetrically between InAs and GaSb layers, along the growth direction, as an electron barrier. This configuration significantly increases the electron–hole overlap under bias and consequently increases absorption while decreasing the dark current.

Table 6.1 illustrates some flat-band energy band diagrams and describes superlattice-based infrared detectors that make use of unipolar barriers, including: double heterojunction (DH), dual-band nBn, DH with graded-gap junction, and complementary barrier structures. As can be seen, these structures are based on either the nBn/pBp/XBn architecture or different double-heterostructure designs.

Table 6.1 Type-II superlattice barrier detectors.

Flat-band energy diagrams	Examples	Description	Refs.
Double heterostructure		**Double heterojunction (DH) photodiode** The first LWIR InAs/InGaSb SL DH photodiode grown on GaSb substrates with a photoresponse out to 10.6 μm. The active region consisting of n-type 39-Å InAs/16-Å $Ga_{0.65}In_{0.35}Sb$ SL (2×10^{16} cm^{-3}) is surrounded by barriers made from p-GaSb and n-GaSb.	35
nB$_p$p barrier		**p-π-M-n photodiode structure** The "M" structure is inserted between the π and n regions of a typical p-π-n structure. The T2SL part is chosen to have nominally 13 ML InAs and 7 ML GaSb for a cutoff wavelength of around 11 μm. The M structure is designed with 18 ML InAs/3 ML GaSb/5 ML AlSb/3 ML GaSb for a cutoff wavelength of approximately 6 μm. The p-π-M-n structure consists of a 250-nm-thick GaSb:Be p$^+$ buffer ($p \sim 10^{18}$ cm^{-3}) followed by a 500-nm-thick InAs/GaSb:Be p$^+$ ($p \sim 10^{18}$ cm^{-3}) superlattice, a 2000-nm-thick slightly p-type doped InAs:Be/GaSb region (π-region $p \sim 10^{18}$ cm^{-3}), a M-structure barrier, a 500-nm-thick InAs:Si/GaSb n$^+$ ($n \sim 10^{18}$ cm^{-3}) region, and topped with a thin InAs:Si n$^+$ doped ($n \sim 10^{18}$ cm^{-3}) contact layer.	28, 30–32
Dual-band nBn		**MWIR/LWIR nBn detector** In this dual-band SL nBn detector, the LWIR SL and MWIR SL are separated by an AlGaSb unipolar barrier. The dual-band response is achieved by changing the polarity of the applied bias (see Fig. 6.15). The advantage of this structure is design simplicity and compatibility with commercially available ROICs. The concerns are associated with low hole mobility and lateral diffusion.	36

(continued)

Table 6.1   (*Continued*)

Flat-band energy diagrams	Examples	Description	Refs.
DH with graded-gap junction	N-grown-junction in wider band gap SL P-graded-gap "W" SL structure p-type-II SL active layer E_c E_v	**Shallow-etch mesa isolation (SEMI) structure** This is an n-on-p graded-gap W photodiode structure in which the energy gap is increased in a series of steps from that of the lightly p-type absorbing region to a value typically two to three times larger. The hole-blocking unipolar barrier is typically made from a four-layer InAs/GaInSb/InAs/AlGaInSb SL. The wider gap levels off about 10 nm short of the doping-defined junction and continues for another 0.25 μm into the heavily n-doped cathode before the structure is terminated by an n^+-doped InAs top cap layer. Individual photodiodes are defined using a shallow etch that typically terminates only 10–20 nm past the junction, which is sufficient to isolate neighboring pixels while leaving the narrow-gap absorber layer buried 100–200 nm below the surface.	37, 38
Complementary barrier	InAs/AlSb SL hole barrier p-type LW InAs/GaSb SL absorber MW InAs/GaSb SL electron barrier n-type InAsSb emitter GaSb buffer GaSb substrate E_c E_v	**Complementary-barrier infrared detector (CBIRD)** This device consists of a lightly-doped p-type InAs/GaSb SL absorber sandwiched between an n-type InAs/AlSb hole barrier (hB) SL and wider InAs/GaSb electron barrier (eB). The barriers are designed to have approximately zero conduction and valence band offset with respect to the SL absorber. A heavily doped n-type InAsSb layer adjacent to the eB SL acts as the bottom contact layer. The N–p junction between the hB InAs/AlSb SL and the absorber SL reduces SRH-related dark current and trap-assisted tunneling. The LWIR CBIRD superlattice detector performance is close to the "Rule 07" trend line. For a detector having a cutoff wavelength of 9.24 μm, a value of $R_0A > 10^5$ Ωcm^2 at 78 K was measured.	9, 39
	SL p⁺-layer contact layer (5/8ML InAs/GaSb)x38; 130 nm p SL electron blocking (eB) (7/4ML GaSb/AlSb)x45; 149 nm B SL absorber (p-type, 10¹⁶ cm⁻³) (14/7ML InAs/GaSb)x300; 1940 nm i SL hole blocking (hB) (16/4ML InAs/AlSb)x45; 275 nm B SL n⁺-contact layer (9/4ML InAs/GaSb)x200; 800 nm n GaSb substrate	**pBiBn detector structure** This is another variation of the DH CBIRD structure. In this design a pin photodiode is modified such that the unipolar eB and hB layers are sandwiched between the p-contact layer and the absorber, and the n-contact layer and the absorber, respectively. This design facilitates a significant reduction in the electric field drop across the narrow-gap absorber region (most of the electric field drops across the wider-bandgap eB and hB layers), leading to a very small depletion region in the absorber layer, and hence reduction in the SRH, BBT, and TAT current components.	40, 41

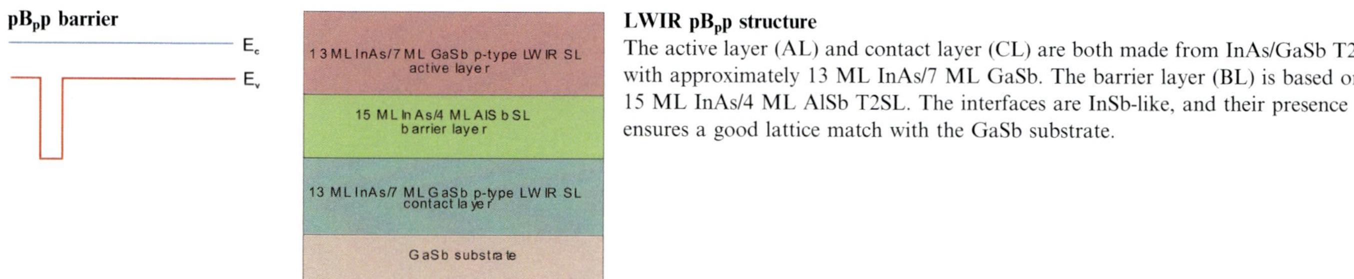

LWIR pB$_p$p structure

13, 42

The active layer (AL) and contact layer (CL) are both made from InAs/GaSb T2SLs with approximately 13 ML InAs/7 ML GaSb. The barrier layer (BL) is based on a 15 ML InAs/4 ML AlSb T2SL. The interfaces are InSb-like, and their presence ensures a good lattice match with the GaSb substrate.

The first LWIR InAs/InGaSb SL double-heterostructure photodiode (see Table 6.1) grown on GaSb substrates with a photoresponse out to 10.6 μm was described by Johnson et al. in 1996.[35] In this structure the active SL region was surrounded by barriers made from p-GaSb and n-GaSb binary compounds. More recently, these barriers have also been fabricated using different types of superlattices.

The realization of dual-band detection capabilities with an nBn design (see Table 6.1) is schematically shown in Fig. 6.15.[4,36] Under forward bias (defined as negative voltage applied to the top contact), the photocarriers are collected from the SL absorber with the λ_2 cutoff wavelength. When the device is under reverse bias (defined as positive voltage applied to the top contact), the photocarriers are collected from the SL absorber with the λ_1 cutoff wavelength, while those from the absorber with the λ_2 cutoff wavelength are blocked by the barrier. Thus, a two-color response is obtained under two different bias polarities.

Hood et al.[43] have modified the nBn concept to make the superior pBn LWIR device (see Fig. 6.3). In this structure, the p–n junction can be located at the interface between the heavily doped p-type contact material and the more lower-doped barrier, or within the lower-doped barrier itself. Similar

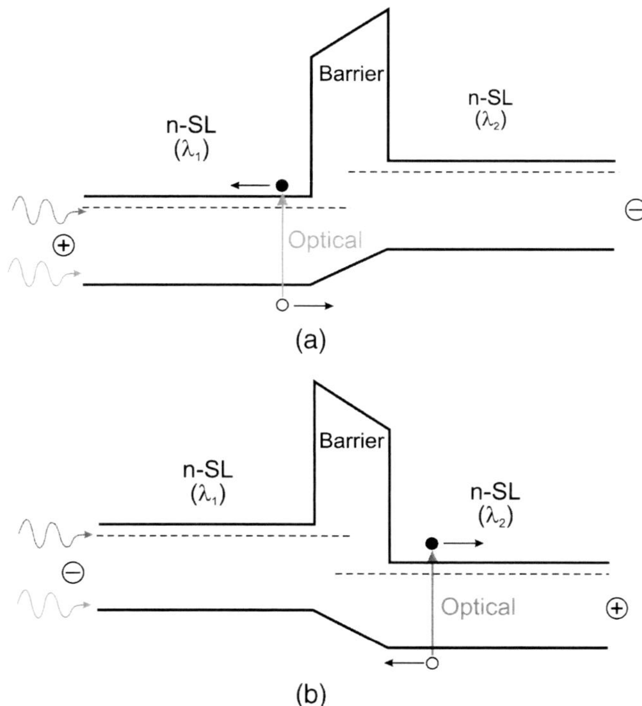

Figure 6.15 Schematic band diagram for the dual-band nBn detector under (a) forward and (b) reverse bias.

to the nBn structure, the pBn structure still reduces GR currents associated with SRH centers (the depletion region exists primarily in the barrier and does not appreciably penetrate the narrow-bandgap n-type absorber). In addition, the electric field in the barrier improves the response of the detector by sweeping from the active layer those photogenerated carriers that reach the barrier before they can recombine.

The p-π-M-n structure (see Table 6.1) is similar to the standard p-π-n structure; however, in comparison with the latter, the electric field in the depletion region of the p-π-M-n structure is reduced. As a result, the GR current is negligible in comparison with the diffusion current from the active region, and the tunneling contribution is reduced because the barrier between the p- and the n-regions is spatially broadened.

A variation of the DH detector structure is a structure with a graded bandgap in the depletion region (see Table 6.1). The graded-gap region is inserted between the absorber and the hole barrier to reduce tunnelling and GR processes. A similar structure that was developed by The Naval Research Laboratory and Teledyne enables the shallow-etch mesa isolation (SEMI) structure for surface leakage current reduction. The junction is placed in the wider-gap portion of the transition-graded-gap W layer. A shallow-mesa etch just through the junction but not into the active layer isolates the diode but still leaves a wide-bandgap surface for ease of passivation. A modest reverse bias allows efficient collection from the active layer, similar to planar double-heterojunction p-on-n HgCdTe photodiodes.[44]

However, the culminating feature is the use of a pair of complementary barriers, namely, an electron barrier and a hole barrier formed at different depths in the growth sequence (see Table 6.1). Such a structure is known as a complementary-barrier infrared device (CBIRD) and was invented by Ting and others at the Jet Propulsion Laboratory. An electron barrier appears in the conduction band, and a hole barrier in the valence band. The two barriers complement one another to impede the flow of dark current. The absorber region, where the bandgap is smallest, is p-type, and the top contact region is n-type, making an n-on-p polarity for the detector element. In sequence from the top, the first three regions are composed of superlattice material: the n-type cap, the p-type absorber, and the p-type electron barrier. The highly doped n-InAsSb layer below the electron barrier is an alloy. Further towards the bottom are a GaSb buffer layer and the GaSb substrate.

The introduction of device designs containing unipolar barriers has taken the LWIR CBIRD superlattice detector performance close to the "Rule 07" trend line, which provides a heuristic predictor of the state-of-the-art HgCdTe photodiode performance.[45] The barriers prove to be very effective in suppressing the dark current. Figure 6.16 compares the J–V characteristics of a CBIRD device to a homojunction device made with the same absorber superlattice. Lower dark current results in a higher RA product.

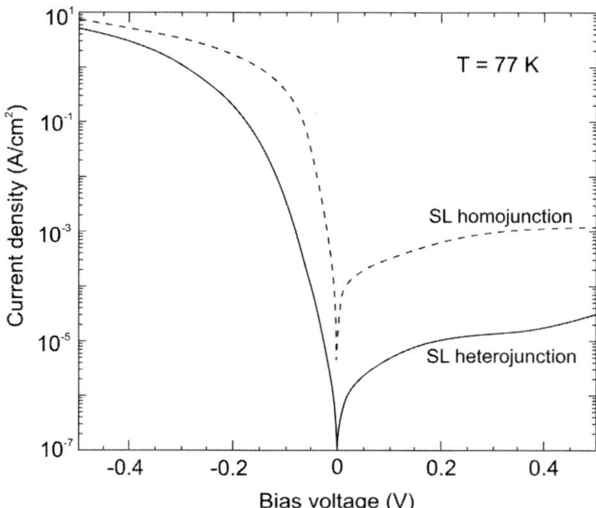

Figure 6.16 Dark *J–V* characteristics for a LWIR CBIRD detector and a superlattice homojunction at 77 K (reprinted from Ref. 4).

Usually, R_0A values are plotted for devices with near-zero-bias turn-on. However, since the detector is expected to operate at a higher bias, a more relevant quantity is the effective resistance–area product. In the case of a detector having a cutoff wavelength of 9.24 μm, the value of $R_0A > 10^5$ Ωcm^2 at 78 K was measured, as compared with about 100 Ωcm^2 for an InAs/GaSb homojunction having the same cutoff. For good photoresponse, the device must be biased to typically –200 mV; the estimated internal quantum efficiency is greater than 50%, while the RA_{eff} remains above 10^4 Ωcm^2.[9]

Recently, considerable progress has been achieved at SCD in the development of LWIR pB$_p$p T2SL barrier detectors. These detectors enable diffusion-limited behavior with dark currents comparable to the HgCdTe "Rule 07" and with high quantum efficiency. The active and contact layers of these devices are both made from InAs/GaSb T2SLs with approximately 13 ML InAs/7 ML GaSb, while the barrier layer is based on a 15 ML InAs/4 ML AlSb T2SL. The individual InAs layers are terminated with indium, and the AlSb or GaSb layers are terminated with antimony, in order to create InSb-like interfaces with the correct amount of strain for lattice matching between the T2SL and the GaSb substrate.

As is shown in Fig. 6.17(a), the dark current in a standard LWIR n-on-p diode based solely on InAs/GaSb T2SLs is higher in the lower temperature region in comparison with pB$_p$p T2SL barrier detector, whose schematic profile of band edges is shown in the inset of the figure. The barrier device is diffusion limited down to 77 K, while the p-n diode is GR-limited at this temperature, with a dark current over 20× larger. The dark current in pB$_p$p

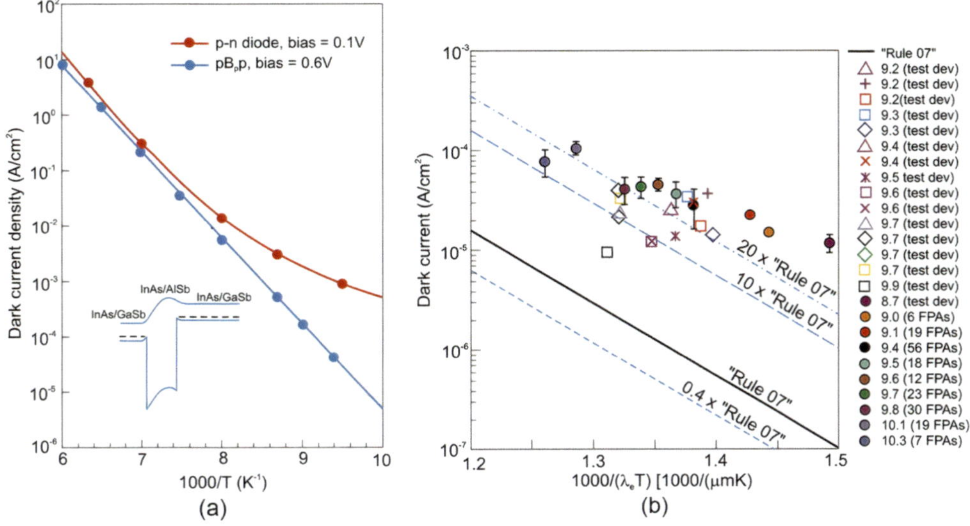

Figure 6.17 Dark current density of the pB$_p$p T2SL barrier detector (area $= 100 \times 100$ μm^2): (a) in comparison with a p-n diode (adapted from Ref. 13), (b) "Rule 07" plot for barrier structures with different thicknesses of active layers. Range of bandgap wavelengths: $9.0 < \lambda_e < 10.3$ μm. Solid line shows HgCdTe "Rule 07" with uncertainty factors of 0.4, 10, and 20 (dashed lines) (adapted from Ref. 46).

devices, with a thickness of active layers between 1.5 and 6.0 μm, is within one order of magnitude of the HgCdTe "Rule 07" [see Fig. 6.17(b)].

Theoretical predictions of spectral response curves based on **k·p** and optical transfer matrix methods for both antireflection coating (ARC) and no antireflection coating are in good agreement with experimental data (see Fig. 6.18). The detector structures include a mirror on the contact layer to reflect 80% of the light back for a second pass. The inset in Fig. 6.18 shows the typical bias dependence of the quantum efficiency for a 100×100 μm^2 test device without antireflection coating. The signal reaches its full value until a positive bias of $\sim$0.6 V is applied. This bias is needed to overcome the electrostatic barrier to minority carriers caused by negative space charge in the depleted p-type barrier. LWIR T2SL pB$_p$p devices have a BLIP temperature of $\sim$100 K at F/2 optics.

For an equivalent detector based on the InAs/InAsSb T2SL, the predicted average quantum efficiency is only about two-thirds of the InAs/GaSb value. This estimation can be attributed to the smaller absorption coefficient of the InAs/InAsSb T2SL near the cutoff wavelength.[47]

Recently, Martyniuk et al.[48] have presented the performance of the T2SL mesa-type InAs/GaSb nBn detectors on GaAs substrates with a bulk AlGaSb barrier operated under higher-operation-temperature conditions ($>$200 K) with 50% cutoff wavelength $\sim$5.1 μm at 230 K. The detector structure

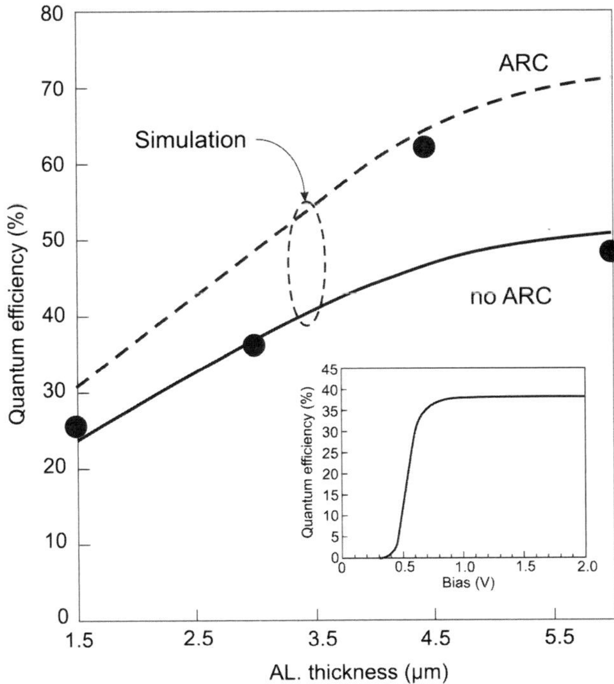

Figure 6.18   Simulated and measured (dots) quantum efficiency values at 77 K for LWIR pB$_p$p InAs/GaSb T2SL devices with active layer thicknesses from 1.5 to 6.0 µm and a cutoff wavelength of ~9.5 µm. A mirror on the contact layer reflects 80% of the light back for a second pass. *Inset:* Example of quantum efficiency versus bias (adapted from Ref. 47).

(see Fig. 6.19) consists of two 10 ML InAs/10 ML GaSb:Te n-type doped contact layers. Between contact layers, the non-intentionally doped absorber (2 µm) and non-intentionally doped Al$_{0.2}$Ga$_{0.8}$Sb barrier (100 nm) were grown. The total device thickness was estimated at ~3.3 µm. The interfacial misfit dislocation array and GaSb buffer layers were introduced to accommodate the 7.8% lattice mismatch between the detector structure and GaAs substrate. The 1.1-mm-thick GaAs substrate was converted into an immersion lens to limit the influence of the defects occurring during growth on GaAs substrate and to increase detectivity (~10^{10} cmHz$^{1/2}$/W at 230 K) under reverse bias 200 mV and ~3×10^9 cmHz$^{1/2}$/W at 300 K to under 400 mV. The presented results are better than for nBn architectures with the same and slightly higher cutoff wavelength grown on GaAs without immersion lens and GaSb substrates.

6.6 Barrier Detectors versus HgCdTe Photodiodes

There are currently two broad IR detector materials, namely, HgCdTe ternary alloys and III-V semiconductors. In the mid-1990s the U.S. Government placed a research emphasis on III-V materials as an alternative technology option to

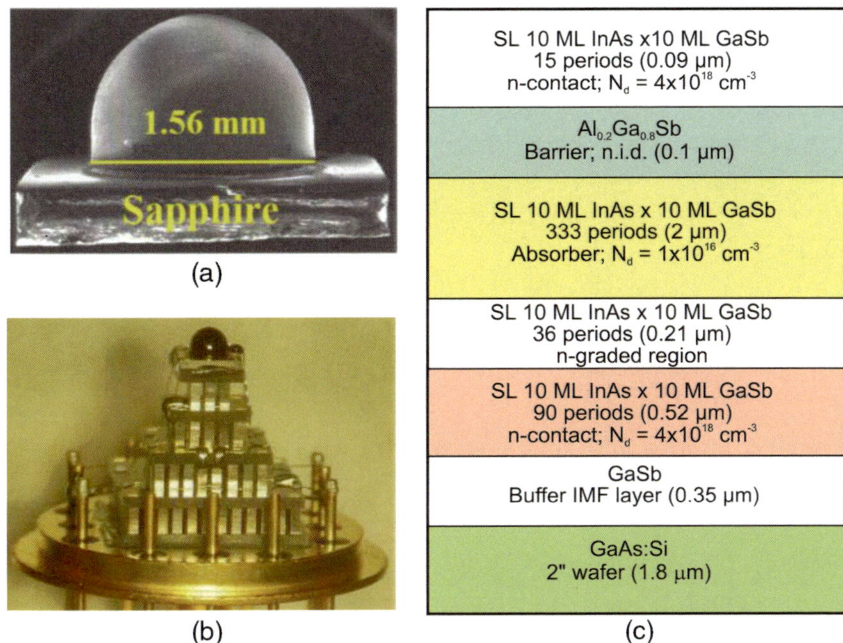

Figure 6.19 (a) SEM image of the GaAs immersion lens, (b) fully operational device before housing with four-stage thermoelectrical cooler, and (c) MWIR T2SL nBn detector structure with GaAs substrate (reprinted from Ref. 48 with permission from IEEE).

HgCdTe to attain its stated goal of inexpensive large-area IR FPAs. This concept typically involves the fabrication of III-V superlattices to tailor the bandgap of the material to detect the desired IR radiation.

The bandgap structure and physical properties of III-V compounds are found to be remarkably similar to HgCdTe with the same bandgap. It is interesting to compare the performance of detectors composed of III-V material systems with HgCdTe photodiodes operating in both MWIR and LWIR spectral ranges.

In general, the dark current density generated in a p–n junction is the sum of the diffusion current of the active region and the depletion layer region, and is given by

$$I_{dark} = qG_{diff}V_{diff} + qG_{dep}V_{dep}, \tag{6.1}$$

where q is the electron charge, G_{diff} is the thermal generation rate in the diffusion region, $V_{diff} = At$ is the volume of the active region (A is the detector area, and t is the thickness of active region), G_{dep} is the generation rate in space charge region, and $V_{dep} = Aw$ is the volume of the depletion region (w is the width of depletion region).

The diffusion current can be estimated as

$$I_{diff} = \frac{qn_i^2 At}{N_{dop}\tau_{diff}},$$ (6.2)

were N_{dop} is the majority-carrier density in the absorption region (extrinsic regime), τ_{diff} is the diffusion carrier lifetime, and n_i is the intrinsic carrier concentration.

The depletion layer current can be given by a simplification formula:

$$I_{dep} = \frac{qn_i Aw}{2\tau_{SRH}}.$$ (6.3)

To obtain Eq. (6.3) it is assumed that a trap is located at the intrinsic level (electron and hole concentration are equal n_i), the and SRH lifetime is τ_{SRH}.

From last two equations two important conclusions can be drawn:

- For a diffusion-limited photodiode, the dark current is inversely proportional to the product $N_{dop} \times \tau_{diff}$; and
- The ratio of the diffusion current to the depletion current is

$$\frac{I_{diff}}{I_{dep}} \approx 2\frac{n_i}{N_{dop}}\frac{\tau_{SRH}}{\tau_{diff}}\frac{t}{w}.$$ (6.4)

Equation (6.4) indicates that the diffusion current dominates the depletion current for high intrinsic-carrier concentration (e.g., in materials with a narrow bandgap). The reverse is observed for semiconductors with a larger bandgap. As is shown in the equation, the I_{diff}/I_{dep} ratio depends also on the doping level N_{dop}, the ratios of the respective volumes (t/w), and lifetimes. The last issue requires additional discussion of the GR mechanisms.

6.6.1 The $N_{dop} \times \tau_{diff}$ product as the figure of merit for diffusion-limited photodetectors

The basic properties of the artificial T2SL materials are completely different from those of the constituent layers (see section 3.3). Since the electron effective mass of an InAs/GaSb SL is larger compared to HgCdTe alloy with the same bandgap, diode tunnelling currents in the SL can be reduced compared to the HgCdTe alloy. As a result, the doping density of a T2SL is on the order of 1×10^{16} cm^{-3}, which is considerably higher than the doping level in HgCdTe (typically about 10^{15} cm^{-3}).

Taking into account both the experimental and theoretical lifetime data shown in Figs. 6.20 and 6.21, we can present semi-empirical curves that describe the lifetime behavior with doping concentration. Figure 6.20 shows the fit data for the major infrared absorbing materials.

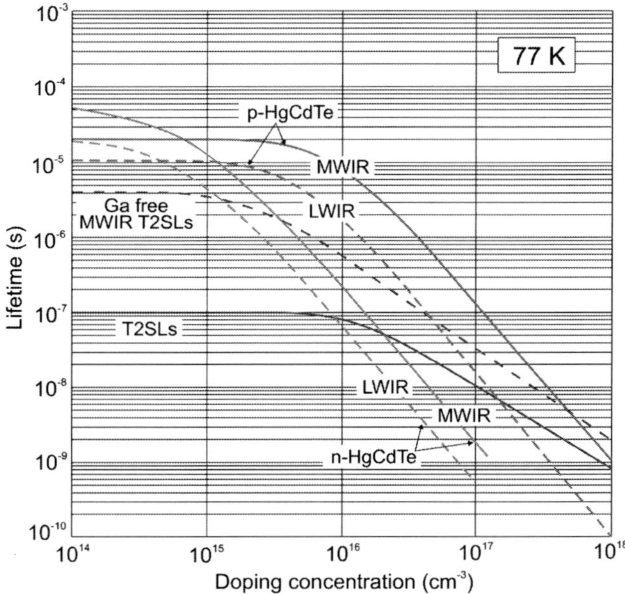

Figure 6.20   The lifetime data fit versus doping concentration for different infrared material systems at 77 K.

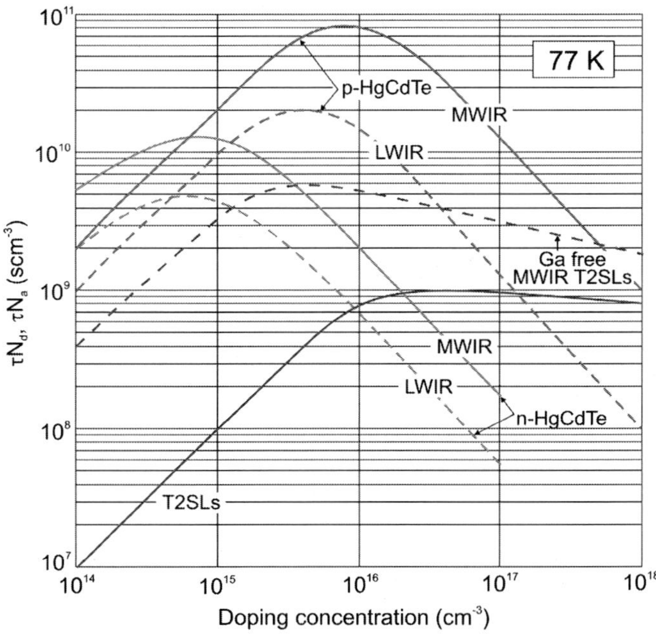

Figure 6.21   The $N_{dop} \times \tau_{diff}$ product versus doping concentration for diffusion-limited MWIR and LWIR photodiodes operated at 77 K fabricated with HgCdTe and T2SL material systems.

As seen from Eq. (6.2), the figure of merit of a diffusion-limited photodiode is the $N_{dop} \times \tau_{diff}$ product, so the highest lifetime is required with the highest doping. The transition between SRH lifetime (for low doping concentrations) to an Auger lifetime (for higher doping) introduces a bell function of the $N_{dop} \times \tau_{diff}$ product for HgCdTe diffusion-limited photodiodes, as illustrated in Fig. 6.21.

For p-on-n HgCdTe photodiodes, the optimum doping concentration is about 10^{15} cm^{-3} and corresponds to the so-called "Rule 07"[45] describing the empirically dark current of the Auger 1 and SRH limitation. In the case of n-on-p photodiodes, the optimum doping is at a higher level—below 10^{16} cm^{-3}.

Figure 6.21 also compares the $N_{dop} \times \tau_{diff}$ product for HgCdTe photodiodes with equivalent T2SL photodiodes. As is shown, the optimum doping concentration for InAs/GaSb T2SLs is higher (about 2×10^{16} cm^{-3}) than for HgCdTe material system. As a result, the higher doping compensates for the shorter lifetime, resulting in relatively low diffusion dark current of InAs/GaSb T2SL photodiodes.

Another situation is observed for Ga-free T2SLs due to considerably higher carrier lifetime; in this case, the optimal doping concentration is in the middle at 10^{15} cm^{-3}.

6.6.2 Dark current density

The dark diffusion current density generated in an absorber region is given by

$$J_{dark} = qGt, \tag{6.5}$$

where q is the electron charge, G is the thermal generation rate in the base region, and t is the thickness of the active region.

Assuming that the thermal generation is a sum of Auger 1 and SRH mechanisms in an n-type absorber region, i.e.,

$$G = \frac{n_i^2}{N_d \tau_{A1}} + \frac{n_i^2}{(N_d + n_i)\tau_{SRH}}, \tag{6.6}$$

and that $\tau_{A1} = 2\tau_{A1}^i/[1 + (N_d/n_i)^2]$, the dark current density is

$$J_{dark} = \frac{qN_d t}{2\tau_{A1}^i} + \frac{qn_i^2 t}{(N_d + n_i)\tau_{SRH}}. \tag{6.7}$$

The influence of generation in a depletion region ($J_{dep} = qn_i w/\tau_{SRH}$, were w is the depletion width) can be omitted in both III-V barrier detectors and HgCdTe photodiodes. Here, the contribution of J_{dep} is taken into consideration only for a 5-μm-cutoff InAsSb photodiode.

In the above equations, n_i is the intrinsic carrier concentration, N_d is the absorber donor doping, and τ^i_{A1} is the intrinsic Auger 1 lifetime. In estimation of the SRH generation term of Eq. (6.6), it is assumed that the energy level of the recombination center approaches the midgap, and then the SRH recombination rate approaches the maximum.

For the dark current density of the p-type absorber region of detectors, we can obtain a similar equation:

$$ J_{dark} = \frac{qN_a t}{2\tau^i_{A7}} + \frac{qn_i^2 t}{(N_a + n_i)\tau_{SRH}}, \tag{6.8} $$

where N_a is the absorber acceptor doping, τ^i_{A7} is the intrinsic Auger 7 lifetime, and $\tau_{A7} = 2\tau^i_{A7}/[1 + (N_a/n_i)^2]$.

In the farther estimations we have chosen III-V barrier detectors and HgCdTe photodiodes with cutoff wavelengths of 5 and 10 μm. The most commonly fabricated detector structures used homo- (n^+-on-p) and heterojunction (p-on-n) photodiodes. In both photodiodes, the lightly doped narrow-gap absorbing region ["base" of the photodiode: p(n)-type carrier concentration of about 3×10^{15} cm^{-3} (5×10^{14} cm^{-3})], determines the dark current and photocurrent. To receive high quantum efficiency, the thickness of an active detector layer should be equal, at least, to the cutoff wavelength. Typical quantum efficiency without antireflection coating is about 70%.

The main technological challenge for the fabrication of III-V barrier detectors is the growth of thick-active-detector-region SL structures without degrading the material quality. This is especially important when using T2SL structures. Using high-quality SL materials thick enough to achieve acceptable quantum efficiency is crucial to the success of the technology. For these reasons, at the present stage of the technological development, a typical thickness of the active region is about 3 μm and, as a rule, this thickess is independent of the detector cutoff wavelength. The influence of considerable surface leakage attributed to the discontinuity in the periodic crystal structure caused by mesa delineation is eliminated in barrier detectors.

Equation (6.7) demonstrates that the contribution of Auger 1 generation varies as N_d, whereas the SRH generation varies as $1/N_d$. As a result, the minimum dark current density depends on the absorber doping concentration and on the value of the SRH lifetime. This dependence is shown in Fig. 6.22 after Kinch et al.[49] for a MWIR nBn InAsSb barrier detector. To approach BLIP performance, the detector with a 4.8-μm cutoff wavelength operating at 160 K with F/3 optics requires a generic IR material SRH lifetime of about 1 μs and an optimized absorber doping of $\sim 10^{16}$ cm^{-3}. A value of 0.6 μs, relatively independent of temperature, for τ_{SR} has been suggested in Ref. 50.

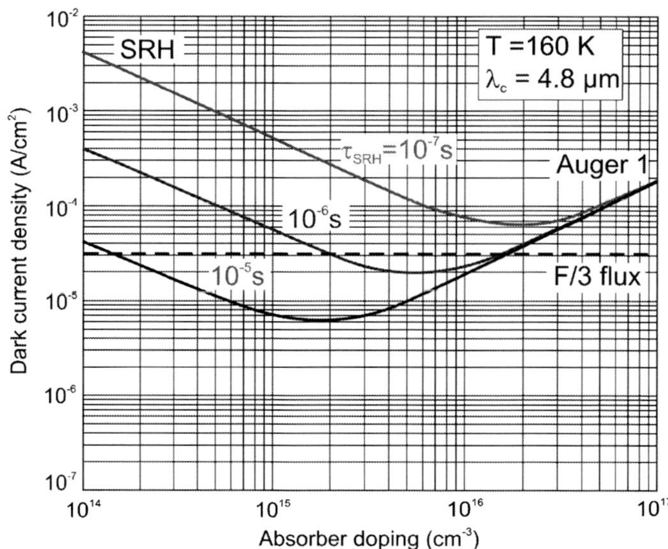

Figure 6.22 Dependence of absorber dark current density on doping concentration for various values of SRH lifetime at an operating temperature of 160 K and a cutoff wavelength of 4.8 μm, operating at an *F*/3 background flux (adapted from Ref. 49).

Numerical estimations of dark current were performed utilizing the commercial software APSYS by Crosslight Software Inc.[51] Specific parameters are listed in Table 6.2, but other relations used in device modeling are given in Rogalski's monograph[44] and Ting's et al. review paper.[4]

It is well known that at present, the performance of InAs/GaSb T2SL detectors is limited by the fairly fast SRH transition time between the conduction band and valence band states of the absorber material. To our knowledge, there are no systematic studies of the carrier lifetime and diffusion length dependence on temperature for T2SL materials. Several groups have estimated SRH carrier lifetimes ranging from several tens of nanoseconds to 157 ns.[52] For these reasons, we assume a carrier lifetime of 200 ns in both MWIR and LWIR compositions.

Pixel reduction is mandatory to increase the detection and identification range of infrared imaging systems. A pitch of 15 μm is being used in current infrared array production. In our estimations we assume a detector area of 15 × 15 μm^2.

Figure 6.23 compares the predicted dependence of dark current density on operating temperature for different types of detectors with a cutoff wavelength of 5 μm. In comparison with the InAsSb photodiode (with a built-in depletion region), the benefits of the nBn structure is clear—it allows for operation at considerably higher temperatures. However, HgCdTe photodiodes enable higher operating temperatures than the InAsSb nBn detector by ~20 K. The best MWIR III-V devices are heavily doped when the Auger lifetime is

Table 6.2 Material parameters of active regions.

Type of detector	Material	Cutoff wavelength [μm]	Doping of active region N_d, N_a [cm^{-3}]	Thickness of active region [μm]	τ_{SRH} [μs]	Detector area [μm^2]	Quantum efficiency	$F_1 F_2$	τ_{int} [ms]
nBn	InAsSb	5	1×10^{16}	3	0.40	15×15	0.70	0.28	10
nBn	InAs/GaSb	10	1×10^{16}	3	0.20	15×15	0.50	0.28	1
n-on-p	HgCdTe	5	3×10^{15}	5	-	15×15	0.70	0.20	10
n-on-p	HgCdTe	10	3×10^{15}	10	-	15×15	0.70	0.20	1
p-on-n	HgCdTe	5	5×10^{14}	5	-	15×15	0.70	0.20	10
p-on-n	HgCdTe	10	5×10^{14}	10	-	15×15	0.70	0.20	1

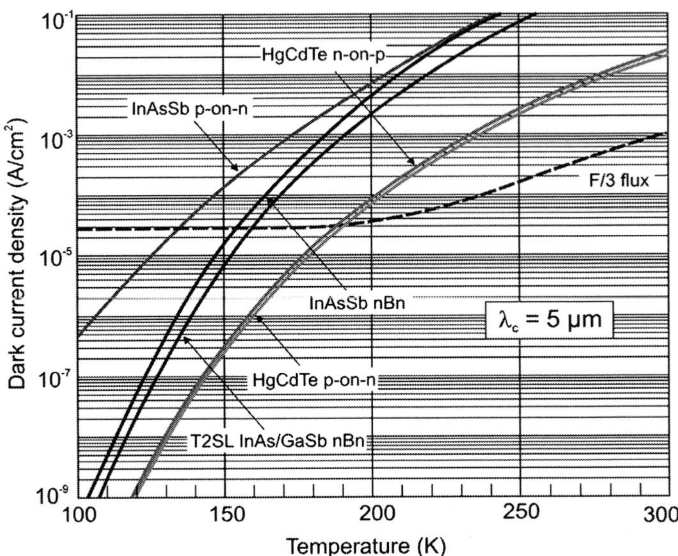

Figure 6.23 Dark current density versus temperature for an InAsSb photodiode and an nBn detector, and HgCdTe photodiodes with a cutoff wavelength of 5 μm (adapted from Ref. 51).

significantly reduced. 300-K data from Bewley et al.[53] show that Auger coefficients of SL devices are 5–20× lower than those of HgCdTe. To attain their full potential, the detector developers need to realize Auger-limited devices at doping in the 10^{15} cm^{-3} range. It should be stated that III-V MWIR detector architectures fall short of the ultimate performance possible with *F*/3 optics, namely, operation at about 150 K. MWIR HgCdTe systems that operate at *F*/3 optics at 160 K are commercially available.

Theoretically, LWIR T2SL materials have lower fundamental dark currents than HgCdTe. However, their performance has not achieved theoretical values. This limitation appears to be due to two main factors: relatively high background concentrations (about 10^{16} cm^{-3}) and a short minority-carrier lifetime (typically tens of nanoseconds). Up until now, nonoptimized carrier lifetimes limited by the SRH recombination mechanism have been observed. The minority-carrier diffusion length is in the range of several micrometers. Improving these fundamental parameters is essential to realizing the predicted performance of T2SL photodiodes.

To date, LWIR T2SL photodiodes perform slightly worse than HgCdTe. For example, Fig. 6.24 compares the predicted dependence of dark current density on operating temperature for InAs/GaSb nBn and HgCdTe photodiodes with a cutoff wavelength of 10 μm. The BLIP performance with *F*/1 optics for a HgCdTe photodiode is achieved at about 130 K—about 15 K higher than for InAs/GaSb nBn detector.

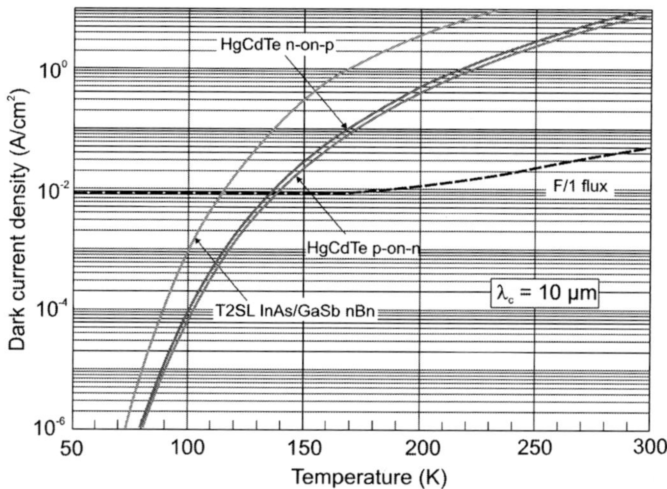

Figure 6.24 Dark current density versus temperature for an InAs/GaSb T2SL nBn detector and a HgCdTe n-on-p photodiode with a cutoff wavelength of 10 μm (adapted from Ref. 51).

6.6.3 Noise equivalent difference temperature

The detector sensitivity can be also expressed by the noise equivalent difference temperature (*NEDT*). This parameter is a figure of merit for thermal imagers (see section 1.5.2).

NEDT can be determined knowing the dark current density J_{dark}, the background flux (system optics) Φ_B, and the integration time τ_{int}, according to the relations[54]

$$NEDT = \frac{1 + (J_{dark}/J_\Phi)}{\sqrt{N_w}\left(\frac{1}{\Phi_B}\frac{d\Phi_B}{dT}\right)},$$ (6.9)

$$J_\Phi = q\eta\Phi_B,$$ (6.10)

$$\sqrt{N_w} = \frac{(J_{dark} + J_\Phi)\tau_{int}}{q},$$ (6.11)

where N_w is the well capacity of the readout, $J_\Phi = \eta\Phi_B A$ is the background flux current, and A is the detector area. Here η represents the overall quantum efficiency of the detector, including the internal quantum efficiency. The optics transmission and cold shield efficiency are assumed to be unity.

Taking into account that $(d\Phi_B/dT)/\Phi_B = C$ is the scene contrast, Eq. (6.9) assumes the form of

$$NEDT = \frac{1 + (J_{dark}/J_\Phi)}{\sqrt{N_w}\,C}.$$ (6.12)

The contrast in the MWIR band at 300 K is 3.5–4% compared to 1.7% for the LWIR band. Values of N are typically in the range of 1×10^6 to 1×10^7 electrons for a 15-μm-pixel design with available node capacities for current CMOS ROIC designs. In our estimation we assume 1×10^7 electrons.

Equation (6.12) demonstrates that if the value of the I_{dark}/I_ϕ ratio increases and/or the value of η decreases, more integration time and a faster speed of the optics are required. Thus, inefficient detectors can be utilized in faster-optics and slower-frame-rate systems.

Figure 6.25 shows the temperature dependence of the $NEDT$ for barrier detectors and HgCdTe photodiodes with a cutoff wavelength of 5 μm and 10 μm. The comparison of both detector technologies indicates that theoretical performance limits for HgCdTe photodiodes are more favorable than for barrier detectors in a temperature range above 150 K in a mid-wavelength range and above 80 K in a long-wavelength spectral range. In a low-temperature range the figures of merit for both material systems provide a similar performance because they are predominantly limited by the readout circuits.

6.6.4 Comparison with experimental data

Here we evaluate the current state of the technology of III-V barrier detectors by examining the dark current density in the mid-wavelength and long-wavelength spectral ranges as a function of a cutoff wavelength. Using the HgCdTe benchmark known as "Rule 07," we compare the experimental data for barrier detectors with a simple empirical relationship that describes the dark current behavior of HgCdTe photodiodes using temperature and wavelength.[45]

Figure 6.26 collects values of dark current density in MWIR barrier detectors published in literature for comparison with "Rule 07." The cutoff wavelength was taken as the point of a 50% response. The empirical data are

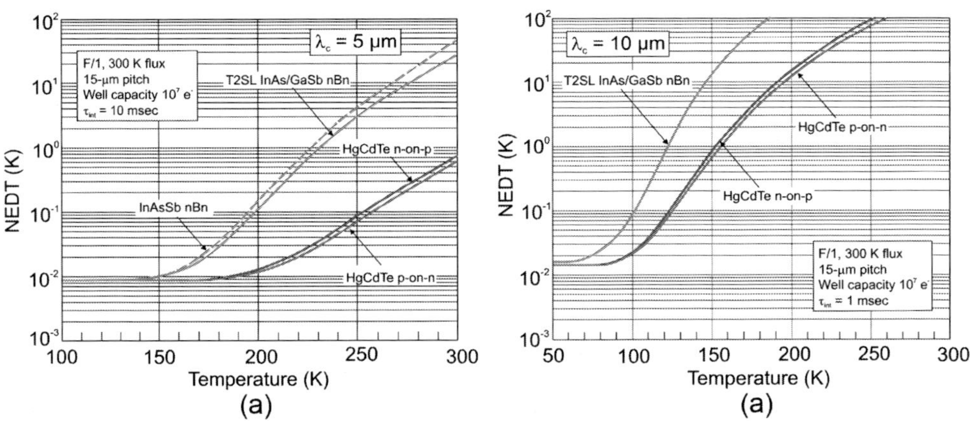

Figure 6.25 Temperature dependence of the *NEDT* for barrier detectors and HgCdTe photodiodes with cutoff wavelengths of (a) 5 μm and (b) 10 μm (adapted from Ref. 51).

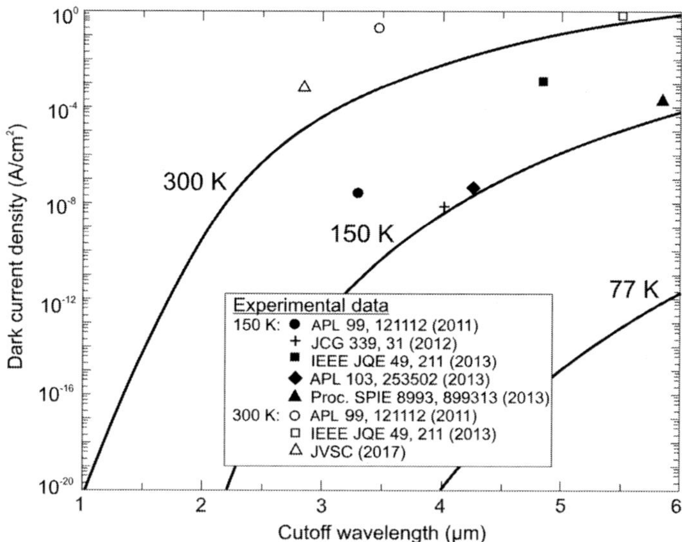

Figure 6.26 Collected values of dark current density in MWIR barrier detectors for comparison with "Rule 07" of HgCdTe photodiodes calculated for 77, 150, and 300 K (adapted from Ref. 51).

confined to 150 K and 300 K. For the barrier devices, the characteristics at zero bias are not relevant; therefore, a reverse bias about 150 mV has been chosen to extract the photogenerated minority carriers.

At liquid nitrogen temperature, the experimental data of MWIR barrier devices show considerable greater leakage current than "Rule 07" by many orders of magnitude. Results close to "Rule 07" are reported at a cutoff wavelength close to 4 µm at 150 K. These best-quality devices are fabricated using an InAsSb active region that is lattice matched to GaSb substrates. As we can see, the overall trend moves closer to "Rule 07" as the wavelength increases. This trend is especially observed in the LWIR region—it is generally more difficult to control dark current at shorter wavelengths.

A similar collection of experimental data for nonbarrier (homojunction) and barrier (heterojunction) LWIR T2SLs devices operating at 78 K has been gathered by Rhiger (see Fig. 6.27).[55] The nonbarrier dark currents are generally higher, with the best approaching "Rule 07" to within a factor of about 8. The barrier devices clearly show lower dark currents on average, and some are close to the "Rule 07" curve for a cutoff wavelength of ≥ 9 µm. It can be concluded that although the minority-carrier lifetime in the SL detector active region is not limited by an Auger mechanism, the diffusion current does not differ so greatly from "Rule 07."

In theoretical evaluations of *NEDT* for nBn InAsSb detector we assume a detector design similar to that shown in Fig. 6.9(a). Figure 6.11 shows the temperature dependence of *NEDT* at optics *F*/3.2 for a 15-µm-pitch

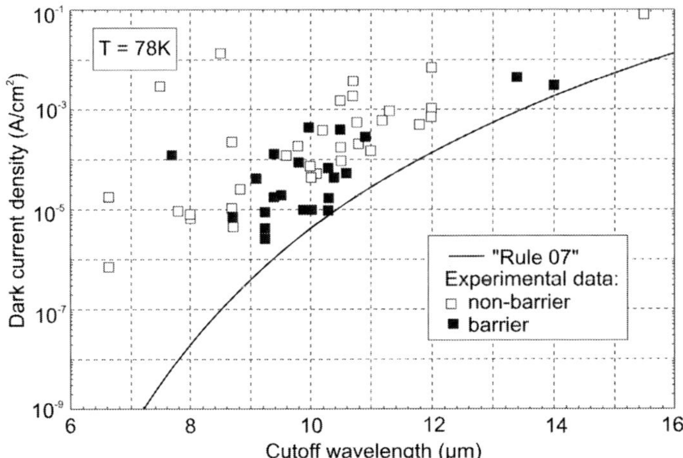

Figure 6.27   The 78-K dark current densities plotted against cutoff wavelength for T2SL nonbarrier and barrier detectors reported in the literature since late 2010. The solid line indicates the dark current density calculated using the empirical "Rule 07" model (adapted from Ref. 55).

$InAs_{0.91}Sb_{0.09}$/B-AlAsSb barrier detector operated in the blue part of the MWIR window of the atmosphere (3.4–4.2 µm). The *NEDT* is 20 mK at 10-ms integration time. As we can see, a good agreement between experimental data and the theoretical prediction has been achieved. A strong increase in *NEDT* above 170 K is consistent with the estimated BLIP temperature of 175 K.

To compare the performance of MWIR III-V barrier detector arrays with state-of-the-art HgCdTe technology, we chose the best-quality MWIR arrays—Hawk detectors with 640 × 512 pixels, 16-µm pitch, and an *F*/4 radiation shield. The 50% cutoff wavelength is 5.5 µm at 80 K. These N⁺-p(As) heterostructure photodiodes of $Hg_{1-x}Cd_xTe$, $x = 0.3$ and $x = 0.2867$, are optimized for HOT conditions. The devices were grown by metalorganic vapor phase epitaxy, MOCVD on GaAs substrates. The N⁺-region is doped with iodine at a level of 10^{16} cm⁻³, and an absorbing p-layer about 3-µm-thick having a smaller energy gap is doped with arsenic at a level of 10^{15} cm⁻³.

The Hawk array demonstrates a high-quality image at temperatures of 160–190 K. Although the acceptable 210-K image is slightly grainier than the 160-K image, the 210-K image is very useable. Figure 6.28 shows *NEDT* versus temperature with results from 2011 (standard) as well as results after an improvement in device technology.[56] The new results predict better detector performance together with extended useful range of operating temperatures.

For standard production arrays, the *NEDT* remains constant up to 150–160 K and doubles by 185 K (see Fig. 6.28). After improving the device technology, the near-BLIP achieved at 150 K by the standard process has been raised around 30 K to 180 K with the expectation of background-dominated performance to well above 200 K. To explain the temperature

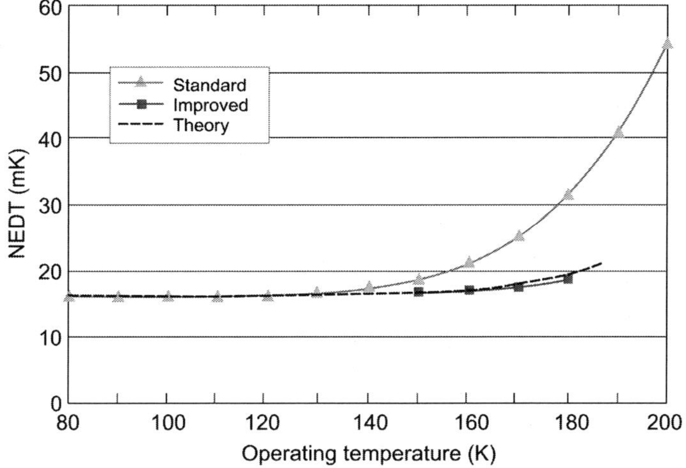

Figure 6.28 *NEDT* performance as a function of operating temperature (adapted from Ref. 56).

dependence of *NEDT* for the Hawk detector we estimated the dark current using a model given by DeWames and Pellegrino.[57] With the publication of their model, the deliberate introduction of recombination centers emerged as a new technique in HOT detector engineering.

At low temperatures, the performance of MWIR and LWIR FPAs is usually limited by the readout circuits (by storage capacity of the ROIC). In this case,[58]

$$NEDT = (\tau C \eta_{BLIP} \sqrt{N_w})^{-1},\qquad(6.13)$$

where N_w is the number of photogenerated carriers integrated for one integration time τ_{int} and is defined as

$$N_w = \eta A \tau_{int} \Phi_B.\qquad(6.14)$$

The percentage of BLIP η_{BLIP} is simply the ratio of photon noise to composite FPA noise:

$$\eta_{BLIP} = \left(\frac{N^2_{photon}}{N^2_{photon} + N^2_{FPA}}\right)^{1/2}.\qquad(6.15)$$

Equations (6.13) through (6.15) demonstrate that the charge-handling capacity of the readout, the integration time linked to the frame time, and the dark current of the sensitive material become the major issues of IR FPAs. The *NEDT* is inversely proportional to the square root of the integrated charge; therefore, the greater the charge the higher the performance. The well charge capacity is the maximum amount of the charge that can be stored in the storage capacitor of each cell. The size of the unit cell is limited to the dimensions of the detector element in the array.

The distinction between integration time and FPA frame time must be noted. At high backgrounds it is often impossible to handle the large number of carriers generated over a frame time compatible with standard video rates. Off-FPA frame integration can be used to attain a level of sensor sensitivity that is commensurate with the detector-limited D^* and not the charge-handling-limited D^*.

Figure 6.29 shows the theoretical *NEDT* versus charge handling capacity for different FPAs, assuming that the integration capacitor is filled to half of the maximum capacity (to preserve dynamic range) under nominal operating conditions in two spectral bandpasses: 3.4–4.8 μm and 7.8–10 μm. We can see that the measured sensitivities agree with the expected values for FPAs fabricated with different material systems, including barrier detectors and T2SLs.

6.7 Multicolor Barrier Detectors

Different detector architectures are involved in multicolor infrared applications related to remote sensing and object identification. Gautam et al.[60] have demonstrated a three-color heterojunction bandgap-engineered T2SL InAs/GaSb SWIR, MWIR, and LWIR detection. The most popular is a vertical

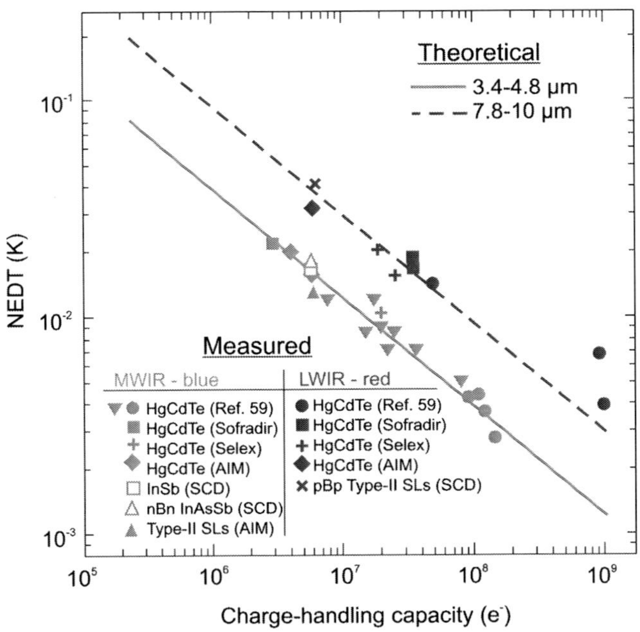

Figure 6.29 *NEDT* versus charge-handling capacity.

dual-band detector design with two 'back-to-back' photodiodes separated by a common ground contact layer.[61]

Utilization of barrier architectures (nBn or pBp) for dual-band T2SL barrier detectors introduces ease of bandgap tuning for different wavelength bands. Schematic operation of a dual-band nBn structure is explained in Section 6.4 (see Fig. 6.15). The device structure consists of two n-type (p-type)-doped InAs/GaSb T2SL absorption regions with designed operating wavelengths separated by a conduction band (valence band) T2SL barrier. Both barriers should be characterized by negligibly small valence band (conduction band) discontinuity with respect to both n-type (p-type) absorption regions. The absorption regions (channels) can be addressed consequently by changing the polarity of the applied bias.

Dual-band barrier technology benefits from a relatively easy growth procedure and mature III-V fabrication technology. In addition, these detectors are expected to demonstrate lower dark current levels compared to the vertical photovoltaic detector design due to elimination of the depletion region in the detector architecture.

Krishna et al. have demonstrated two-color barrier detectors based on nBn or pBp architectures.[62–64] For example, the structure of LW/LWIR dual-band pBp InAs/GaSb detector is schematically presented in Fig. 6.30. First, a quaternary AlGaAsSb etch stop layer (ESL) was grown lattice matched to the GaSb substrate. Next, a thick p-type GaSb contact layer was grown between the ESL and T2SL contact region to ensure a smooth surface for growth of the detector structure. Only p-type Be-doped GaSb layers were used for the top and bottom contacts. Also, the absorber layers and barrier were p-type doped with berillum as the dopant material.

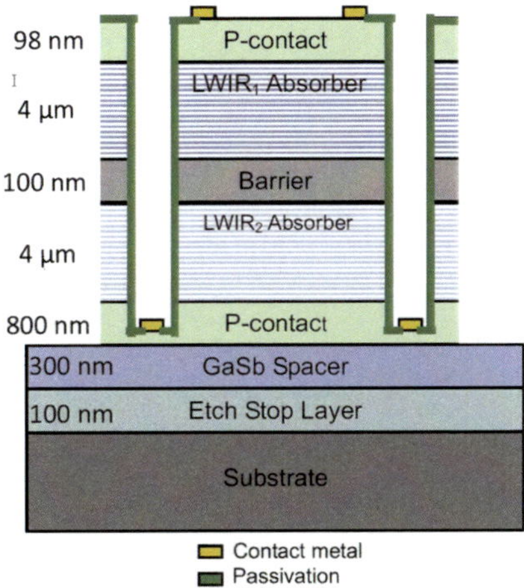

Figure 6.30 Schematic view of a dual-band LW/LWIR InAs/GaSb T2SL detector with the pBp architecture (reprinted from Ref. 63 with permission from AIP Publishing).

The dual-band pBp detector fabrication procedure was initiated by a 9-μm-deep etching using Cl_2-based inductively couple plasma (ICP) at 200 °C substrate temperature and with a SiO_2 etch mask, from the top contact layer to the middle of the bottom contact layer. The device fabrication was completed by depositing 50 nm Ti/50 nm Pt/350 nm Au to form ohmic contacts, followed by SU-8 passivation.

Figure 6.31 shows representative spectral response curves of the dual-band pBp detector at 78 K. The photo-carriers are collected from the "red" $LWIR_1$ channel (top absorbing region) when negative voltage is applied on the top contact. Positive voltage applied to the top contact results in collection of photo-carriers from the "blue" $LWIR_2$ channel (bottom absorbing region). The cutoff wavelengths are 9.2 μm ("blue" channel) and ~12 μm ("red" channel).

The dark current–voltage characteristics of the dual-band pBp T2SL detectors with variable areas measured at 78 K are shown in Fig. 6.32. The positive bias voltage range above 15 mV corresponds to the transport of minority electrons in the "blue" $LWIR_2$ absorption region, while the negative bias voltage range below 15 mV corresponds to the transport in the "red" $LWIR_1$ absorption region. The best signal-to-noise ratios are received at the applied biases +100 mV and –200 mV. The bulk-limited effective resistance–area product RA_{eff} (at 78 K) at +100 mV and –200 mV are 7.7×10^5 Ωcm^2 and 1.2×10^2 Ωcm^2, respectively.

Analysis of the experimental data presented in Fig. 6.32 indicates that the surface leakage component of dark current limits the performance of

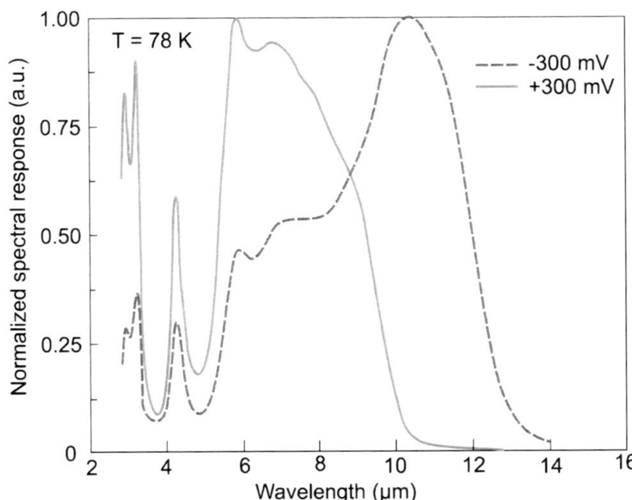

Figure 6.31 Normalized spectral response curves of the dual-band InAs/GaSb T2SL detector measured at 78 K (adapted from Ref. 64).

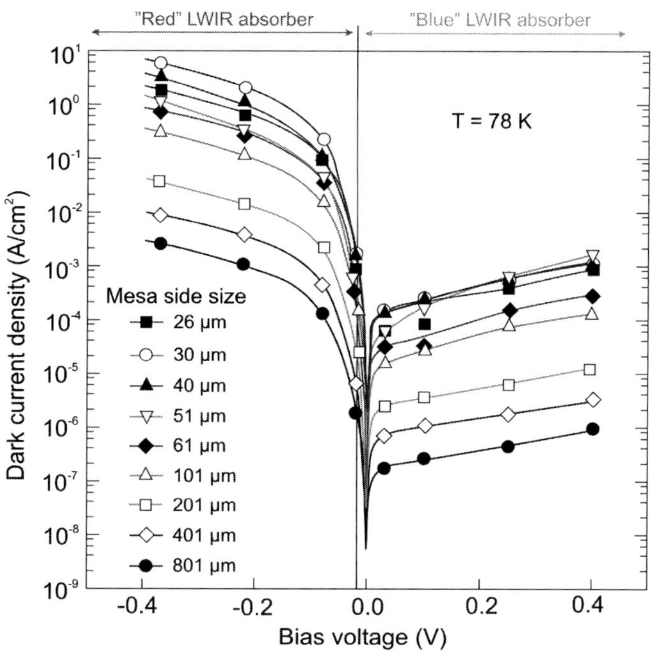

Figure 6.32 Dark current density of pBp LW/LWIR T2SL detectors with variable areas measured at 78 K under different applied bias (adapted from Ref. 64).

small-size LW/LWIR detectors. Due to the fabrication specifics, the lower absorber of the LW/LWIR detector is partially etched, while the upper absorber is fully delineated. It is expected that the surface current contribution to the total dark current affects the "red" absorber of the LW/LWIR structure more significantly than the "blue" absorber. The surface passivation of p-type LWIR material is a key area for continued development.

References

1. A. M. White, "Infrared detectors," U.S. Patent 4,679,063 (22 September 1983).
2. P. C. Klipstein, "Depletionless photodiode with suppressed dark current and method for producing the same," U.S. Patent 7,795,640 (2 July 2003).
3. S. Maimon and G. Wicks, "nBn detector, an infrared detector with reduced dark current and higher operating temperature," *Appl. Phys. Lett.* **89**, 151109 (2006).
4. D. Z.-Y. Ting, A. Soibel, L. Höglund, J. Nguyen, C. J. Hill, A. Khoshakhlagh, and S. D. Gunapala, "Type-II superlattice infrared detectors," in *Semiconductors and Semimetals*, Vol. **84**, edited by S. D. Gunapala, D. R. Rhiger, and C. Jagadish, Elsevier, Amsterdam, pp. 1–57 (2011).

5. G. R. Savich, J. R. Pedrazzani, D. E. Sidor, and G. W. Wicks, "Benefits and limitations of unipolar barriers in infrared photodetectors," *Infrared Physics & Technol.* **59**, 152–155 (2013).

6. M. Reine, J. Schuster, B. Pinkie, and E. Bellotti, "Numerical simulation and analytical modeling of InAs nBn infrared detectors with p-type barriers," *J. Electron. Mater.* **42**(11), 3015–3033 (2013).

7. M. Reine, J. Schuster, B. Pinkie, and E. Bellotti, "Numerical simulation and analytical modeling of InAs nBn infrared detectors with n-type barrier layers," *J. Electron. Mater.* **43**(8), 2915–2934 (2014).

8. P. Klipstein, "'XBn' barrier photodetectors for high sensitivity operating temperature infrared sensors," *Proc. SPIE* **6940**, 69402U (2008) [doi: 10.1117/12.778848].

9. D. Z. Ting, C. J. Hill, A. Soibel, J. Nguyen, S. A. Keo, M. C. Lee, J. M. Mumolo, J. K. Liu, and S. D. Gunapala, "Antimonide-based barrier infrared detectors," *Proc. SPIE* **7660**, 76601R (2010) [doi: 10.1117/12.851383].

10. P. Klipstein, O. Klin, S. Grossman, N. Snapi, I. Lukomsky, D. Aronov, M. Yassen, A. Glozman, T. Fishman, E. Berkowicz, O. Magen, I. Shtrichman, and E. Weiss, "XB*n* barrier photodetectors based on InAsSb with high operating temperatures," *Opt. Eng.* **50**(6), 061002 (2011) [doi: 10.1117/1.3572149].

11. G. R. Savich, J. R. Pedrazzani, D. E. Sidor, S. Maimon, and G. W. Wicks, "Use of unipolar barriers to block dark currents in infrared detectors," *Proc. SPIE* **8012**, 80122T (2012) [doi: 10.1117/12.884075].

12. P. Martyniuk and A. Rogalski, "HOT infrared photodetectors," *Opto-Electron. Rev.* **21**, 240–258 (2013).

13. P. C. Klipstein, "XB$_n$n and XB$_p$p infrared detectors," *J. Cryst. Growth* **425**, 351–256 (2015).

14. J. B. Rodriguez, E. Plis, G. Bishop, Y. D. Sharma, H. Kim, L. R. Dawson, and S. Krishna, "nBn structure based on InAs/GaSb type-II strained layer superlattices," *Appl. Phys. Lett.* **91**, 043514 (2007).

15. P. Klipstein, D. Aronov, E. Berkowicz, R. Fraenkel, A. Glozman, S. Grossman, O. Klin, I. Lukomsky, I. Shtrichman, N. Snapi, M. Yassem, and E. Weiss, "Reducing the cooling requirements of mid-wave IR detector arrys," SPIE Newsroom, 2011 [doi: 10.1117/2.1201111.003919].

16. M. Razeghi, S. P. Abdollahi, E. K. Huang, G. Chen, A. Haddadi, and B. M. Nquyen, "Type-II InAs/GaSb photodiodes and focal plane arrays aimed at high operating temperatures," *Opto-Electr. Rev.* **19**, 261–269 (2011).

17. M. Razeghi, "Type II superlattice enables high operating temperature," *SPIE Newsroom*, 2011 [doi: 10.1117/2.1201110.003870].

18. G. R. Savich, J. R. Pedrazzani, D. E. Sidor, S. Maimon, and G. W. Wicks, "Dark current filtering in unipolar barrier infrared detectors," *Appl. Phys. Lett.* **99**, 121112 (2011).

19. A. P. Craig, M. Jain, G. Wicks, T. Golding, K. Hossain, K. McEwan, C. Howle, B. Percy, and A. R. J. Marshall, "Short-wave infrared barriode detectors using InGaAsSb absorption material lattice matched to GaSb," *Appl. Phys. Lett.* **106**, 201103 (2015).

20. G. R. Savich, D. E. Sidor, X. Du, G. W. Wicks, M. C. Debnath, T. D. Mishima, M. B. Santos, T. D. Golding, M. Jain, A. P. Craig, and A. R. J. Marshall, "III-V semiconductor extended short-wave infrared detectors," *J. Vac. Sci. & Tech. B* **35**(2), 02B105 (2017).

21. A. Khoshakhlagh, S. Myers, E. Plis, M. N. Kutty, B. Klein, N. Gautam, H. Kim, E. P. G. Smith, D. Rhiger, S. M. Johnson, and S. Krishna, "Mid-wavelength InAsSb detectors based on nBn design," *Proc. SPIE* **7660**, 76602Z (2010) [doi: 10.1117/12.850428].

22. E. Weiss, O. Klin, S. Grossmann, N. Snapi, I. Lukomsky, D. Aronov, M. Yassen, E. Berkowicz, A. Glozman, P. Klipstein, A. Fraenkel, and I. Shtrichman, "InAsSb-based XB_nn bariodes grown by molecular beam epitaxy on GaAs," *J. Crystal Growth* **339**(1), 31–35 (2012).

23. P. Martyniuk and A. Rogalski, "Modeling of InAsSb/AlAsSb nBn HOT detector's performance limits," *Proc. SPIE* **8704**, 87041X (2013) [doi: 10.1117/12.2017721].

24. A. I. D'souza, E. Robinson, A. C. Ionescu, D. Okerlund, T. J. de Lyon, R. D. Rajavel, H. Sharifi, N. K. Dhar, P. S. Wijewarnasuriya, and C. Grein, "MWIR InAsSb barrier detector data and analysis," *Proc. SPIE* **8704**, 87041V (2013) [doi: 10.1117/12.2018427].

25. P. C. Klipstein, Y. Gross, A. Aronov, M. ben Ezra, E. Berkowicz, Y. Cohen, R. Fraenkel, A. Glozman, S. Grossman, O. Kin, I. Lukomsky, T. Markowitz, L. Shkedy, I. Sntrichman, N. Snapi, A. Tuito, M. Yassen, and E. Weiss, "Low SWaP MWIR detector based on XBn focal plane array," *Proc. SPIE* **8704**, 87041S (2013) [doi: 10.1117/12.2015747].

26. Y. Lin, D. Donetsky, D. Wang, D. Westerfeld, G. Kipshidze, L. Shterengas, W. L. Sarney, S. P. Svensson, and G. Belenky, "Development of bulk InAsSb alloys and barrier heterostructures for long-wavelength infrared detectors," *J. Electron. Mater.* **44**(10), 3360–3366 (2015).

27. E. H. Aifer, J. G. Tischler, J. H. Warner, I. Vurgaftman, W. W. Bewley, J. R. Meyer, C. L. Canedy, and E. M. Jackson, "W-structured type-II superlattice long-wave infrared photodiodes with high quantum efficiency," *Appl. Phys. Lett.* **89**, 053519 (2006).

28. B.-M. Nguyen, M. Razeghi, V. Nathan, and G. J. Brown, "Type-II M structure photodiodes: an alternative material design for mid-wave to long wavelength infrared regimes," *Proc. SPIE* **6479**, 64790S (2007) [doi: 10.1117/12.711588].

29. C. L. Canedy, H. Aifer, I. Vurgaftman, J. G. Tischler, J. R. Meyer, J. H. Warner, and E. M. Jackson, "Antimonide type-II W photodiodes with

long-wave infrared R_0A comparable to HgCdTe," *J. Electron. Mater.* **36**, 852–856 (2007).

30. B.-M. Nguyen, D. Hoffman, P.-Y. Delaunay, and M. Razeghi, "Dark current suppression in type II InAs/GaSb superlattice long wavelength infrared photodiodes with M-structure," *Appl. Phys. Lett.* **91**, 163511 (2007).

31. B.-M. Nguyen, D. Hoffman, P.-Y. Delaunay, E. K. Huang, M. Razeghi, and J. Pellegrino, "Band edge tunability of M-structure for heterojunction design in Sb based type II superlattice photodiodes," *Appl. Phys. Lett.* **93**, 163502 (2008).

32. M. Razeghi, H. Haddadi, A. M. Hoang, E. K. Huang, G. Chen, S. Bogdanov, S. R. Darvish, F. Callewaert, and R. McClintock, "Advances in antimonide-based Type-II superlattices for infrared detection and imaging at center for quantum devices," *Infrared Phys. & Technol.* **59**, 41–52 (2013).

33. E. K. Huang, D. Hoffman, B.-M. Nguyen, P.-Y. Delaunay, and M. Razeghi, "Surface leakage reduction in narrow band gap type-II antimonide-based superlattice photodiodes," *Appl. Phys. Lett.* **94**, 053506 (2009).

34. O. Salihoglu, A. Muti, K. Kutluer, T. Tansel, R. Turan, Y. Ergun, and A. Aydinli, "'N' structure for type-II superlattice photodetectors," *Appl. Phys. Lett.* **101**, 073505 (2012).

35. J. L. Johnson, L. A. Samoska, A. C. Gossard, J. L. Merz, M. D. Jack, G. H. Chapman, B. A. Baumgratz, K. Kosai, and S. M. Johnson, "Electrical and optical properties of infrared photodiodes using the InAs/Ga$_{1-x}$In$_x$Sb superlattice in heterojunctions with GaSb," *J. Appl. Phys.* **80**, 1116–1127 (1996).

36. A. Khoshakhlagh, J. B. Rodriguez, E. Plis, G. D. Bishop, Y. D. Sharma, H. S. Kim, L. R. Dawson, and S. Krishna, "Bias dependent dual band response from InAs/Ga(In)Sb type II strain layer superlattice detectors," *Appl. Phys. Lett.* **91**, 263504 (2007).

37. E. H. Aifer, H. Warner, C. L. Canedy, I. Vurgaftman, J. M. Jackson, J. G. Tischler, J. R. Meyer, S. P. Powell, K. Oliver, and W. E. Tennant, "Shallow-etch mesa isolation of graded-bandgap W-structured type II superlattice photodiodes," *J. Electron. Mater.* **39**, 1070–1079 (2010).

38. I. Vurgaftman, E. H. Aifer, C. L. Canedy, J. G. Tischler, J. R. Meyer, and J. H. Warner, "Graded band gap for dark-current suppression in long-wave infrared W-structured type-II superlattice photodiodes," *Appl. Phys. Lett.* **89**, 121114 (2006).

39. D. Z.-Y. Ting, C. J. Hill, A. Soibel, S. A. Keo, J. M. Mumolo, J. Nguyen, and S. D. Gunapala, "A high-performance long wavelength superlattice complementary barrier infrared detector," *Appl. Phys. Lett.* **95**, 023508 (2009).

40. E. A. DeCuir, G. P. Meissner, P. S. Wijewarnasuriya, N. Gautam, S. Krishna, N. K. Dhar, R. E. Welser, and A. K. Sood, "Long-wave type-II

superlattice detectors with unipolar electron and hole barriers," *Opt. Eng.* **51**(12), 124001 (2012) [doi: 10.1117/1.OE.51.12.124001].

41. N. Gautam, S. Myers, A. V. Barve, B. Klein, E. P. Smith, D. Rhiger, E. Plis, M. N. Kutty, N. Henry, T. Schuler-Sandyy, and S. Krishna, "Band engineering HOT midwave infrared detectors based on type-II InAs/GaSb strained layer superlattices," *Infrared Physics & Techol.* **59**, 72–77 (2013).

42. P. C. Klipstein, E. Avnon, Y. Benny, R. Fraenkel, A. Glozman, S. Grossman, O. Klin, L. Langoff, Y. Livneh, I. Lukomsky, M. Nitzani, L. Shkedy, I. Shtrichman, N. Snapi, A. Tuito, and E. Weiss, "InAs/GaSb type II superlattice barrier devices with a low dark current and a high quantum efficiency," *Proc. SPIE* **9070**, 90700U (2014) [doi: 10.1117/12.2049825].

43. A. D. Hood, A. J. Evans, A. Ikhlassi, D. L. Lee, and W. E. Tennant, "LWIR strained-layer superlattice materials and devices at Teledyne Imaging Sensors," *J. Elect. Mater.* **39**, 1001–1006 (2010).

44. A. Rogalski, *Infrared Detectors*, 2nd edition, CRC Press, Boca Raton, Florida (2010).

45. W. E. Tennant, D. Lee, M. Zandian, E. Piquette, and M. Carmody, "MBE HgCdTe Technology: A very general solution to IR detection, described by 'Rule 07,' a very convenient heuristic," *J. Elect. Mat.* **37**, 1406 (2008).

46. P. C. Klipstein, E. Avnon, D. Azulai, Y. Benny, R. Fraenkel, A. Glozman, E. Hojman, O. Klin, L. Krasovitsky, L. Langof, I. Lukomsky, M. Nitzani, I. Shtrichman, N. Rappaport, N. Snapi, E. Weiss, and A. Tuito, "Type II superlattice technology for LWIR detectors," *Proc. SPIE* **9819**, 98190T (2016) [doi: 10.1117/12.2222776].

47. P. Klipstein, "Physics and technology of antimonide heterostructure devices at SCD," *Proc. SPIE* **9370**, 937020 (2015) [doi: 10.1117/12. 2082938].

48. P. Martyniuk, W. Gawron, D. Stępień, D. Benyahia, A. Kowalewski, K. Michalczewski, and A. Rogalski, "Demonstration of mid-wavelength type-II superlattice InAs/GaSb single pixel barrier detector with GaAs immersion lens," *IEEE Electron Dev. Lett.* **37**(1), 64–65 (2016).

49. M. A. Kinch, H. F. Schaake, R. L. Strong, P. K. Liao, M. J. Ohlson, J. Jacques, C.-F. Wan, D. Chandra, R. D. Burford, and C. A. Schaake, "High operating temperature MWIR detectors," *Proc. SPIE* **7660**, 76602V (2010) [doi: 10.1117/12.850965].

50. J. F. Klem, J. K. Kim, M. J. Cich, S. D. Hawkins, T. R. Fortune, and J. L. Rienstra, "Comparison of nBn and nBp mid-wave barrier infrared photodetectors," *Proc. SPIE* **7608**, 76081P (2010) [doi: 10.1117/12. 842772].

51. P. Martyniuk and A. Rogalski, "Performance comparison of barrier detectors and HgCdTe photodiodes," *Opt. Eng.* **53**(10), 106105 (2014) [doi: 10.1117/1.OE.53.10.106105].

52. D. Zuo, P. Qiao, D. Wasserman, and S. L. Chuang, "Direct observation of minority carrier lifetime improvement in InAs/GaSb type-II superlattice photodiodes via interfacial layer control," *Appl. Phys. Lett.* **102**, 141107 (2013).

53. W. W. Bewley, J. R. Lindle, C. S. Kim, M. Kim, C. L. Canedy, I. Vurgaftman, and J. R. Meyer, "Lifetime and Auger coefficients in type-II W interband cascade lasers," *Appl. Phys. Lett.* **93**, 041118 (2008).

54. M. A. Kinch, *Fundamentals of Infrared Detector Materials*, SPIE Press, Bellingham, 2007 [doi: 10.1117/3.741688].

55. D. R. Rhiger, "Performance comparison of long-wavelength infrared type II superlattice devices with HgCdTe," *J. Elect. Mater.* **40**, 1815–1822 (2011).

56. P. Knowles, L. Hipwood, N. Shorrocks, I. M. Baker, L. Pillans, P. Abbott, R. Ash, and J. Harji, "Status of IR detectors for high operating temperature produced by MOVPE growth of MCT on GaAs substrates," *Proc. SPIE* **8541**, 854108 (2012).

57. R. R. DeWames and J. Pellegrino, "Electrical characteristics of MOVPE grown MWIR $N^+p(As)$ HgCdTe hetero-structure photodiodes build on GaAs substrates," *Proc. SPIE* **8353**, 83532P (2012) [doi: 10.1117/12.921093].

58. L. J. Kozlowski and W. F. Kosonocky, "Infrared detector arrays," in *Handbook of Optics*, Chap. 23, edited by M. Bass, E. W. Van Stryland, D. R. Williams, and W. L. Wolfe, McGraw-Hill, Inc., New York (1995).

59. L. J. Kozlowski, "HgCdTe focal plane arrays for high performance infrared cameras," *Proc. SPIE* **3179**, 200–211 (1997) [doi: 10.1117/12.276226].

60. N. Gautam, M. Naydenkov, S. Myers, A. V. Barve, E. Plis, T. Rotter, L. R. Dawson, and S. Krishna, "Three color infrared detector using InAs/GaSb superlattices with unipolar barriers," *Appl. Phys. Lett.* **98**, 121106 (2011).

61. A. Rogalski, J. Antoszewski, and L. Faraone, "Third-generation infrared photodetector arrays," *J. Appl. Phys.* **105**, 091101 (2009).

62. A. Khoshakhlagh, J. B. Rodriguez, E. Plis, G. D. Bishop, Y. D. Sharma, H. S. Kim, L. R. Dawson, and S. Krishna, "Bias dependent dual band response from InAs/Ga(In)Sb type II strain layer superlattice detectors," *Appl. Phys. Lett.* **91**(26), 263504 (2007).

63. E. Plis, S. Myers, D. Ramirez, E. P. Smith, D. Rhiger, C. Chen, J. D. Phillips, and S. Krishna, "Dual color longwave InAs/GaSb type-II strained layer superlattice detectors," *Infrared Phys.& Technol.* **70**, 93–98 (2015).

64. E. Plis, S. A. Myers, D. A. Ramirez, and S. Krishna, "Development of dual-band barrier detectors," *Proc. SPIE* **9819**, 981911 (2016) [doi: 10.1117/12.2228166].

Chapter 7
Cascade Infrared Photodetectors

In a conventional photodiode, the responsivity and diffusion length are closely coupled and an increase in the absorber thickness beyond the diffusion length may not result in the desired improvement in the signal to noise ratio (*SNR*). This effect is particularly pronounced at high temperatures, where diffusion lengths are typically reduced. Only charge carriers that are photogenerated at a distance shorter than the diffusion length from a junction can be collected. In HOT detectors the absorption depth of LWIR radiation is longer than the diffusion length. Therefore, only a limited fraction of the photogenerated charge contributes to the quantum efficiency.

To avoid the limitation imposed by the reduced diffusion length and to effectively increase the absorption efficiency, innovative detector designs based on multistage detection and currently termed as cascade infrared detectors (CIDs) have been introduced in the last decade. CIDs contain multiple discrete absorbers, where each one is shorter or narrower than the diffusion length. In this discrete CID absorber architecture, the individual absorbers are sandwiched between engineered electron and hole barriers to form a series of cascade stages. The photogenerated carriers travel over only one cascade stage before they recombine in the next stage, and every individual cascade stage can be significantly shorter than the diffusion length, while the total thickness of all of the absorbers can be comparable or even longer than the diffusion length.

In this case, the *SNR* and the detectivity will continue to increase with an increasing number of discrete absorbers, resulting in improved device performance at elevated temperatures compared to a conventional p-n photodiode. In addition, the flexibility to vary the number and thicknesses of the discrete absorbers results in the ability to tailor the CID designs for optimized performance in meeting specific applications.

7.1 Multistage Infrared Detectors

Different types of multistage IR detectors have been proposed and are now grouped into two main classes: (1) so-called intersubband (IS) unipolar quantum cascade IR detectors (QCIDs), and (2) interband (IB) ambipolar CIDs. Intersubband QCIDs have evolved from the research on quantum cascade lasers (QCLs) and have been built for about 15 years.[1–6] A schematic comparison between the band structure of a photoconductive quantum well infrared photodetector (QWIP) and a photovoltaic QCID is shown in Fig. 7.1. The QWIP structure is polarized in order to make the electrons circulate in the external circuit and to record the variation. The active detector region consists of identical QWs separated by thicker barriers. Electrons are excited from the quantum wells by either photoemission (red arrows) or by thermionic emission (black arrows). In contrast, QCIDs are usually designed to be photovoltaic detectors. They consist of several identical periods made of one active doped well and some other coupled wells. The photoexcited electrons are transported from one active well to the next one by phonon emission through cascaded levels. Figure 7.1(b) shows the conduction band of one period. An incident photon induces an electron to go from the ground state E_1

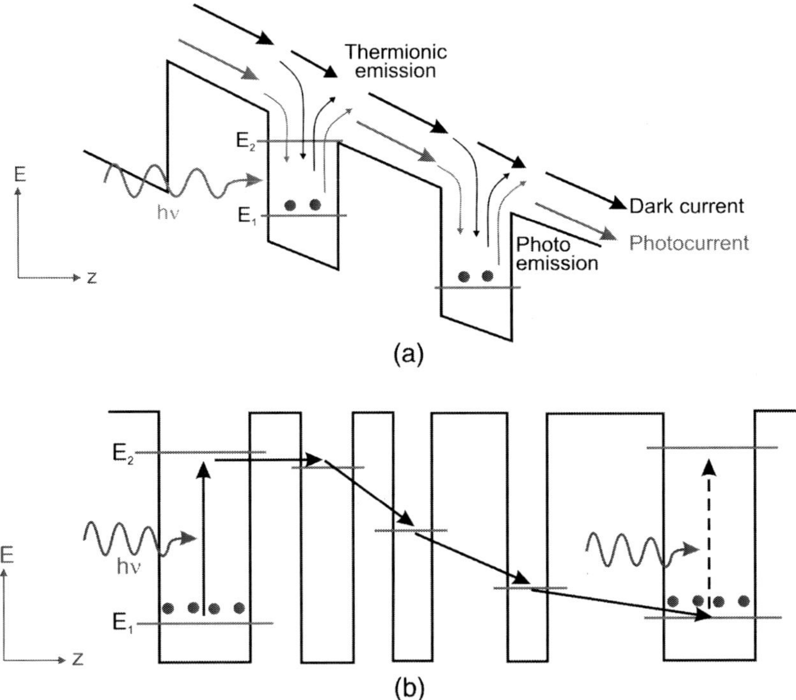

Figure 7.1 Schematic conduction band diagram of (a) a QWIP and (b) a QCID. In the QWIP, electron transport is accomplished by an external voltage bias, whereas an internal potential ramp ensures carrier transport in the QCID (adapted from Ref. 4).

to the excited level E_2. The electron is next transferred to the right-hand QWs through longitudinal optical phonon relaxations, and finally to the fundamental subband of the next period. The detector period is repeated N times in order to increase the detectivity.

To describe the performance of IS QCIDs, it is convenient to use the formalism originally developed for QWIPs.[7] A theoretical model is presented in Refs. 1, 2, and 5.

The detectivity of a QCID, including Johnson noise and electrical shot noise components, is determined by[1]

$$D^* = \frac{\eta \lambda q}{hc} \left(\frac{4kT}{NR_0A} + \frac{2qI_{dark}}{N} \right)^{-1/2}, \tag{7.1}$$

where R_0A is the resistance at zero bias times the detector area, corresponding to one period of the QCID; T is the detector temperature; N is the number of periods; and I_{dark} is the dark current. Equation (7.1) shows that the *SNR* is $\propto \sqrt{N}$.

The IS QCID technology has been proven in the wavelength range from the near IR to the terahertz (THz) region and the attained detectivity is presented in Fig. 7.2. At present, well-established semiconductor material systems and processing methods are available. Early QCIDs have been demonstrated in the near-IR using InGaAs/AlAsSb, in the mid-IR using

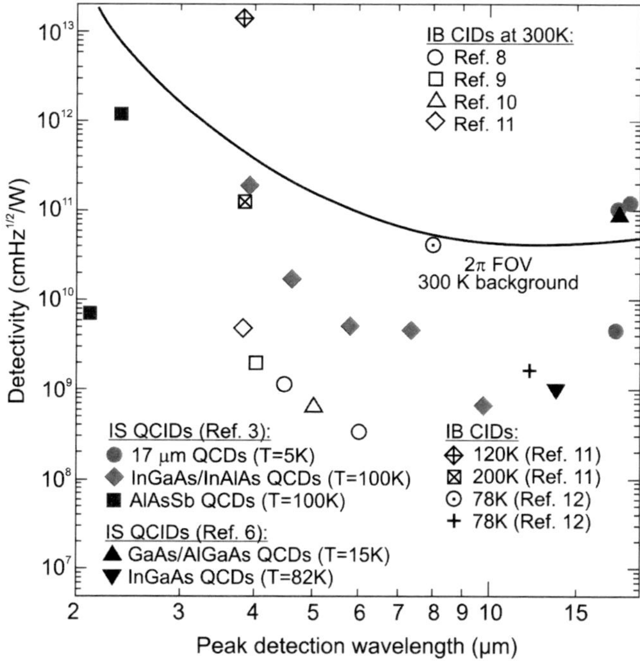

Figure 7.2 Detectivity as function of wavelength for different types of CIDs.

InGaAs/InAlAs, and in long-IR up to the THz region using GaAs/AlGaAs materials. These detectors have been cryogenically cooled.[3,4]

7.2 Type-II Superlattice Interband Cascade Infrared Detectors

It has been recently demonstrated that bipolar devices based on type-II InAs/GaSb IB SL absorbers[8–17] are good candidates for detectors operating at near room temperature. These IB cascade detectors combine the advantages of IB optical transitions with the excellent carrier transport properties of the IB cascade laser structures. The thermal generation rate at any specific temperature and cutoff wavelength in these devices is usually orders of magnitude smaller than that for corresponding IS QCIDs, and devices with good performance have been recently demonstrated. The operating temperature of IB cascade detectors is considerable higher in comparison with IS cascade detectors (see Fig. 7.2).

Hinkley and Yang[14] have shown that multiple-stage architecture is useful for improving the sensitivity of HOT detectors, where the quantum efficiency is limited by short diffusion length. In the case of HgCdTe photodiodes at room temperature, the absorption depth for LWIR radiation ($\lambda > 5$ µm) is longer than the diffusion length. Therefore, only a limited fraction of the photogenerated charge contributes to the quantum efficiency. Calculations considering the example of an uncooled 10.6-µm photodiode show that the ambipolar diffusion length is less than 2 µm, while the absorption depth is ~13 µm. This reduces the quantum efficiency to ~15% for a single pass of radiation through the detector.

A similar situation occurs in the case of HOT T2SL interband cascade IR detectors (IB CIDs). For detector designs where the absorber lengths in each stage are equal, the multiple-stage architecture offers the potential for significant detectivity improvement when $\alpha L \leq 0.2$, where α is the absorption coefficient, and L is the diffusion length.[14] This theoretical prediction has been confirmed by experimental data, as is shown in Fig. 7.2.

7.2.1 Principle of operation

The operation concept behind IB cascade photodetectors is similar to that described by Piotrowski and Rogalski (see Fig. 7.3).[18,19] Earlier attempts to

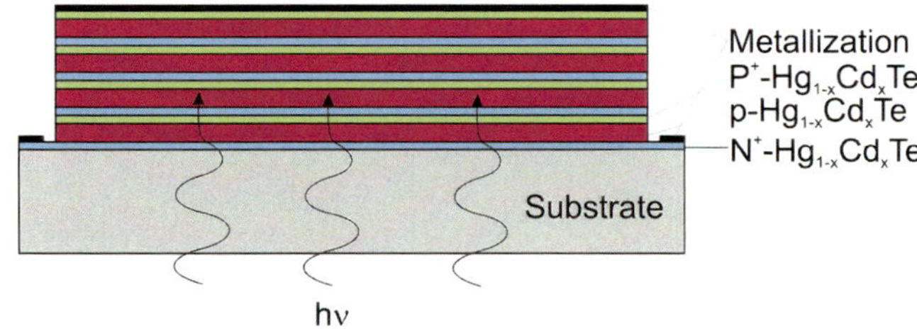

Figure 7.3   Backside-illuminated four-cell stacked HgCdTe photovoltaic detector.

realize this type of device using HgCdTe were carried out by utilizing tunnel junctions to electrically connect the conduction band of one absorber to the valence band of an adjacent absorber in a way similar to multi-junction solar cells. Each cell is composed of a p-type-doped narrow-gap absorber and heavily doped N^+ and P^+ heterojunction contacts. The incoming radiation is absorbed only in absorber regions, while the heterojunction contacts collect the photogenerated charge carriers. Such devices are capable of achieving high quantum efficiency, large differential resistance, and fast response. A practical problem is associated with electrical conductivity through the adjacent N^+ and P^+ layers; however, this is generally overcome by employing tunnel currents through the N^+ and P^+ interface.

The T2SL material system is a natural candidate for realizing multiple-stage IB devices.[20] Figure 7.4 shows the general design structure for a T2SL cascade detector.[17,21] Each stage is composed of an n-period InAs/GaSb T2SL sandwiched between an AlSb/GaSb QW electron barrier and an InAs/Al(In)Sb QW hole barrier.

Because the design of IB CIDs is relatively complicated, involving many interfaces and strained thin layers, their growth by MBE is challenging. Detector designs exhibit key differences in their approach to construct the relaxation and tunneling regions, as well as the contact layers. These designs are described in detail in Ref. 20. Here we focus on the high-quality devices presented by Tian and Krishna.[11]

Tian and Krishna have proposed a cascade detector structure whose operation is shown schematically in Fig. 7.5.[11] The incoming photons are absorbed in thin InAs/GaSb T2SLs sandwiched between the electron-relaxation and the interband-tunneling regions that serve also as the hole (eR) and the electron barriers (eB), respectively. The barriers act as a means

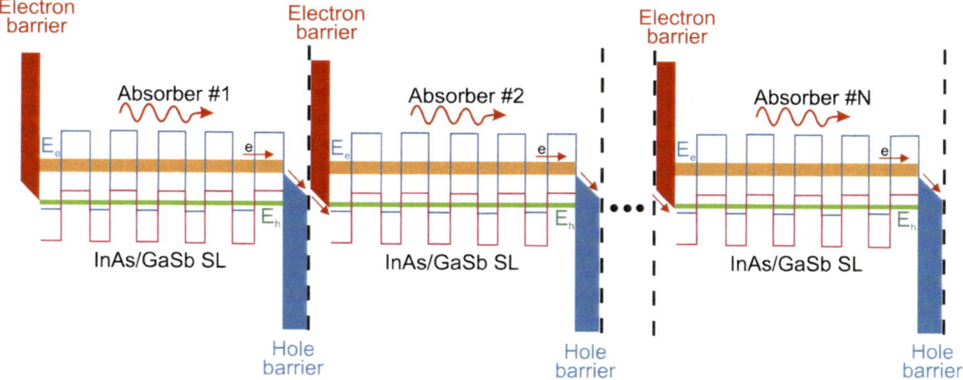

Figure 7.4 Schematic illustration of an IB CID device with multiple stages. Each stage is composed of a SL absorber sandwiched between electron and hole barriers. E_e and E_h denote the energy for electron and hole minibands, respectively. The energy difference $(E_e - E_h)$ is the bandgap E_g of the SL (adapted from Ref. 21).

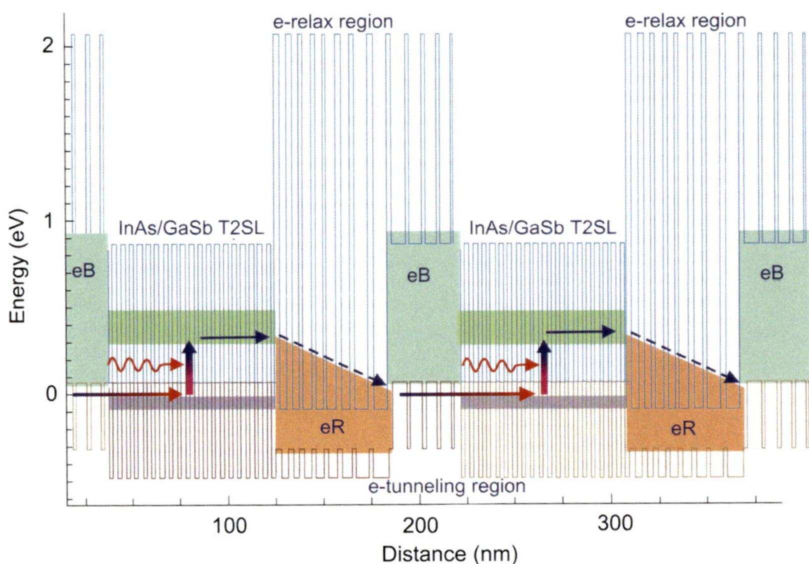

Figure 7.5 Schematic illustration of the interband cascade type-II InAs/GaSb superlattice photodetector. The photons are absorbed into the absorbers, generating electron–hole pairs. The electrons diffuse into the eR region and then transport into the valence band of the next stage through ultrafast LO-phonon-assist intersubband relaxation and interband tunneling (adapted from Ref. 11).

for suppressing leakage current. The electron-relaxation region is designed to facilitate the extraction of photogenerated carriers from the conduction miniband of the absorber and transport them ideally (with little or no resistance) to the valence band of the absorber in the next stage. The energy levels of coupled InAs/AlSb multi-QWs in the conduction band form a six-step staircase, with energy-ladder separations comparable to the longitudinal optical (LO)-phonon energy. The uppermost energy level of the relaxation region staircase is close to the conduction miniband in the InAs/GaSb SL, and the bottom energy level is positioned below the valence band edge of the adjacent GaSb layer, allowing the interband tunneling of extracted carriers to the next stage. The eB region consists of GaSb/AlSb QWs with estimated electron barrier thickness and height (relative to the conduction miniband minimum of the InAs/GaSb T2SLs absorber) of 45 nm and 0.72 eV, respectively.

7.2.2 MWIR interband cascade detectors

The 5-stage detector structures are MBE-grown on Zn-doped 2-inch (001) GaSb substrates. The absorbers are composed of lightly p-doped ($\sim 5 \times 10^{15}$ cm^{-3}) InAs/GaSb T2SLs with InSb interfacial layers to balance strain of the lattice-mismatched InAs. The V/III beam equivalent flux ratios for Sb/Ga and As/In are set as 4.0 and 3.2 respectively. The detectors consist of a 0.5-µm p-type GaSb buffer layer, a 5-stage interband cascade absorber structure, and finally a 45-nm-thick n-type InAs top contact layer. The individual absorbers consist of 30, 60, and 90 periods of 7 MLs InAs/8 MLs GaSb (ML stands for monolayer) T2SLs, which correspond to total absorber thicknesses of 0.73, 1.45, and 2.16 µm, respectively. Single-pixel detectors with circular mesa sizes ranging from 25 to 400 µm in diameter were fabricated. A 200-nm-thick SiN_x film is then deposited for sidewall passivation and electrical isolation. The top and bottom contacts are formed by e-beam-evaporated Ti/Au. No antireflection coating is applied on top of the mesa.

Figure 7.6 shows the representative temperature-dependent dark current characteristics of a 90-period MWIR interband cascade device. The low-temperature J–V curves for IB cascade detectors are relatively steep [see Fig. 7.6(a)], suggesting contributions from tunneling components. At higher temperatures, the dark current is much less sensitive to the operation bias volatge and is diffusion limited.

Additional insight into the dark current characteristics gives the Arrhenius plot of the dark current density at -10 mV as well as the measured zero-bias-resistance–area product R_0A [see Fig. 7.6(b)]. The extracted activation energy E_A at higher operating temperatures is about 0.302 eV, which is very close to the effective bandgap in the InAs/GaSb T2SL absorber, confirming that the dark current at higher temperatures is mostly due to the diffusion component. The R_0A product of the device exceeds 1.25×10^7 Ωcm^2 at 120 K, is 2470 Ωcm^2 at 200 K, and is 3.93 Ωcm^2 at room temperature, which is the highest R_0A product reported in T2SL detectors. The dark current density is as low as

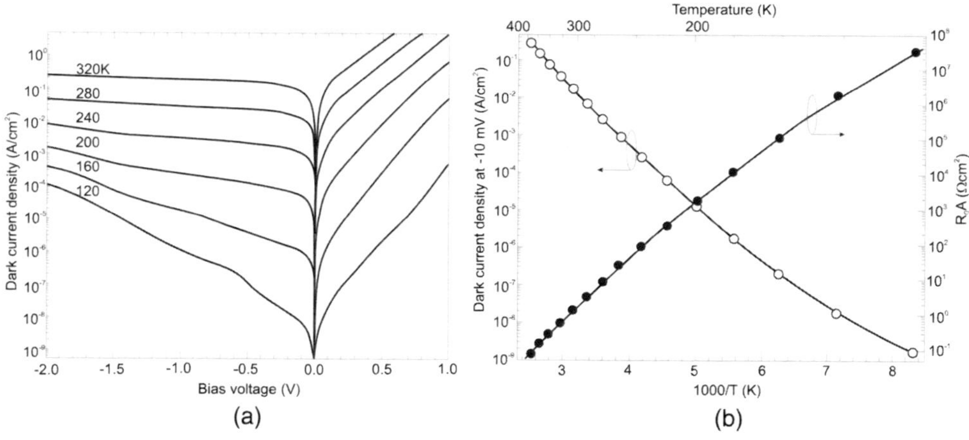

Figure 7.6 Dark current characteristics of a 5-stage MWIR InAs/GaSb T2SL IB 90-period cascade detector: (a) current–volatge characteristics, (b) Arrhenius plots of the electrical performance (adapted from Ref. 11).

1.28×10^{-7} A/cm^2 at 160 K, and the extracted R_0A is 9.42×10^4 Ωcm^2, which are both slightly better than the HgCdTe "Rule 07."[22]

The spectral responsivity of MWIR T2SL cascade detectors is about 0.3 A/W at room temperature and wavelength 4 μm, and has been observed up to 380 K. Figure 7.7 presents the Johnson-noise-limited detectivity spectra at various temperatures, extracted from the measured responsivity spectra and R_0A product for the above-described detector structure. The Johnson-limited D^* reaches 1.29×10^{13} Jones at 3.8 μm and 120 K, and 9.73×10^{11} Jones at 200 K. The BLIP performance is 180 K at 4 μm for 5-stage/junction devices with 70% absorption quantum efficiency, which correspond to 14% external quantum efficiency.

In the presented design, the total thickness of the absorber is about 1 μm, and theoretically, the absorption quantum efficiency could be increased by increasing the number of stages. However, the conversion quantum efficiency is lower than that of the absorption quantum efficiency by a factor of N.

Using the equation for spectral responsivity, $R_i = (\lambda\eta/hc)qg$, (where h, c, and g are Planck's constant, speed of light, and photoconductive gain, respectively) and the experimental data for $R_i \approx 0.3$ A/W, we can estimate room-temperature conversion quantum efficiency at a wavelength of 4 μm as $\eta g \approx 9\%$. Since the device has five stages, one would estimate a gain of 1/5 (0.20). This leads to an absorption quantum efficiency of 45%. The absorption quantum efficiency would be increased by increasing the number

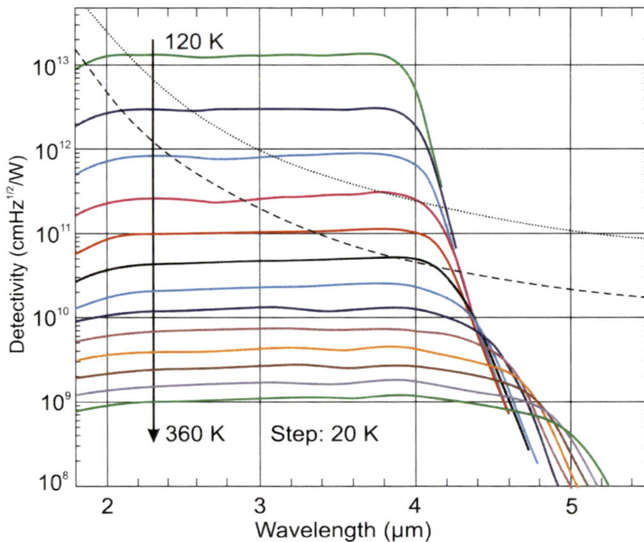

Figure 7.7 Johnson-noise-limited detectivity spectra of a 5-stage MWIR InAs/GaSb T2SL IB 90-period cascade detector at various temperatures (adapted from Ref. 11). The dashed lines represent the BLIP D^* for a photovoltaic detector with an external quantum efficiency of 70%, and the dotted lines are the BLIP D^* for 5-stage devices with absorption quantum efficiency of 70%, both under 300 K background with 2π FOV.

of stages provided that the absorbers are distributed closely in real space and are not very thick, ensuring equal absorption of the photon flux in each of the stages (total thickness of all stages should be comparable to the diffusion length).

The transport of photoexcited carriers is very fast and occurs over a distance in each cascade stage that is much shorter than a typical diffusion length ($\sim$50–200 nm depending on wavelength). Therefore, the lateral diffusion transport may not be significant over such a short distance; thus, the deeply etched mesa structures for confining photoexcited carriers may not be necessary in QCIDs in contrast to conventional photodiodes. Moreover, significant wave function overlap of energy states in the multiple-QW region (relaxation region) causes the intersubband relaxation time (e.g., optical-phonon scattering time $\approx$1 ps) to be much shorter than the interband recombination time ($\sim$1 ns, or $\sim$0.1 ns at high temperatures with significant Auger recombination). Consequently, the photoexcited electrons in the active region are transferred to the bottom of the energy ladder with very high efficiency. This mechanism enables the quick and efficient removal of carriers after photoexcitation.

Figure 7.8 presents the experimentally measured response time of a cascade detector versus temperature at zero bias [see Fig. 7.8(a)] and versus bias voltage for three temperatures of operation: 225 K, 293 K, and 380 K [see Fig. 7.8(b)]. These results confirm short response time of interband cascade detectors. At zero bias, in a temperature range of 225–280 K, the response time increases with increasing temperature from about 1 ns to 5 ns. It stabilizes at about 5 ns for further increase in temperature to about 360 K, when it decreases reaching a value of $\sim$2 ns at 380 K.

The negative bias is beneficial for a cascade detector response time [see Fig. 7.8(b)]. The negative correlation between the response time and the

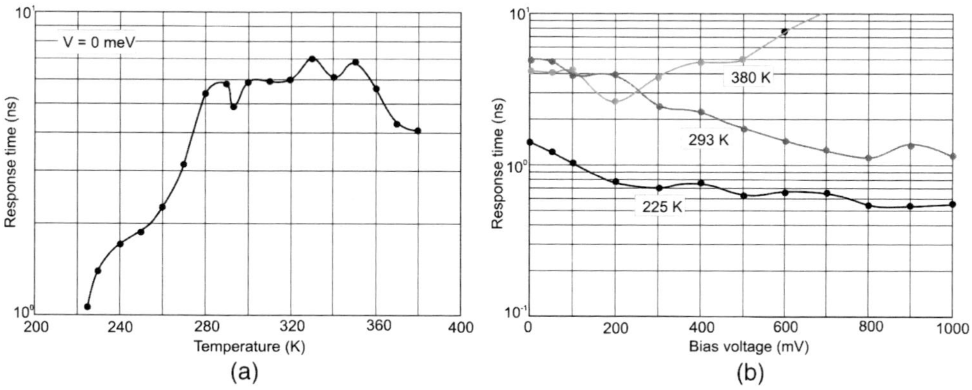

Figure 7.8 Response time of a T2SL MWIR cascade detector: (a) at zero bias voltage versus temperature and (b) versus bias voltage at 225 K, 293 K, and 380 K (adapted from Ref. 15).

applied voltage for temperatures of 225 and 293 K is probably related to the drift component decreasing with increasing bias as the electric field increases across the absorber region. In this context, the behavior of the response time as a function of bias above a bias of 200 mV for the detector operated at 380 K is not identified to the full extent as the time response increases with increasing voltage. The authors believe that under this condition the separation between the quantized energy level in the GaSb QW of the tunneling region and the valence band in the transport region does not match the LO-phonon energy in AlSb, which is responsible for tunneling of holes by the phonon-assisted process. In addition, ambipolar mobility is reduced, which in turn influences the detector's time response.

Recently, the first 5-stage MWIR interband cascade detector 320×256 FPA was demonstrated with a pixel size of 24×24 μm and a pitch of 30 μm.[23] This device demonstrated BLIP performance above 150 K (300 K, 2π FOV).

7.2.3 LWIR interband cascade detectors

Recently, preliminary studies of LWIR and VLWIR interband cascade infrared detectors with cutoff wavelengths up to 16 μm at 78 K have been presented.[12,16,17]

Figure 7.9 shows an exemplary device structure of two-stage LWIR device grown by MBE. The absorber layers have thicknesses of 620.0 nm and

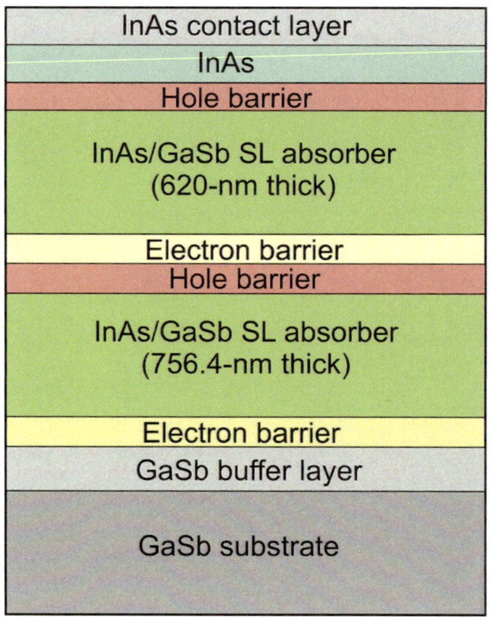

Figure 7.9 Device structure for two-stage LWIR IB CIDs.

756.4 nm, with each SL period composed of 36.3 Å of InAs and 21.9 Å of GaSb. The dipper absorber is made thicker in order to achieve current matching. In each SL period, a 1.9-Å-thick InSb layer is intentionally inserted into both the InAs-on-GaSb and the GaSb-on-InAs layers as the interface strain-balancing layer. In order to make electrons the minority carriers, half of the GaSb layers in the SL absorbers are p-doped with a doping density of 3.5×10^{16} cm^{-3}. The electron and hole barriers in each of these devices have identical designs. After the structure growth, square mesa devices with edge lengths ranging from 200 to 1000 µm are fabricated by using conventional contact UV lithography and wet etching. For passivation, two layers consisting of 170 nm of Si$_3$N$_4$ followed by 137 nm of SiO$_2$ were used.

Dark current characteristics versus bias voltage at different temperatures are shown in Fig. 7.10. The activation energy is estimated to be 102 meV (see insert of the figure) in comparison with the corresponding bandgap energy at 78 K equal to 135 meV. These data imply that the detector is neither diffusion limited nor dominated by the GR process (activation energy is larger than $E_g/2$). It is suggested that the deviation from the diffusion limit is probably related to the nonuniform doping that is applied to the absorber region, which creates an electric field that could affect the SRH GR in the absorber layer. The unintentional electrostatic barrier leads also to lower collection efficiencies due to the inefficiency of hole transport in the absorbers. It is suspected that the absorbers are n-type, especially at high temperatures. In that case the carrier transport is less efficient compared to the electrons, and external bias is required to aid the collection of photocarriers.

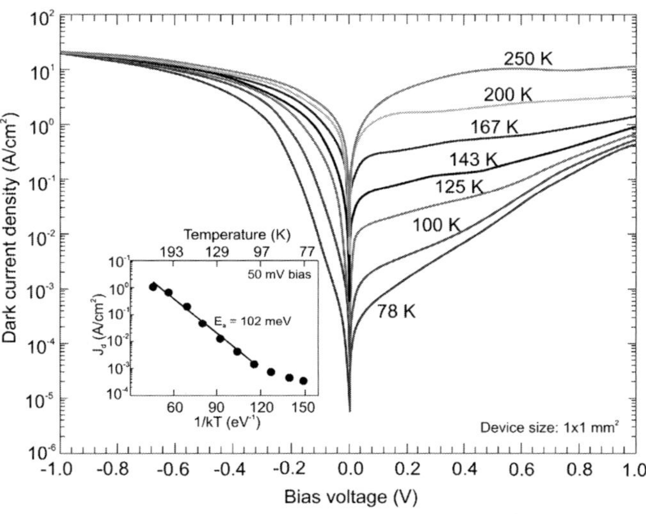

Figure 7.10 Dark current–voltage characteristics at different temperatures for the LWIR detector. The inset shows the fitted activation energy for the Arrhenius plot of the dark current (reprinted from Ref. 12).

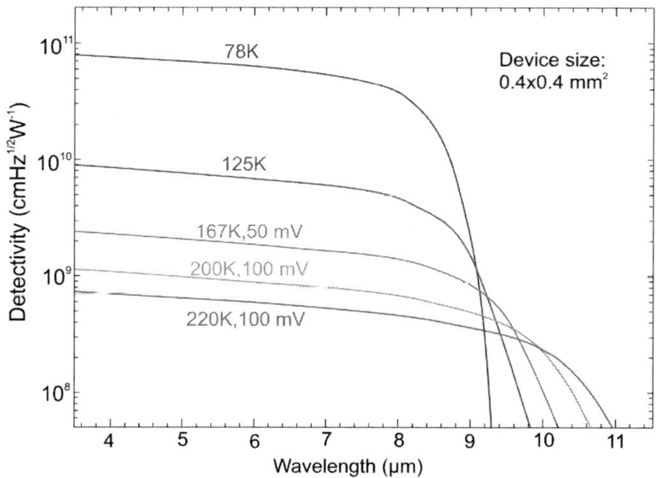

Figure 7.11   Spectral detectivity of a LWIR interband QCID at temperatures up to 220 K (reprinted from Ref. 12).

Figure 7.11 shows spectral Johnson-limited detectivities of LWIR interband QCIDs at different temperatures. At 78 K, the R_0A product value of 115 Ωcm^2 has been achieved, which corresponds to a detectivity of 3.7×10^{10} cm $Hz^{1/2}$ at 8 μm.

To improve the performance of LWIR T2SL QCIDs, further modifications in device technology and design are required, such as shorter absorbents and better bandgap alignments between the absorbers and the unipolar barriers, p-type doping of absorbers, and correction in processing.

7.3 Performance Comparison with HgCdTe HOT Photodetectors

At present, HgCdTe is the most widely used variable-gap semiconductor for IR photodetectors, including uncooled operation. However, the junction resistance of HgCdTe photodiodes operated in the LWIR region is very low due to high thermal generation. For example, small-sized, uncooled, 10.6-μm photodiodes (50×50 μm²) exhibit less than 1Ω zero bias junction resistance, which is well below the series resistance of a diode. Consequently, the performance of conventional devices is very poor, and these devices are not usable for practical applications.

Figure 7.12 compares the R_0A product of HgCdTe photodiodes with room-temperature experimental data for interband CIDs fabricated with type-II InAs/GaSb SL absorbers. It is evident that at the present early stage of the CID technology the experimentally measured R_0A values at room temperature are higher than those for state-of-the-art HgCdTe photodiodes. However, their quantum efficiencies are low, typically below 10%, resulting in

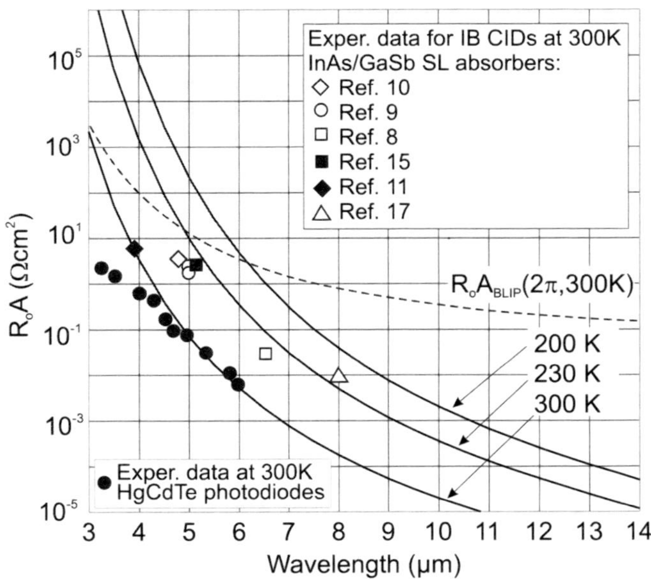

Figure 7.12 R_0A product of HgCdTe photodiodes (solid lines) in comparison with room-temperature experimental data for IB CIDs with type-II InAs/GaSb SL absorbers (adapted from Ref. 15).

lower detectivity for interband T2SL cascade detectors in comparison to HgCdTe photodiodes.

Figure 7.13 shows the performance of optically immersed two-stage thermoelectrically (2TE) cooled HgCdTe devices. Without optical immersion, MWIR photovoltaic detectors are sub-BLIP devices with performance close to the GR limit. However, well-designed optically immersed devices approach the BLIP limit when thermoelectrically cooled with two-stage Peltier coolers. The situation is less favorable for >8-µm LWIR photovoltaic detectors, which exhibit detectivities below the BLIP limit by an order of magnitude. Typically, the devices are used at zero bias. The attempts to use Auger-suppressed nonequilibrium devices were not successful due to large $1/f$ noise extending to $\sim$100 MHz in the extracted photodiode.

Lenses for monolithically immersed HgCdTe detectors are formed directly from transparent GaAs substrates. Due to immersion, the apparent optical detector area increases by a factor of n^2 for hemispherical lenses, where n is the refraction coefficient. Using GaAs lenses, the expected increase in detectivity is $n^2 \approx 10$. The detectivity of nonimmersed HgCdTe detectors is estimated to be about one order of magnitude below the values shown in Fig. 7.13. Therefore, the performance of IB CIDs is comparable to that obtained for Peltier-cooled HgCdTe devices.

Figure 7.14 gathers the experimentally measured response times of a T2SL interband CID and HgCdTe photodetectors (mainly photodiodes) of different designs and operated in a temperature range between 220 K and room

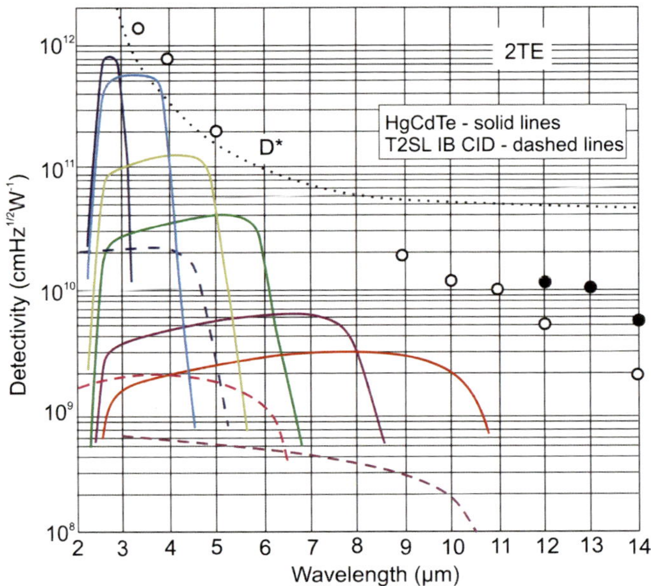

Figure 7.13 Typical spectral detectivity curves of HgCdTe immersed detectors and with two-stage TE coolers (solid lines). The best experimental data (white dots) are measured for detectors with a FOV of 36 deg. BLIP detectivity is calculated for FOV $= 2\pi$. The black dots are measured for detectors with four-stage TE coolers. Spectral detectivity curves for three T2SL IB CIDs (without immersion) are also shown (dashed lines) for comparison (figure adapted from Ref. 15 courtesy of Vigo Systems).

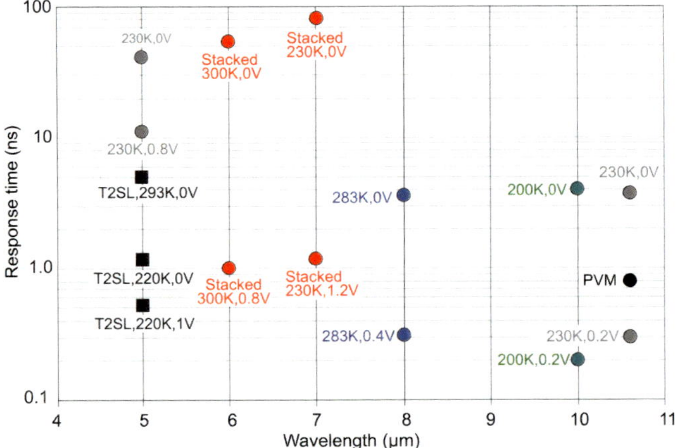

Figure 7.14 Response time versus wavelength for zero-biased and reverse-biased (as indicated) HgCdTe photodiodes and a T2SL MWIR cascade detector operating between 220 and 300 K. (Stacked – double-stacked photovoltaic detector; PVM – multiple-heterojunction photovoltaic detector) (reprinted from Ref. 15).

temperature. Most of the zero-biased LWIR photodiodes are characterized by response times below 10 ns. The device response time decreases under reverse bias, achieving a value below 1 ns. In this comparison, the response time of T2SL cascade detectors is comparable or even shorter than that of HgCdTe photodetectors.

Interband QCIDs have complicated cascade detector architectures. Their design is particularly pronounced for high-temperature operation, where diffusion lengths are considerably reduced. Currently, their performance is comparable to HgCdTe performance. However, due to the strong covalent bonding of III-V semiconductors, QCIDs can be operated at temperatures up to 400 °C, which is not achievable for their HgCdTe counterparts.

It is expected that better understanding of quantum cascade device physics and other aspects related to their design and material properties will enable improvement of high-performance HOT detectors. In addition, the discrete architecture of QCIDs provides a great deal of flexibility for manipulating carrier transports to achieve high-speed operation, which determines the maximum bandwidth. The possibility of monolithically integrating them with active components, for instance lasers, offers entirely new avenues for telecommunication systems based on quantum devices.

References

1. A. Gomez, M. Carras, A. Nedelcu, E. Costard, X. Marcadet, and V. Berger, "Advantages of quantum cascade detectors," *Proc. SPIE* **6900**, 69000J (2008) [doi: 10.1117/12.754215].
2. F. R. Giorgetta, E. Baumann, M. Graf, Q. Yang, C. Manz, K. Köhler, H. E. Beere, D. A. Ritchie, E. Linfield, A. G. Davies, Y. Fedoryshyn, H. Jäckel, M. Fischer, J. Faist, and D. Hofstetter, "Quantum cascade detectors," *IEEE J. Quantum Electron.* **45**, 1039–1052 (2009).
3. D. Hofstetter, F. R. Giorgetta, E. Baumann, Q. Yang, C. Manz, and K. Köhler, "Mid-infrared quantum cascade detectors for applications in spectroscopy and pyrometry," *Appl. Phys. B* **100**, 313–320 (2010).
4. A. Buffaz, M. Carras, L. Doyennette, A. Nedelcu, P. Bois, and V. Berger, "State of the art of quantum cascade photodetectors," *Proc. SPIE* **7660**, 76603Q (2010) [doi: 10.1117/12.853525].
5. Buffaz, A. Gomez, M. Carras, L. Doyennette, and V. Berger, "Role of subband occupancy on electronic transport in quantum cascade detectors," *Phys. Rev. B* **81**, 075304 (2010).
6. J. Q. Liu, S. Q. Zhai, F. Q. Liu, S. M. Liu, L. J. Wang, J. C. Zhang, N. Zhuo, and Z. G. Wang, "Quantum cascade detectors in very long wave infrared," *2014 Conference on Optoelectronic and Microelectronic Materials & Devices (COMMAD)*, 14–17 Dec. 2014, Perth, Australia, pp. 127–129.

7. H. Schneider and H. C. Liu, *Quantum Well Infrared Photodetectors*, Springer, Berlin (2007).

8. R. Q. Yang, Z. Tian, Z. Cai, J. F. Klem, M. B. Johnson, and H. C. Liu, "Interband-cascade infrared photodetectors with superlattice absorbers," *J. Appl. Phys.* **107**, 054514 (2010).

9. Z. Tian, R. T. Hinkey, R. Q. Yang, D. Lubyshev, Y. Qiu, J. M. Fastenau, W. K. Liu, and M. B. Johnson, "Interband cascade infrared photo-detectors with enhanced electron barriers and p-type superlattice absorbers," *J. Appl. Phys.* **111**, 024510 (2012).

10. N. Gautam, S. Myers, A. V. Barve, B. Klein, E. P. Smith, D. R. Rhiger, L. R. Dawson, and S. Krishna, "High operating temperature interband cascade midwave infrared detector based on type-II InAs/GaSb strained layer superlattice," *Appl. Phys. Lett.* **101**, 021106 (2012).

11. Z.-B. Tian and S. Krishna, "Mid-infrared interband cascade photo-detectors with different absorber designs," *IEEE J. Quant. Electron.* **51**(4), 4300105 (2015).

12. H. Lotfi, L. Lu, H. Ye, R. T. Hinkey, L. Lei, R. Q. Yang, J. C. Keay, T. D. Mishima, M. B. Santos, and M. B. Johnson, "Interband cascade infrared photodetectors with long and very-long cutoff wavelengths," *Infrared Phys. & Technol.* **70**, 162–167 (2015).

13. J. V. Li, R. Q. Yang, C. J. Hill, and S. L. Chung, "Interbad cascade detectors with room temperature photovoltaic operation," *Appl. Phys. Lett.* **86**, 101102 (2005).

14. R. T. Hinkey and R. Q. Yang, "Theory of multiple-stage interband photovoltaic devices and ultimate performance limit comparison of multiple-stage and single-stage interband infrared detectors," *J. Appl. Phys.* **114**, 104506 (2013).

15. W. Pusz, A. Kowalewski, P. Martyniuk, W. Gawron, E. Plis, S. Krishna, and A. Rogalski, "Mid-wavelength infrared type-II InAs/GaSb super-llatice interband cascade photodetectors," *Opt. Eng.* **53**(4), 043107 (2014) [doi: 10.1117/1.OE.53.4.043107].

16. H. Lotfi, L. Lin, L. Lu, R. Q. Yang, J. C. Keay, M. B. Johnson, Y. Qiu, D. Lubyshev, J. M. Fastenau, and A. W. K. Liu, "High-temperature operation of interband cascade infrared photodetectors with cutoff wavelengths near 8 μm," *Opt. Eng.* **54**(6), 063103 (2015) [doi: 10.1117/1 .OE.54.6.063103].

17. L. Lei, L. Li, H. Ye, H. Lotfi, R. Q. Yang, M. B. Johnson, J. A. Massengale, T. D. Mishima, and M. B. Santos, "Long wavelength interband cascade infrared photodetectors operating at high tempera-tures," *J. Appl. Phys.* **120**, 193102 (2016).

18. J. Piotrowski and A. Rogalski, *High-Operating Temperature Infrared Photodetectors*, SPIE Press, Bellingham, Washington (2007) [doi: 10.1117/ 3.717228].

19. J. Piotrowski, P. Brzozowski, and K. Jóźwikowski, "Stacked multi-junction photodetectors of long-wavelength radiation," *J. Electron. Mater.* **32**, 672–676 (2003).
20. P. Martyniuk, J. Antoszewski, M. Martyniuk, L. Faraone, and A. Rogalski, "New concepts in infrared photodetector design," *Appl. Phys. Rev.* **1**, 041102 (2014).
21. H. Lotfi, R. T. Hinkey, L. Li, R. Q. Yang, J. F. Klem, and M. B. Johnson, "Narrow-bandgap photovoltaic devices operating at room temperature and above with high open-circuit voltage," *Appl. Phys. Lett.* **102**, 211103 (2013).
22. W. E. Tennant, D. Lee, M. Zandian, E. Piquette, and M. Carmody, "MBE HgCdTe Technology: A very general solution to IR detection, described by 'Rule 07,' a very convenient heuristic," *J. Elect. Mat.* **37**, 1406 (2008).
23. Z.-B. Tian, S. E. Godoy, H. S. Kim, T. Schuler-Sandy, J. A. Montoya, and S. Krishna, "High operating temperature interband cascade focal plane arrays," *Appl. Phys. Lett.* **105**, 051109 (2014).

Chapter 8
Coupling of Infrared Radiation with Detectors

There are different methods of light coupling in a photodetector to enhance quantum efficiency.[1] A notable example of a method described for thin film solar cells[2,3] can be applied to infrared photodetectors. In general, these absorption enhancement methods can be divided into four categories that use either optical concentration, antireflection structures, optical path increase, or light localization, as shown in Fig. 8.1.

8.1 Standard Coupling

Semiconductor materials used for photodetectors have large values of refractive index and thus large values of reflection coefficient at the device surface. This reflection is minimized using antireflection structures. The simplest way to enhance absorption is to use a retroreflector to double pass infrared radiation. In thin devices the quantum efficiency can be significantly enhanced using interference phenomena to set up a resonant cavity within the photodetector.[1,4] Various optical resonator structures are used. In the simplest method, interference occurs between the waves reflected at the rear, highly reflective surface and at the front surface of the semiconductor. The thickness of the semiconductor is selected to set up the standing waves in the structure with peaks at the front and nodes at the back surface. The quantum efficiency oscillates with thickness of the structure, with the peaks at a thickness corresponding to an odd multiple of $\lambda/4n$, where n is the refractive index of the semiconductor. The gain in quantum efficiency increases with n. With the use of interference effects, a strong and highly nonuniform absorption can be achieved, even for long-wavelength radiation with a low absorption coefficient.

Another possible way to improve the performance of IR photodetectors is to increase the apparent "optical" size of the detector in comparison with the actual physical size using a suitable concentrator that compresses

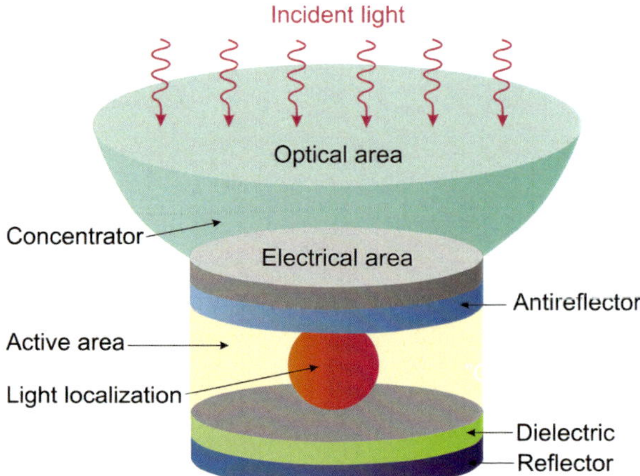

Figure 8.1 Different methods of absorption enhancement in a photodetector use an optical concentrator, an antireflection structure, structures for optical path increase (cavity enhancement), and light localization structures.

impinging IR radiation. The concentration efficiency can be then defined as the ratio between the optical area and the electrical area, minus absorption and scattering losses.

This must be achieved without reduction of the acceptance angle, or at least, with limited reduction to angles required for the fast optics of IR systems. Various types of suitable optical concentrators can be used, including optical cones, conical fibers, and other types of reflective, diffractive, and refractive optical concentrators.[5]

An efficient way to achieve an effective concentration of radiation uses a variety of immersion lenses. These lenses can be roughly divided into refractive, reflective, and diffractive elements, although hybrid solutions are also possible. Examples of these lenses are shown in Fig. 8.2. Microlenses monolithically integrated with detectors are typically used in CCD and CMOS active pixel imagers for visible application, and concentrate the incoming light into the photosensitive region when they are accurately

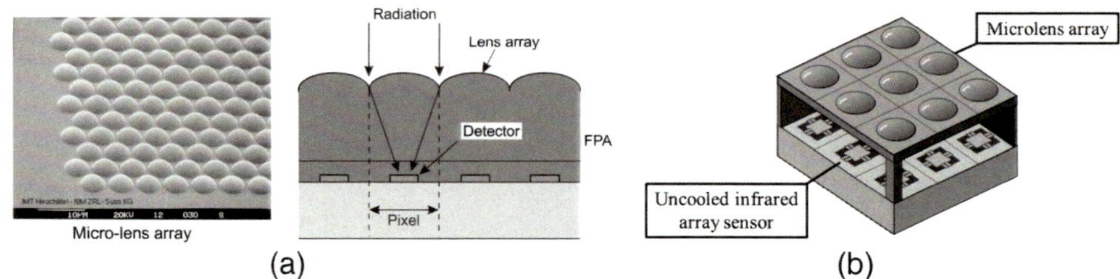

Figure 8.2 Microlenses for infrared arrays sensors: (a) micrograph of and cross-sectional drawing of microlensed FPA, and (b) diagram showing the concept of an uncooled infrared array sensor with microlenses.

deposited over each pixel [see Fig. 8.2(a)]. When the fill factor is low and microlenses are not used, the light falling elsewhere is either lost or, in some cases, creates artifacts in the imagery by generating electrical currents in the active circuitry. An example of the concept of an uncooled infrared array sensor with microlenses is shown in Fig. 8.2(b).[6]

The operating principle of a hemispherical immersion lens is shown in Fig. 8.3. The detector is located at the center of curvature of the immersion lens. The lens produces an image of the detector. No spherical or coma aberration exists (aplanatic imaging). Due to immersion, the apparent linear size of the detector increases by a factor of n. The image is located at the detector plane. The use of a hemispheric immersion lens in combination with an objective lens of an optical imaging system is shown in Fig. 8.3(b). The immersion lens plays the role of a field lens, which increases the FOV of the optical system.

The limit to compression is determined by the Lagrange invariant ($A\Omega$ product) and the sine condition for an aplanatic system.[5] In air the physical and the apparent size of the detector are related by the equation

$$n^2 A_e \sin \theta' = A_o \sin \theta, \tag{8.1}$$

where n is the lens refractive index, A_e and A_o are the physical and apparent size of the detector, respectively, and θ and θ' are the marginal ray angle before refraction at the lens and at the image, respectively. Therefore,

$$\frac{A_o}{A_e} = n^2 \frac{\sin \theta'}{\sin \theta}. \tag{8.2}$$

For hemispherical lenses, the marginal ray angles are 90 deg. Therefore, the area gain is n^2. Larger gain can be obtained for a hyperhemisphere used as an aplanatic lens.[5] This results in an apparent increase in linear detector size by a factor of n^2. In this case, the image plane is shifted.

An alternative approach uses a compound parabolic concentrator (CPC, also called a Winston collector or Winston cone).[7,8] QinetiQ has developed a

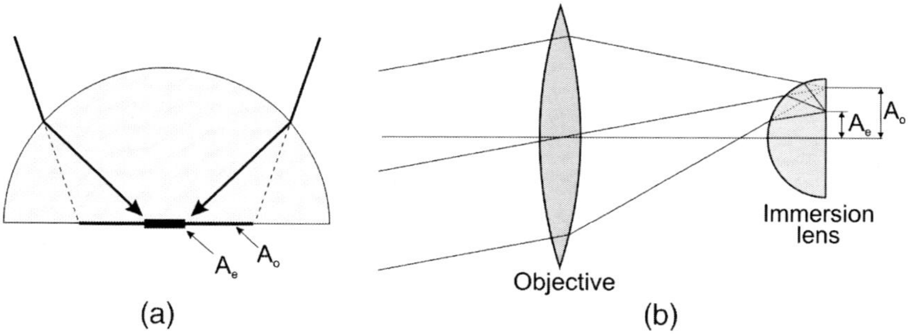

Figure 8.3 (a) Operating principle of optical immersion and (b) ray tracing for an optical system with an objective lens and an immersion lens (reprinted from Ref. 5).

micromachining technique involving dry etching to fabricate the cone concentrators for detector and luminescent devices.[9,10]

As mentioned above, the simplest way to couple radiation with the detector active region is to place a mirror at the backside of detector, thus doubling the optical path through the active region. At present, however, much more sophisticated methods are available, including the use of different cavities with reflective walls, as well as surface structuring. These methods belong to the optical trapping approaches. The advent of nanophotonics enabled optical trapping, such as subwavelength optical localization utilizing plasmonic nanocomposites. Some metal–dielectric structures ensure the possibility of light localization on a level much smaller than the operating wavelength.

Advances in optoelectronics-related materials science, such as metamaterials and nanostructures, have opened doors for new nonclassical approaches to device design methodologies, which are expected to offer enhanced performance along with reduced product cost for a wide range of applications. Surface plasmons are widely recognized in the field of surface science following the pioneering work of Ritchie in the 1950s.[11] The relative ease of manipulating surface plasmons provides an opportunity for their applications to photonics and optoelectronics for scaling down optical and electronic devices to nanometric dimensions. For the first time, it is possible to reliably control light at the nanoscale. Additionally, plasmonics takes advantage of the very large (and negative) dielectric constant of metals to compress the wavelength and enhance electromagnetic fields in the vicinity of metal conductors. Coupling light into semiconductor materials remains a challenging and active research topic. Micro- and nanostructured surfaces have become widely used design tools for increasing light absorption and enhancing the performance of broadband detectors without employing antireflection coatings.

The following sections presents in some detail the mentioned methods of photodetector enhancement.

8.2 Plasmonic Coupling

New solutions arising with the use of plasmonic structures open avenues for photodetectors development.[12–16] The goal of infrared plasmonics is to increase the absorption in a given volume of a detector's material. As mentioned in Section 1.4, smaller volumes provide lower noise, while higher absorption results in a stronger output signal. This leads to miniaturized detector structures with length scales that are much smaller than those being currently achieved. The choice of plasmonic material has significant implications for the ultimate utility of any plasmonic device or structure. The structures and materials that support surface plasmon excitations may play a key role in next-generation optical interconnects and sensing technologies.

8.2.1 Surface plasmons

A plasmon is a quantized electron density wave in a conducting material. Bulk plasmons are longitudinal excitations, whereas surface plasmons (SPs) can have both longitudinal and transverse components.[12] Light of a frequency below the frequency of the plasmon for that material (the plasma frequency) is reflected, while light above the plasma frequency is transmitted. SPs on a plane surface are nonradiative electromagnetic modes, i.e.; they cannot be generated directly by light nor can they decay spontaneously into photons. However, if the surface is rough or has a grating on it, or is patterned in some way, light around the plasma frequency couples strongly with the surface plasmons, creating what is called a polariton, or a surface plasmon polariton (SPP)—a transverse-magnetic optical surface wave that may propagate along the surface of a metal until energy is lost either via absorption in the metal or radiation into free space.

In its simplest form an SPP is an electromagnetic excitation (coupled electromagnetic field/charge-density oscillation) that propagates in a wave-like fashion along the planar interface between a metal and a dielectric medium, and whose amplitude decays exponentially with increasing distance into each medium from the interface. Thus, the SPP is a surface electromagnetic wave, whose field is confined to the near vicinity of the dielectric–metal interface. This confinement leads to an enhancement of the field at the interface, resulting in an extraordinary sensitivity of the SPP to surface conditions. The intrinsically 2D nature of SPPs prohibits them from directly coupling to light. Usually, a surface metal grating is required for the excitation of SPPs by normally incident light. Moreover, since the electromagnetic field of an SPP decays exponentially with distance from the surface, it cannot be observed in conventional (far-field) experiments unless the SPP is transformed into light by its interaction with a surface grating.

A schematic representation of an electromagnetic wave and surface charges propagating along a metal–dielectric interface is shown in Fig. 8.4. The charge density oscillations and associated electromagnetic fields comprise SPP waves. The local electric field component is enhanced near the surface and decays exponentially with distance in a direction normal to the interface.

The interaction between the surface charge density and the electromagnetic field results in the momentum of the SP mode $\hbar k_{SP}$ being greater than that of a free-space photon of the same frequency $\hbar k_o$ ($k_o = \omega/c$ is the free-space wavevector) [see Fig. 8.4(c)]. Solving Maxwell's equations under the appropriate boundary conditions yields the SP dispersion relation, which is the frequency-dependent SP wave vector:[17]

$$k_{SP} = k_o \left(\frac{\varepsilon_d \varepsilon_m}{\varepsilon_d + \varepsilon_m} \right)^{1/2}. \tag{8.3}$$

The frequency-dependent permittivities of the metal ε_d and the dielectric material ε_m must have opposite signs if SPs are to be possible at such an interface. This condition is satisfied for metals because ε_m is both negative and

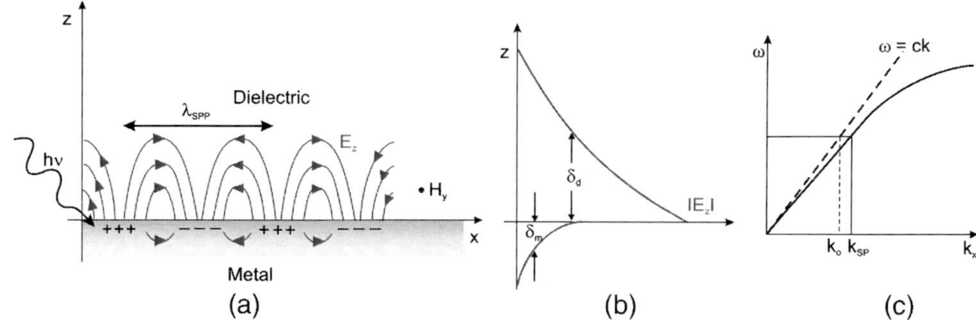

Figure 8.4 (a) Schematic illustration of an electromagnetic wave and surface charges at the interface between the metal and the dielectric material; (b) the local electric field component is enhanced near the surface and decays exponentially with distance in a direction normal to the interface; and (c) dispersion curve for a SP mode.

complex (the latter corresponding to absorption in the metal). The increase in $\hbar k_{SP}$ momentum is associated with the binding of the SP to the surface (the resulting momentum mismatch between light and SPs of the same frequency must be bridged if light is to be used to generate SPs).

In contrast to the propagating nature of SPs along the surface, the field perpendicular to the surface decays exponentially with distance from the surface. This field is said to be evanescent, reflecting the bound, nonradiative nature of SPs, and prevents power from propagating away from the surface.

The SP mode propagates on a flat metal surface with gradually attenuation owing to losses arising from absorption in the metal. The propagation length δ_{SP} can be found from the SP dispersion equation as[18]

$$\delta_{SP} = \frac{1}{2k_{SP}''} = \frac{\lambda}{2\pi} \left(\frac{\varepsilon_m' + \varepsilon_d}{\varepsilon_m' \varepsilon_d} \right)^{3/2} \frac{(\varepsilon_m')^2}{\varepsilon_m''}, \tag{8.4}$$

where k_{SP}'' is the imaging part of the complex SP wavevector, $k_{SP} = k_{SP}' + ik_{SP}''$, and ε_m' and ε_m'' are the real and imaging parts, respectively, of the dielectric function of the metal, $\varepsilon_m = \varepsilon_m' + i\varepsilon_m''$. The propagation length is dependent on the dielectric constant of the metal and the incident wavelength.

SPs were first studied in the visible region. The vast majority of plasmonics research has focused on the shorter-wavelength side of the optical frequency range. In the visible spectrum, silver is the metal with lowest losses, where propagation distances are typically in the range of 10–100 μm. In addition, for a longer incident wavelength, such as the near-infrared telecommunication wavelength 1.55 μm, the propagation length of silver increases towards 1 mm. For a relatively well-absorbing metal such as aluminium, the propagation length is 2 μm at a wavelength of 500 nm.[14]

It appears that common metals such as gold or silver have plasmon resonances in the blue or deep-ultraviolet wavelength ranges. Recently, an increased research effort has been directed to the infrared range. However, when moving from the visible to the infrared range, metal films with arrays of holes that ordinarily show optical transmission are quite opaque. There are no metals available whose plasmon resonances are in the IR range under 10 μm in wavelength. Moreover, the of integration of plasmonic structure with active detector region is intrinsically incompatible due to the low-quality of metal used in deposition techniques in comparison with the high-quality materials used in epitaxial growth of semiconductors or dielectrics. As a result, many intrinsic plasmonic properties can be masked by the poor metal quality or poor semiconductor–metal interfaces. In addition, wavelength tuneability is difficult to realize since the plasmonic resonance frequency is fixed for a given metal. Thus, other alternatives to metals such as highly doped semiconductors have been proposed; e.g., an InAs/GaSb bilayer structure.[16,19]

Figure 8.5 shows schematically the difference in surface enhancement between visible and infrared absorption. As shown in Fig. 8.5(a), the optical

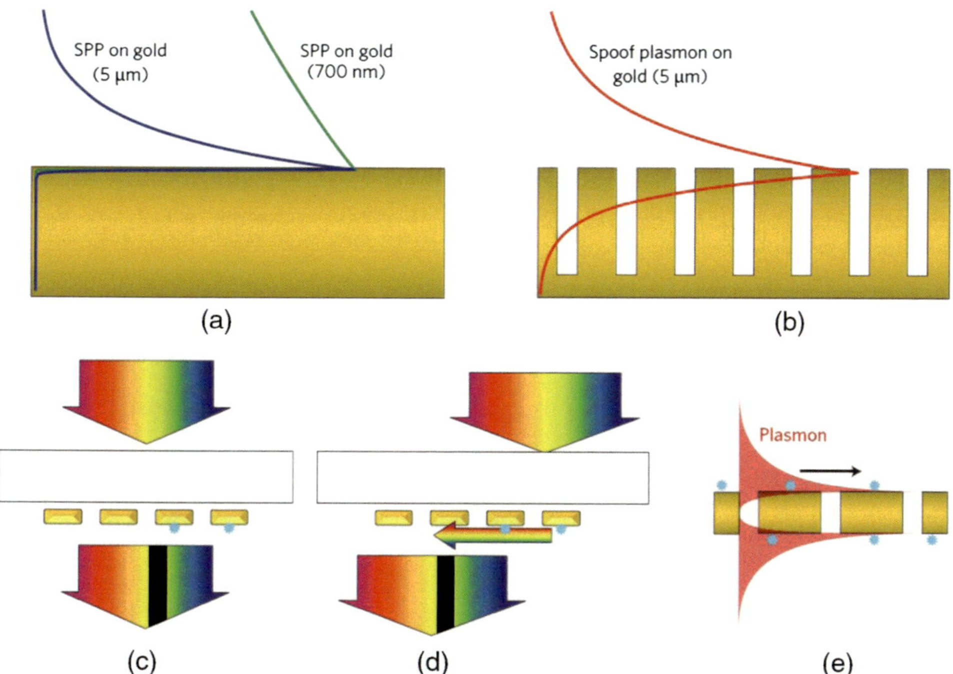

Figure 8.5　Optical fields of SPs and surface-enhanced infrared absorption. (a) The optical field of a SPP is closely bound to the material surface at visible wavelengths (700 nm, green) but weakly bound at mid-IR wavelengths (5 μm, blue). (b) The optical field of spoof plasmons on gold at a wavelength of 5 μm, showing strong confinement (red). (c) Chemicals (light blue dots) on gold islands exhibit better absorption than an unstructured substrate. (d) Use of SPs to increase the absorption enhancement. (e) SP-enhanced infrared absorption in a hole array. Plasmons bound to the surface interact with the molecules deposited on and inside the hole array (reprinted from Ref. 13 with permission from Nature Publishing Group).

field of a SPP is closely bound to the material surface at visible wavelengths but weakly bound at mid-IR wavelengths. Metal surfaces with indentations or holes can give rise to leaky waveguides (spoof plasmonics) or can be used to couple light into dielectric waveguides, as shown in Fig. 8.5(b). This effective approach is based on a deep subwavelength pitch grating in the surface of the metal.[20] This design leads not only to resonance in the grating, but also to extremely tight confinement of light and is used for improving the performance of QWIP and QDIP detectors.[21,22]

Current plasmonic devices at telecommunication and optical frequencies face significant challenges due to losses encountered in the constituent plasmonic materials. These large losses seriously limit the practicality of these metals for many applications. Apart from traditional plasmonic materials (the noble metals), the newer material systems with infrared plasma wavelengths [transition metal nitrides, transparent conducting oxides (TCOs), and silicides] are considered.[16,23,24] TCOs have been shown to be effective plasmonic materials in the infrared region, while transition metal nitrides extend into the visible spectrum. Figure 8.6 shows various classes of materials grouped using two parameters that determine the optical properties of conducting materials: the carrier concentration and carrier mobility. In plasmonics, the carrier concentration has to be high enough to provide a negative real part of the dielectric permittivity. In addition, tunability of dielectric permittivity values with change in carrier concentration is desirable. Lower carrier mobilities could indicate higher damping losses and thus higher material losses.

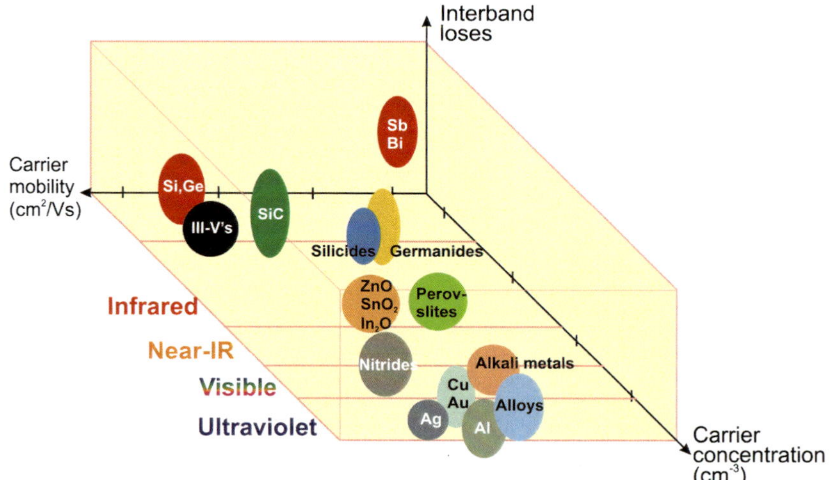

Figure 8.6 Material space for plasmonics and metamaterial applications. Important material parameters such as carrier concentration (maximum doping concentration for semiconductors), carrier mobility, and interband losses form the optimization phase space for various applications. While spherical bubbles represent materials with low interband losses, elliptical bubbles represent those with larger interband losses in the corresponding part of the electromagnetic spectrum (adapted from Ref. 24).

8.2.2 Plasmonic coupling of infrared detectors

Different architectures are used to support SPPs on metalo-dielectric structures. These architectures may involve planar metal waveguides, metal gratings, nanoparticles such as islands, spheres, rods, and antennas, or optical transmission through one or many subwavelength holes in a metal film. However, great challenges remain to fully realize many of the promised potentials.

Figure 8.7 presents three popular geometries that enhance the detector's photoresponse: (a) grating couplers to convert incident light to SPPs that are focused inside a small-scale detector, (b) a particle antenna on a small-scale detector, and (c) metallic photonic crystal structures to enhance the photoresponse. The inclusion of an antenna or resonator enhances the photoresponse or makes the detector wavelength- and polarization specific.

The first architecture involves a nanoscale semiconductor photodetector. Small-area photodetectors benefit from low noise levels, a low junction capacitance, and a possible high-speed operation. However, due to the decrease in the active area of the semiconductor detector under the same optical power density, a lower output is obtained. A nanoantenna in close proximity to the active material of a photodetector allows us to take advantage of the concentrated plasmonic fields.[25] The role of the nanoantenna is to convert free-space plane waves into surface plasmons bound to a patterned metal surface without reflection. The antenna-like structure that couples incident radiation to surface plasmons is used in a technique that is very popular for THz detectors.[26]

Figure 8.7(b) shows an integrating detector nanoscale structure whose photoresponse is enhanced by a local plasmon resonance. Resonant antennas can confine strong optical fields inside a subwavelength volume. By designing the structure in such a way that the region with highly confined optical fields overlaps the active region of the photodetector, strong enhancement of the photocurrent can be achieved using either SPP or localized surface plasmon polariton (LSPP) resonances.

LSPPs are charge oscillations that are bound to a small metal particle or nanostructure. These oscillations can be represented by the displacement of

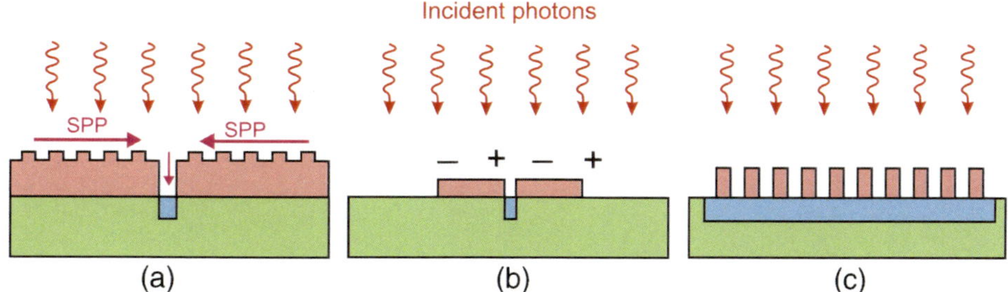

Figure 8.7 Different architectures for plasmon-enhanced detectors: (a) grating couplers that focus the generated SPPs inside a small-scale detector, (b) particle antenna on a small-scale detector, and (c) metallic photonic crystal structures.

charge in the sphere. For example, for a metal sphere in a dielectric, the field inside the metal is given by electrostatic approximation as[27]

$$E_{in} = \frac{3\varepsilon_d E_o}{\varepsilon_m + 2\varepsilon_d},\qquad(8.5)$$

where E_o is the electric field away from the sphere. Ignoring the imaginary contributions to the relative permittivities in Eq. (8.5), it is clear that the field inside the sphere diverges when $\varepsilon_m = -2\varepsilon_d$, which leads to a strong enhancement of the field on the outer surface of the sphere (which is limited in practice by the imaginary part of ε_m). The quality of the resonance is limited by the dispersion of the metal and dielectric, as is clear from the field enhancement denominator in Eq. (8.5).

The third way of enhancing the photodetector photoresponse is shown in Fig. 8.7(c). Photoresponse enhancment is achieved by including a metallic photonic crystal (PC) on the detector area, or arranging the detector structure in a periodic way, forming a PC structure. Integrating a resonant structure into a detector increases the interaction length between the incoming light and the active semiconductor region. This design is interesting for thin film semiconductor detectors with a large absorption length.

Since the absorption coefficient is strongly dependent on wavelength, the wavelength range in which an appreciable photocurrent can be generated is limited for a given detector material. Therefore, broadband absorption is usually inadequate due to quantum efficiency roll-off. Research on PC structures with periodic refractive index modulation has made possible several new ways to control light. Most of such existing devices are realized as 2D PC structures and thus are compatible with standard semiconductor processing.[28-30]

Photonic crystals comprise a regular array of holes (defects) that modify the local refractive index to provide localized modes in the "photonic" band structure (see Fig. 8.8). By removing a single hole, an energy well for photons is formed that is similar to that of electrons in a quantum wire structure. The

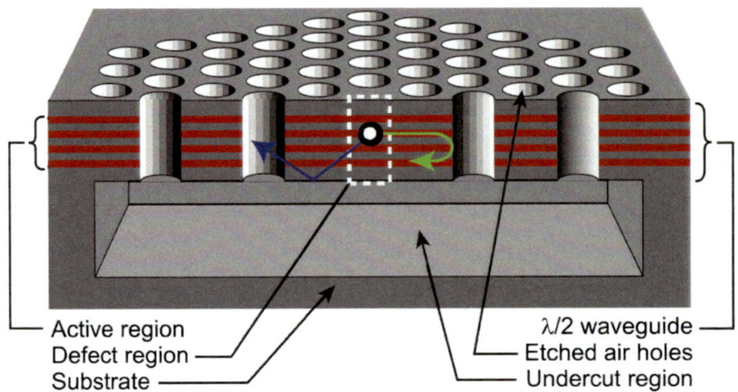

Figure 8.8 Cross-sectional view through the photonic crystal microcavity.

periodic variation in the refractive index gives rise to Bragg scattering of photons, which opens up forbidden energy gaps in the in-plane photon dispersion relation. The PC has a grating effect that "diffracts" the normally incident radiation toward the in-plane direction. In addition, a $\lambda/2$ high-index slab is used to trap photons in the vertical direction by internal reflection at the air–slab interface. As a result, the combination of Bragg reflection from the 2D PC and internal reflection results in a three-dimensionally confined optical mode.

Figure 8.9 is a schematic design of an infrared detector that allows for near-field detection of enhanced evanescent waves transmitted through a structured surface by using a nearby buried quantum detector (with a distance much smaller than the wavelength of the incident electromagnetic field).

An example of metal-PC-integrated detector design is shown in Fig. 8.10 along with a schematic cross-sectional view of the sample structure.[30] The PC is a 100-nm-thick Au film perforated with a 3.6-μm period square array of

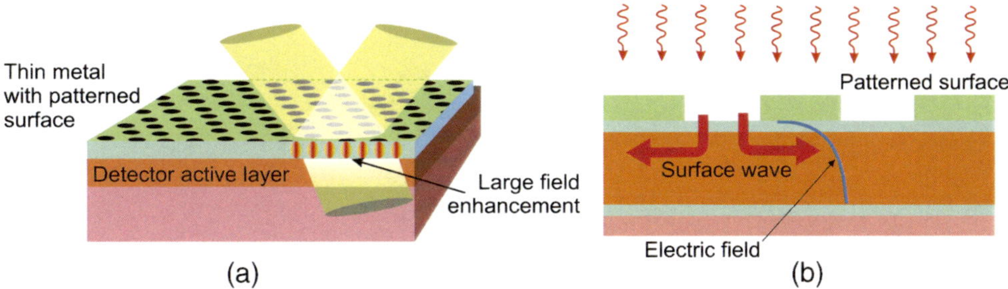

Figure 8.9 Conceptual design of an infrared detector enhanced by surface plasmon polaritons: (a) general view and (b) cross-sectional view.

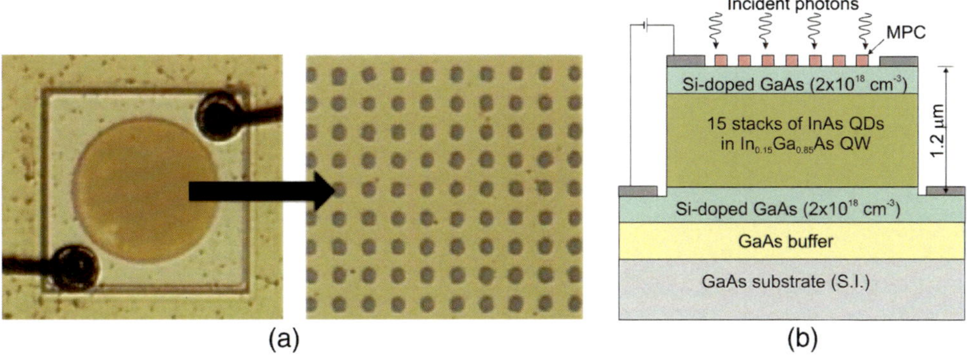

Figure 8.10 (a) Optical microscope images of the metal PC device with $16\times$ magnification revealing the details of the metal PC. The period of the circular holes is 3.6 μm; (b) a schematic cross-sectional view of the metal PC device (reprinted from Ref. 30 with permission from OSA).

circular holes having diameters of 1.65 ± 0.05 μm. This array of circular holes couples to surface plasma waves at 11.3 and 8.1 μm for reverse and forward bias, respectively, where InAs quantum dot infrared photodetectors (QDIP) exhibit the strongest detectivity (up to thirty-fold enhancement).

By confining the quantum dots in a waveguide structure and using a metallic grating coupler, a considerable increase in absorption is observed. Figure 8.11 shows the low-temperature (10 K) photoresponse of the metal PC QDIP and reference devices at −3.0 V and 3.4 V.[30] The arrows in the figure indicate the reference devices that exhibit two rather broad color responses with indistinct peaks for both the −3.0 V and 3.4 V cases. The peak shift with applied voltage has been interpreted in terms of the quantum-confined Stark effect. On the other hand, the metal PC devices have completely different voltage-dependent spectral responsivities in both the peak wavelengths and, especially, the response intensity. The spectral response curves show four peaks at identical wavelengths but varying amplitude for both biases. The peak at 11.3 μm, which is much stronger than that of the reference device, is dominant for reverse bias, while the peak at 8.1 μm is more intense than any other peak for forward bias. The two remaining peaks at 5.8 and 5.4 μm are relatively weak.

The advantage of the above approach is that it can be easily incorporated into the FPA fabrication process of present-day infrared sensors. Holes with 2–3 μm in diameter for a response wavelength range of 8–10 μm can be defined using conventional optical lithography. The introduction of a straightforward modification of a single- or multi-element defect in the PC can selectively increase the response of photons with a specific energy.

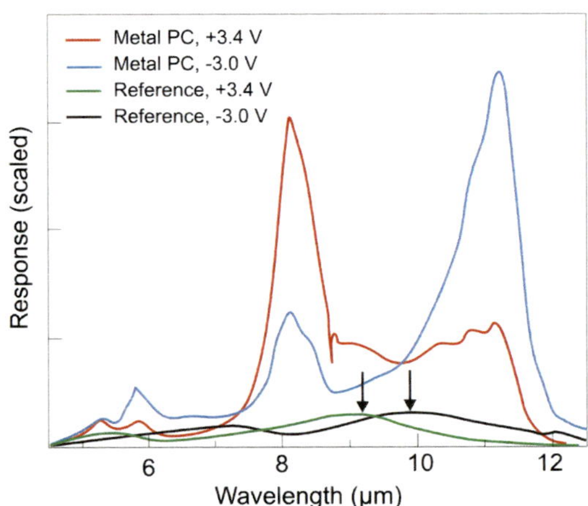

Figure 8.11 Spectral response curves of the reference device (the two spectra at the bottom with arrows indicating the highest peak in each spectrum) and the metal PC device (the other two spectra with higher responsivity) for −3.0 V and 3.4 V at 10 K (reprinted from Ref. 30 with permission from OSA).

Therefore, by changing the dimensions of the defect, the resonance wavelength can be altered, leading to the fabrication of a spectral element in each pixel of the FPA. This approach would have a revolutionary impact on multispectral imaging and hyperspectral imaging detectors.

An important class of 2D PC structures are photonic crystal slabs (PCSs), consisting of a dielectric structure with a periodic modulation in only two dimensions and refractive index guiding in the third. Figure 8.12(a) shows a QWIP fabricated as a PCS structure.[31] The PC structure is underetched by selective removal of the sacrificial AlGaAs layer to create the free-standing PCS. A schematic illustration of the final device is shown in Fig. 8.12(b). The photoresponse of the PCS-QWIP shows a wider response peak but additionally displays several pronounced resonance peaks.

Recently, Qiu et al.[32] have studied the role of the parameters of 2D metallic hole arrays (the periodicity of hole arrays p, hole diameter d, and metal film thickness t) in plasmonic enhancement of InAsSb infrared detectors. To evaluate the transmission performance of the subwavelength hole array, a 3D finite-difference time-domain (FDTD) method has been used. Figure 8.13 shows a cross-sectional view and top view of the 2D hole array (2DHA) with a hole array fabricated above the InAsSb detector active region.

In estimations, the diameter d of the circular hole is fixed at 0.46 μm, and the metal film is made of gold with thickness $t = 20$ nm. The periodicity is

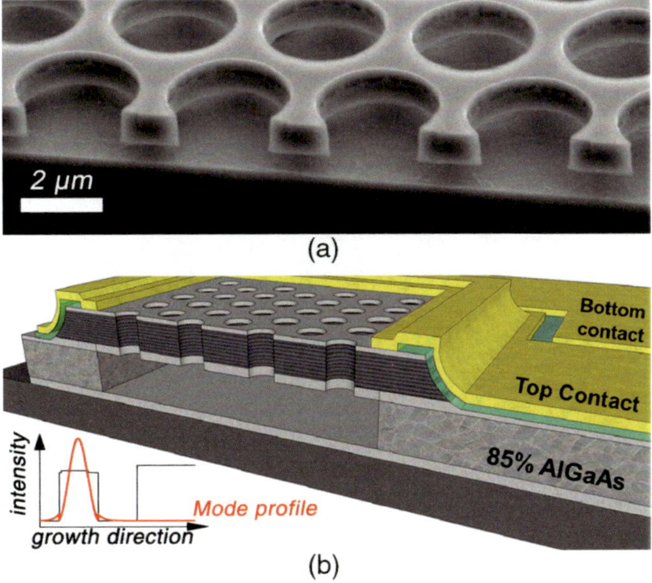

Figure 8.12 PCS-QWIP design: (a) SEM image of a cleaved PCS and (b) cross-section view through the PCS-QWIP structure (reprinted from Ref. 31 with permission from AIP Publishing).

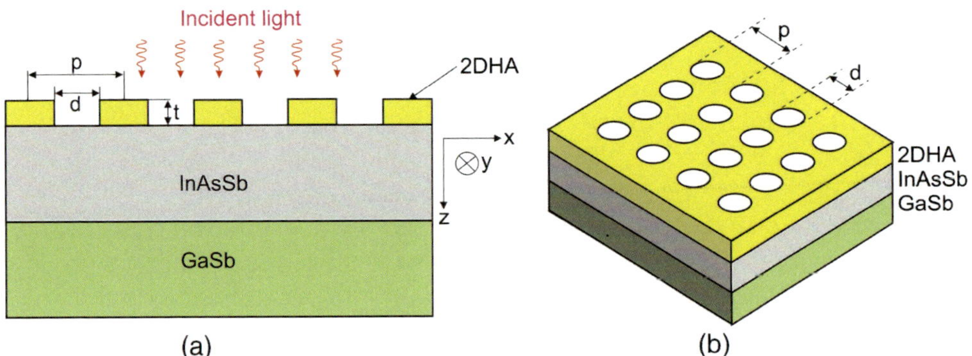

Figure 8.13   (a) A cross-sectional view and (b) top view of a 2D hole array (adapted from Ref. 32).

varied between 0.72 and 1.12 μm, with other parameters remaining unchanged. The wave source is normally incident along positive z direction and polarized in x direction with a wavelength range from 1.5 μm to 6.5 μm.

The transmission efficiency shown in Fig. 8.14 has been calculated for the main peak when either the hole diameter is fixed at $d = 0.46$ μm or the periodicity is fixed at $p = 0.92$ μm. As shown this figure, at resonance wavelengths, the highest transmission efficiency is around 3.85, which indicates that much more light is transmitted than the amount that is directly impinging into the hole area. The transmission efficiency reaches a maximum value when the hole diameter is approximately one-half of the periodicity.

Utilizing either a single-metal or a double-metal plasmon waveguide, Rosenberg et al.[21] have considered a plasmonic photonic crystal resonator for use in mid-infrared photodetectors. Its good frequency and polarization selectivity can be used in hyperspectral and hyperpolarization detectors.

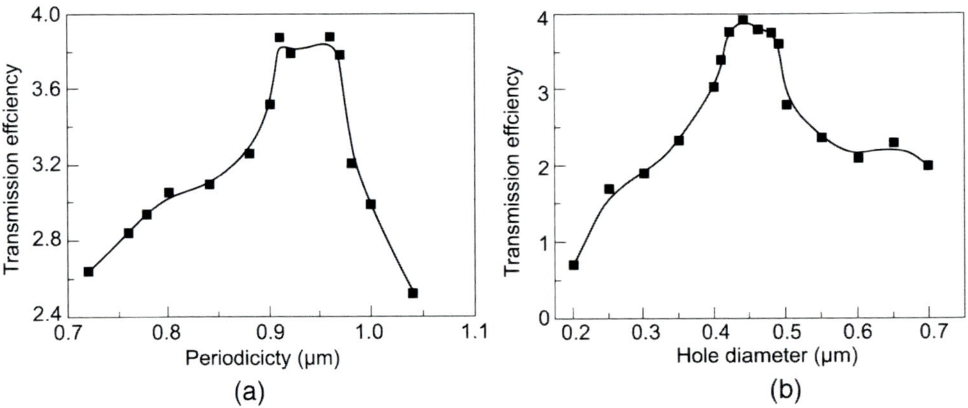

Figure 8.14   Plots of transmission efficiency as a function of (a) periodicity p ($d = 0.46$ μm) and (b) hole diameter d ($p = 0.92$ μm) (adapted from Ref. 32).

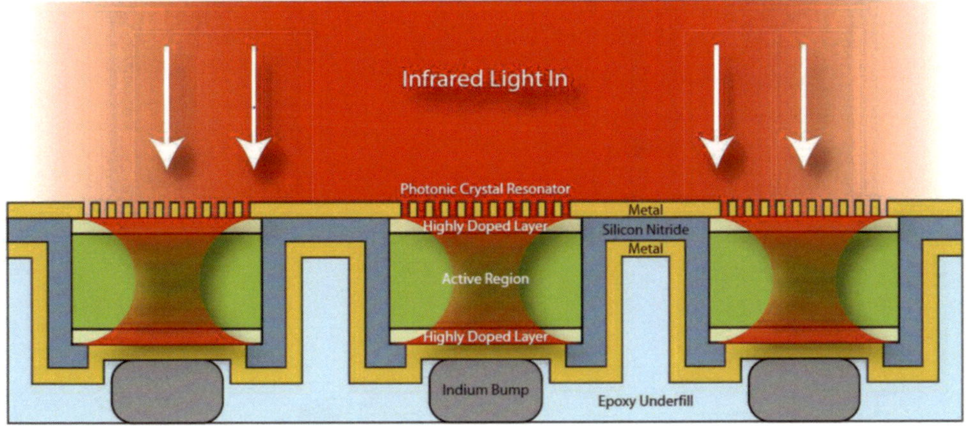

Figure 8.15 A design schematic for a resonant double-metal plasmonic photonic crystal FPA (reprinted from Ref. 21 with permission from OSA).

By suitable scaling of the photonic crystal holes and waveguide width, such a resonator can be optimized for use at any wavelength from the terahertz to the visible bands. Figure 8.15 shows the proposed schematic structure of an FPA with double-metal plasmonic photonic crystal resonators. Only the top metal photonic crystal lithography step differs from the standard fabrication process of hybrid arrays.

8.3 Photon-Trapping Detectors

As Eq. (1.10) indicates, the performance of an infrared detector can be improved by reducing the volume of detector's active region. In this section, we focus our considerations on reducing the detector material volume via the concept of photon trapping (PT). Reduction of the dark current should be achieved without degrading the quantum efficiency. Figure 8.16 shows the effect of volume reduction on quantum efficiency and noise equivalent differential temperature ($NEDT$)[33,34] by a simple first-order model consisting of the Bruggeman effective medium approach[35] when combining HgCdTe (with a composition $x \approx 0.3$) with a void material. The fill factor is calculated as the volume of material remaining divided by the volume of the unit cell. As expected, as the volume is reduced (and fill factor is increased), the quantum efficiency is increased and the $NEDT$ value generally decreases, improving the performance until a critical point when photon collection begins to decrease faster than noise, and hence the overall performance degrades. The modeled trends are observed in measured devices.

Photon-trapping detectors have been demonstrated independently in II-VI[33,34,36] and III-V[36–39] -based epitaxial materials. Subwavelength-sized semiconductor pillar arrays within a single detector have been designed and

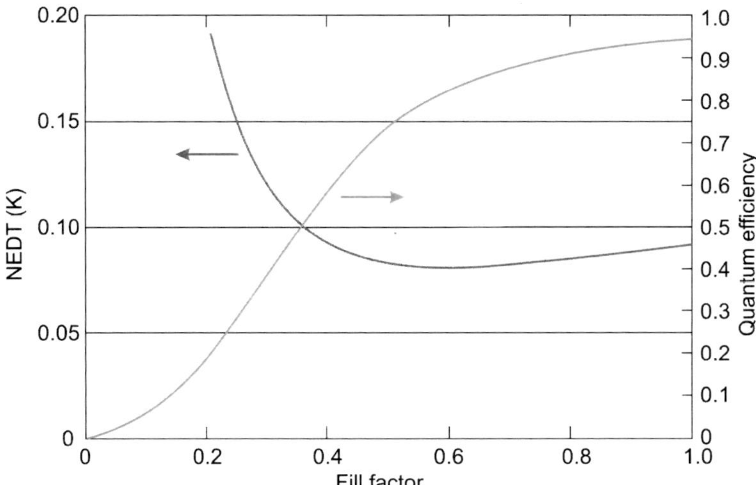

Figure 8.16 Effect of volume reduction on quantum efficiency and *NEDT* (adapted from Ref. 33).

structured as an ensemble of 3D photonic structure units using either a top-down or bottom-up process to significantly increase absorption and quantum efficiency. The sub-element can be of different shapes such as pyramidal, sinusoidal, or rectangular.[37] For example, Fig. 8.17 shows photon-trap structures with pillars and holes of varying volume fill factors. These samples were fabricated from HgCdTe layers on Si grown by MBE with a cutoff wavelength of 5 µm at 300 K.

Theoretical estimations show that PT arrays have a slightly higher optical crosstalk compared to non-PT arrays, but significantly less diffusion crosstalk, thus indicating that PT arrays will have significantly better device performance than non-PT arrays in terms of crosstalk, especially for small pixel pitches.[40] Moreover, the calculation of the modulation transfer function (MTF) from a spot scan of the arrays shows that PT structures have superior resolving capability compared to non-PT structures. Thus, as the detector array technology moves towards pixel-size reduction, the PT approach is an effective means of reducing diffusion crosstalk and an increasing quantum efficiency without employing antireflection coatings.

FDTD simulation of pillar structures indicates resonance between them and confirms that the photon-trapping process via total internal reflection effectively serves as a waveguide to direct incident energy away from the removed regions and into the remaining absorber material. For example, Fig. 8.18 presents the optical generation rate as a function of wavelength from 0.5 to 5.0 µm for a single HgCdTe pillar and the uppermost part of the absorber layer beneath the pillars.[41] At a wavelength of 0.5 µm the optical generation is concentrated at the edge of the pillar, and as the wavelength is increased, the optical generation region gradually extends deeper into the

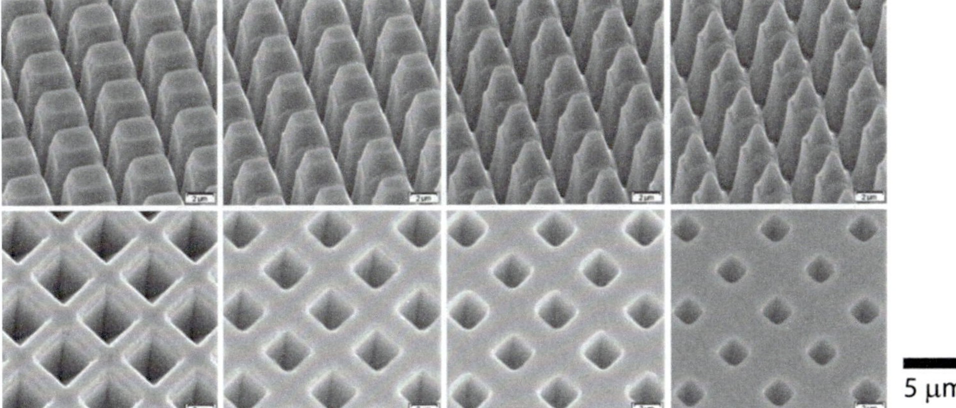

Figure 8.17 Examples of photon-trapping HgCdTe microstructures with test photonic crystal fields with varying fill factors for FTIR demonstration (reprinted from Ref. 33).

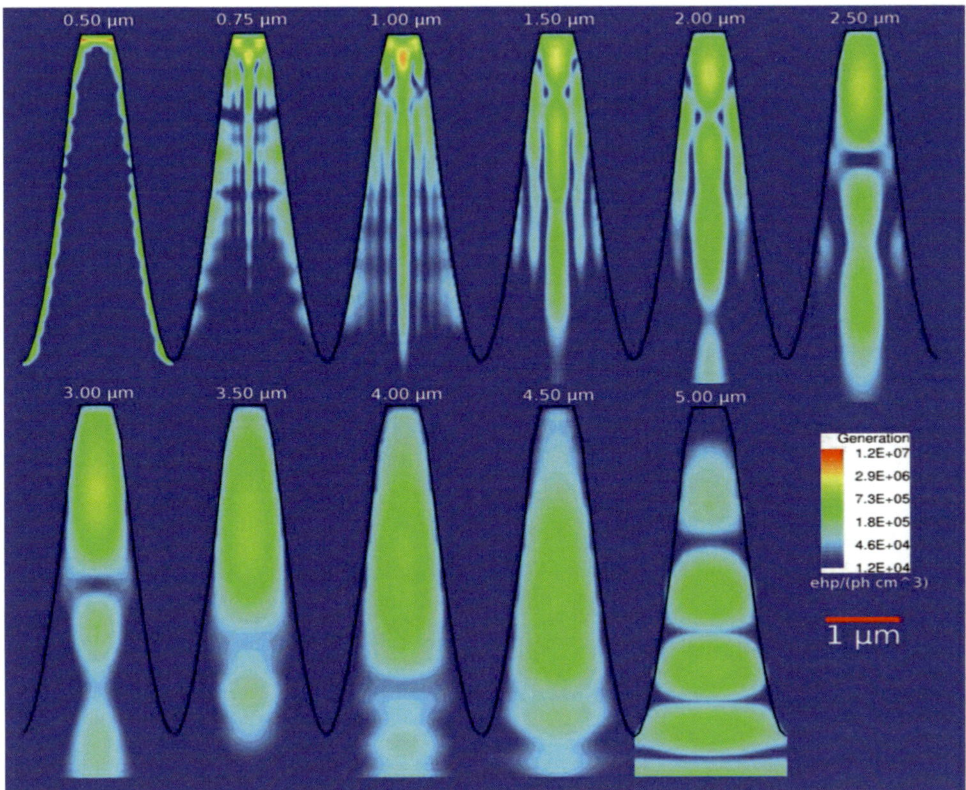

Figure 8.18 Calculated optical generation profile inside the pillar of a HgCdTe photon-trapping structure back-illuminated with planewaves within the wavelength range of 0.5–5.0 μm. This pillar is one of a 2D array of HgCdTe pillars like those shown in Fig. 8.17 (reprinted from Ref. 41).

pillars. At 5.0 μm there is very little optical generation at the tip of the pillar, but there is significant optical generation further into the pillar and into the absorber layer. It has been found that absorption enhancement is weakly dependent on the pillar lattice type, but the lattice period does have a significant impact on the enhancement.[42]

It has been confirmed experimentally that volume reduction leads to improved device performance and consequently to higher operating temperature of detector arrays. Process improvements, such as unique self-aligned contact metal processes and advanced stepper technologies, have been developed to achieve the critical dimensions and lithography alignments required for advanced PT detector designs with a smaller feature within the unit cell. Figure 8.19 shows advanced hexagonal photonic crystal designs with holes in a standard 30-μm mesa with features on a 5-μm pitch. Large-format MBE HgCdTe/Si epitaxial wafer arrays with cutoff wavelengths ranging from 4.3 to 5.1 μm at 200 K exhibit improved performance compared to unpatterned mesas, with a measured *NEDT* of 40 mK and 100 mK at temperatures of 180 K and 200 K, respectively, with good operability.[34]

Utilization of an InAsSb absorber on GaAs substrates instead of a HgCdTe absorber enables fabrication of MWIR low-cost, large-format HOT

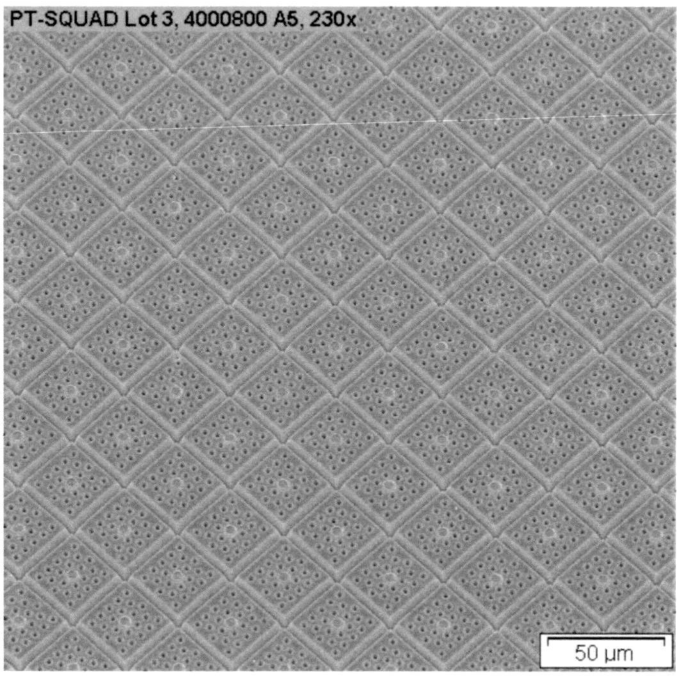

Figure 8.19 MWIR 512 × 512 30-μm-pitch MBE HgCdTe/Si array consisting of photonic crystal holes on a 5-μm pitch in a standard mesa (reprinted from Ref. 34).

FPAs. Souza and co-workers[37-39] have described research efforts to develop visible- to mid-wave (0.5 to 5.0 μm) broadband PT InAsSb-based detectors operating at high temperatures (150–200 K) with low dark current and high quantum efficiency.

The $InAs_{0.82}Sb_{0.18}$ ternary alloy with a 5.25-μm cutoff wavelength at 200 K was grown on a lattice-mismatched GaAs substrate. To compare the detector performance, both $128 \times 128/60$-μm as well as $1024 \times 1024/18$-μm detector arrays consisting of bulk absorber structures as well as photon-trap pyramid structures were fabricated. This innovative detector design was based on pyramidal PT InAsSb structures in conjunction with barrier-based device architecture to suppress both GR dark current as well as the diffusion current through the absorber's reduced volume. The absence of depletion regions in the narrow-gap absorption layer makes the barrier detector immune to dislocations and other defects, possibly allowing growth on lattice-mismatched substrates with a reduced penalty of excess dark current generated by misfit dislocations. The pixel arrays are defined very simply by etching through the contact layer up to the barrier. Figure 8.20(a) shows a 5-μm-cutoff-wavelength nBn detector structure operated at 200 K with an AlAsSb barrier and pyramid-shaped absorbers fabricated in the n-type InAsSb absorber. Based on optical simulation, the engineered pyramidal structures minimize reflection and provide >90% absorption over the entire 0.5- to 5.0-μm spectral range [see Fig. 8.20(b)] while providing up to $3\times$ reduction in absorber volume.

Figure 8.21 shows SEM top and side views of 4.5-μm-height pyramid structures of InAsSb nBn barrier on GaAs substrate. The spacing between adjacent pyramids is smaller than 0.5 μm. The thickness of the InAsSb slab underneath the pyramids is only 0.5 μm. Each die on the 3-inch wafer represents a 1024×1024 pixel, 18-μm- pitch FPA. After pyramid etching the wafer is flipped over and bonded temporarily to another 3-inch handling

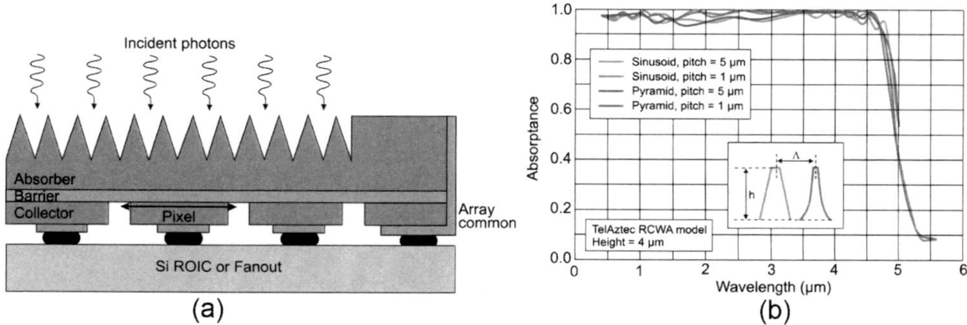

Figure 8.20 Photon trapping nBn detector in the InAsSb/AlAsSb material system: (a) detector architecture with pyramid shaped absorber layer (adapted from Ref. 37), (b) optical simulation of broadband detector response (adapted from Ref. 36).

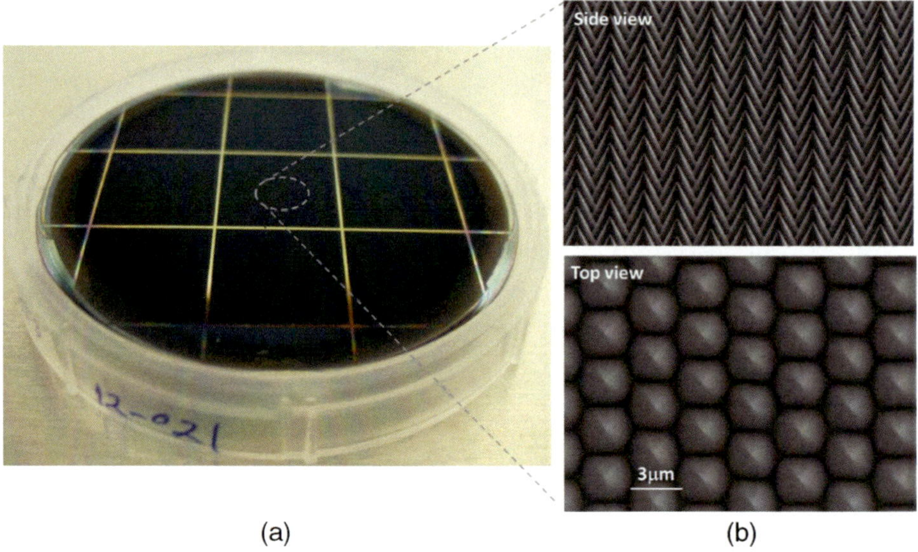

(a)   (b)

Figure 8.21   (a) Photograph of fabricated staggered pyramids on a 3-inch substrate and (b) SEM images of staggered pyramids (reprinted from Ref. 38).

substrate using an epoxy. The entire 600-μm-thick GaAs growth substrate is subsequently removed using a high-etch-rate and high-selectivity ICP dry etch process.

The measured dark current density in the pyramidal structured diodes is reduced by a factor of 3 in comparison with conventional diodes having a bulk absorber [see Fig. 8.22(a)], which is consistent with volume reduction due to the

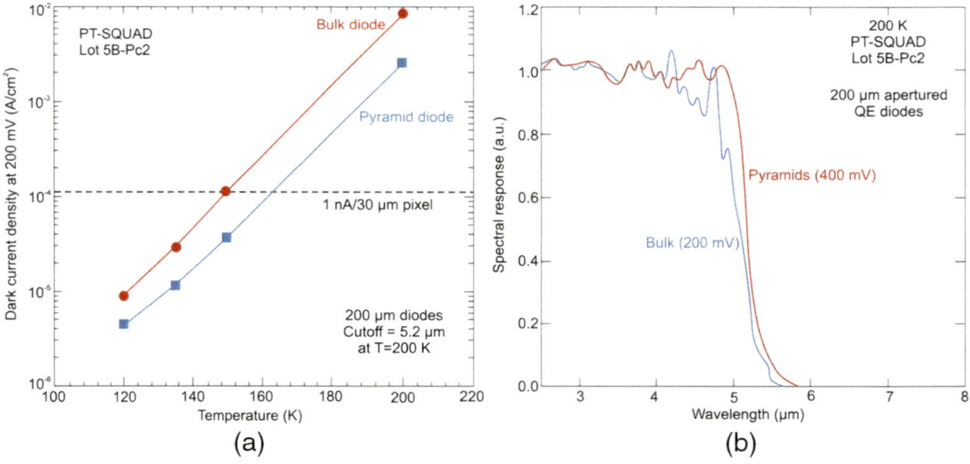

(a)   (b)

Figure 8.22   Comparisons of (a) dark current density between nBn InAsSb bulk detectors and pyramid detectors at different temperatures, and (b) spectral response between bulk- and pyramid-based detectors (adapted from Ref. 38).

creation of the absorber topology. High detectivity ($>1.0 \times 10^{10}$ cm$\sqrt{\text{Hz}}$/W) and high internal quantum efficiency ($>90\%$) have been achieved. The spectral response measurements reveal that staggered pyramids suppress etalon effects and make the spectral response more flat [see Fig. 8.22(b)], as expected from simulation results. Despite the small absorber volume of pyramidal detectors, the internal quantum efficiency is higher than 80%, and detectivity above 1.0×10^{10} cm$\sqrt{\text{Hz}}$/W at 200 K has been estimated over the entire 0.5- to 5.0-μm spectral range.

References

1. Z. Jakšić, *Micro and Nanophotonics for Semiconductor Infrared Detectors*, Springer, Heidelberg (2014).
2. S. J. Fonash, *Solar Cell Device Physics*, Elsevier, Amsterdam (2010).
3. G. Li, R. Zhu, and Y. Yang, "Polymer solar cells," *Nat. Photonics* **6**(3), 153–161 (2012).
4. M. S. Ünlü and S. Strite, "Resonant cavity enhanced photonic devices," *J. Appl. Phys.* **78**, 607–639 (1995).
5. J. Piotrowski and A. Rogalski, *High-Operating-Temperature Infrared Photodetectors*, SPIE Press, Bellingham, Washington (2007) [doi: 10.1117/3.717228].
6. R. Yamazaki, A. Obana, and M. Kimata, "Microlens for uncooled infrared array sensor," *Electronics and Communications in Japan*, **96**(2), 1–8 (2013).
7. R. Winston, "Principles of solar concentrators of a novel design," *Sol. Energy* **16**(2), 89–95 (1974).
8. R. Winston, "Dielectric compound parabolic concentrators," *Appl. Opt.* **15**(2), 291–293 (1976).
9. M. K. Haigh, G. R. Nash, N. T. Gordon, J. Edwards, A. J. Hydes, D. J. Hall, A. Graham, J. Giess, J. E. Hails, and T. Ashley, "Progress in negative luminescent $Hg_{1-x}Cd_xTe$ diode arrays," *Proc. SPIE* **5783**, 376–383 (2005) [doi: 10.1117/12.603292].
10. G. J. Bowen, I. D. Blenkinsop, R. Catchpole, N. T. Gordon, M. A. C. Harper, P. C. Haynes, L. Hipwood, C. J. Hollier, C. Jones, D. J. Lees, C. D. Maxey, D. Milner, M. Ordish, T. S. Philips, R. W. Price, C. Shaw, and P. Southern, "HOTEYE: A novel thermal camera using higher operating temperature infrared detectors," *Proc. SPIE* **5783**, 392–400 (2005) [doi: 10.1117/12.609476].
11. R. H. Ritchie, "Plasma losses by fast electrons in thin flms," *Phys. Rev.* **106**, 874–881 (1957).
12. S. A. Maier, *Plasmonic: Fundamentals and Applications*, Springer, New York (2007).

13. R. Stanley, "Plasmonics in the mid-infrared," *Nature Photonics* **6**, 409–411 (2012).
14. J. Zhang, L. Zhang, and W. Xu, "Surface plasmon polaritons: physics and applications," *J. Phys. D: Appl. Phys.* **45**, 113001 (2012).
15. P. Berini, "Surface plasmon photodetectors and their applications," *Laser Photonics Rev.* **8**, 197–220 (2013).
16. Y. Zhong, S. D. Malagari, T. Hamilton, and D. Wasserman, "Review of mid-infrared plasmonic materials," *J. of Nanophotonics* **9**, 093791 (2015).
17. J. R. Sambles, G. W. Bradbery, and F. Z. Yang, "Optical-excitation of surface-plasmons – an introduction," *Contemp. Phys.* **32**, 173–183 (1991).
18. H. Raether, *Surface Plasmons*, edited by G. Hohler, Springer, Berlin (1988).
19. Debin Li and C. Z. Ning, "All-semiconductor active plasmonic system in mid-infrared wavelengths," *Opt. Express* **19**(15), 147367 (2011).
20. P. Bouchon, F. Pardo, B. Portier, L. Ferlazzo, P. Ghenuche, G. Dagher, C. Dupuis, N. Bardou, R. Haidar, and J.-L. Pelouard, "Total funneling of light aspect ratio plasmonic nanoresonators," *Appl. Phys. Lett.* **98**, 191109 (2011).
21. J. Rosenberg, R. V. Shenoi, S. Krishna, and O. Painter, "Design of plasmonic photonic crystal resonant cavities for polarization sensitive infrared photodetectors," *Opt. Exp.* **18**(4), 3672–3686 (2010).
22. C.-C. Chang, Y. D. Sharma, Y.-S. Kim, J. A. Bur, R. V. Shenoi, S. Krishna, D. Huang, and S.-Y. Lin, "A surface plasmon enhanced infrared photodetector based on InAs quantum dots," *Nano Lett.* **10**, 1704–1709 (2010).
23. G. V. Naik, V. M. Shalaev, and A. Boltasseva, "Alternative plasmonic materials: Beyond gold and silver," *Adv. Mater.* **25**, 3264–3294 (2013).
24. J. B. Khurgin and A. Boltasseva, "Reflecting upon the losses in plasmonics and metamaterials," *MRS Bulletin* **37**, 768–779 (2012).
25. P. Biagioni, J.-S. Huang, and B. Hecht, "Nanoantennas for visible and infrared radiation," *Rep. Prog. Phys.* **75**, 024402 (2012).
26. K. Ishihara, K. Ohashi, T. Ikari, H. Minamide, H. Yokoyama, J.-I. Shikata, and H. Ito, "Therahertz-wave near field imaging with subwavelength resolution using surface-wave-assisted bow-tie aperture," *Appl. Phys. Lett.* **89**, 201120 (2006).
27. U. Kreibig and M. Vollmer, *Optical Properties of Metal Clusters*, Springer, Berlin (1995).
28. K. T. Posani, V. Tripathi, S. Annamalai, N. R. Weisse-Bernstein, S. Krishna, R. Perahia, O. Crisafulli, and O. J. Painter, "Nanoscale quantum dot infrared sensors with photonic crystal cavity," *Appl. Phys. Lett.* **88**, 151104 (2006).
29. K. T. Posani, V. Tripathi, S. Annamalai, S. Krishna, R. Perahia, O. Crisafulli, and O. Painter, "Quantum dot photonic crystal detectors," *Proc. SPIE* **6129**, 612906 (2006) [doi: 10.1117/12.641750].

30. S. C. Lee, S. Krishna, and S. R. J. Brueck, "Quantum dot infrared photodetector enhanced by surface plasma wave excitation," *Opt. Exp.* **17**(25) 23160–23168 (2009).

31. S. Kalchmair, H. Detz, G. D. Cole, A. M. Andrews, P. Klang, M. Nobile, R. Gansch, C. Ostermaier, W. Schrenk, and G. Strasser, "Photonic crystal slab quantum well infrared photodetector," *Appl. Phys. Lett.* **98**, 011105 (2011).

32. S. Qiu, L. Y .M. Tobing, Z. Xu, J. Tong, P. Ni, and D.-H. Zhang, "Surface plasmon enhancement on infrared photodetection," *Procedia Engineering* **140**, 152–158 (2016).

33. J. G. A. Wehner, E. P. G. Smith, G. M. Venzor, K. D. Smith, A. M. Ramirez, B. P. Kolasa, K. R. Olsson, and M. F. Vilela, "HgCdTe photon trapping structure for broadband mid-wavelength infrared absorption," *J. Electron. Mater.* **40**, 1840–1846 (2011).

34. K. D. Smith, J. G. A. Wehner, R. W. Graham, J. E. Randolph, A. M. Ramirez, G. M. Venzor, K. Olsson, M. F. Vilela, and E. P. G. Smith, "High operating temperature mid-wavelength infrared HgCdTe photon trapping focal plane arrays," *Proc. SPIE* **8353**, 83532R (2012) [doi: 10.1117/12.921480].

35. D. A. G. Bruggeman, "Berechnung verschiedener physikalischer Konstanten von heterogenen Substanzen," *Ann. Phys. (Leipzig)* **24**, 636–679 (1935).

36. N. K. Dhar and R. Dat, "Advanced imaging research and development at DARPA," *Proc. SPIE* **8353**, 835302 (2012) [doi: 10.1117/12.923682].

37. A. I. D'souza, E. Robinson, A. C. Ionescu, D. Okerlund, T. J. de Lyon, R. D. Rajavel, H. Sharifi, D. Yap, N. Dhar, P. S. Wijewarnasuriya, and C. Grein, "MWIR $InAs_{1-x}Sb_x$ nCBn detectors data and analysis," *Proc. SPIE* **8353**, 835333 (2012) [doi: 10.1117/12.920495].

38. H. Sharifi, M. Roebuck, T. De Lyon, H. Nguyen, M. Cline, D. Chang, D. Yap, S. Mehta, R. Rajavel, A. Ionescu, A. D'souza, E. Robinson, D. Okerlund, and N. Dhar, "Fabrication of high operating temperature (HOT), visible to MWIR, nCBn photon-trap detector arrays," *Proc. SPIE* **8704**, 87041U (2013) [doi: 10.1117/12.2015083].

39. A. I. D'souza, E. Robinson, A. C. Ionescu, D. Okerlund, T. J. de Lyon, R. D. Rajavel, H. Sharifi, N. K. Dhar, P. S. Wijewarnasuriya, and C. Grein, "MWIR InAsSb barrier detector data and analysis," *Proc. SPIE* **8704**, 87041V (2013) [doi: 10.1117/12.2015083].

40. J. Schuster and E. Bellotti, "Numerical simulation of crosstalk in reduced pitch HgCdTe photon-trapping structure pixel arrays," *Opt. Exp.* **21**(12), 14712 (2013).

41. C. A. Keasler, "Advanced numerical modeling and characterization of infrared focal lane arrays," Ph.D. Thesis, Boston University, Boston, Massachusetts (2012).

42. C. A. Keasler and E. Bellotti, "A numerical study of broadband absorbers for visible to infrared detectors," *Appl. Phys. Lett.* **99**, 091109 (2011).

Chapter 9
Focal Plane Arrays

During the last five decades, different types of detectors have been combined with electronic readouts to make detector arrays. The progress in integrated circuit design and fabrication techniques has resulted in continued rapid growth in the size and performance of these solid-state arrays. In the infrared technique, these devices are based on a readout array connected to an array of detectors.

The term focal plane array refers to an assemblage of individual detector picture elements (pixels) located at the focal plane of an imaging system. Although this definition could include 1D (linear) arrays as well as 2D arrays, it is frequently applied to the latter. Usually, the optics part of an optoelectronic imaging device is limited to focusing the image onto the detector array. These so-called staring arrays are scanned electronically usually using circuits integrated with the arrays. The architecture of detector-readout assemblies has assumed a number of forms that are discussed below.

Infrared FPAs are critical components in many of the military and civilian applications of advanced imaging systems. In this section we focus on the current requirements for extended detector capability to support applications for future generations of infrared sensor systems.

9.1 Trends in Infrared Focal Plane Arrays

Figure 9.1 illustrates the trend in array size over the past 50 years. Imaging IR FPAs have been developing in-line with the ability of silicon integrated circuit (IC) technology to read and process the array signals and to display the resulting image. The progress in IR arrays has also been steady, mirroring the development of dense electronic structures such as dynamic random access memories (DRAMs). FPAs have had nominally the same development rate as DRAM ICs, which have followed Moore's law with a doubling-rate period of approximately 18 months; however, FPAs have been lagging behind DRAMs by about 5–10 years. The 18-month doubling time is evident from the slope of the graph presented in the inset of Fig. 9.1, which shows the log of the number

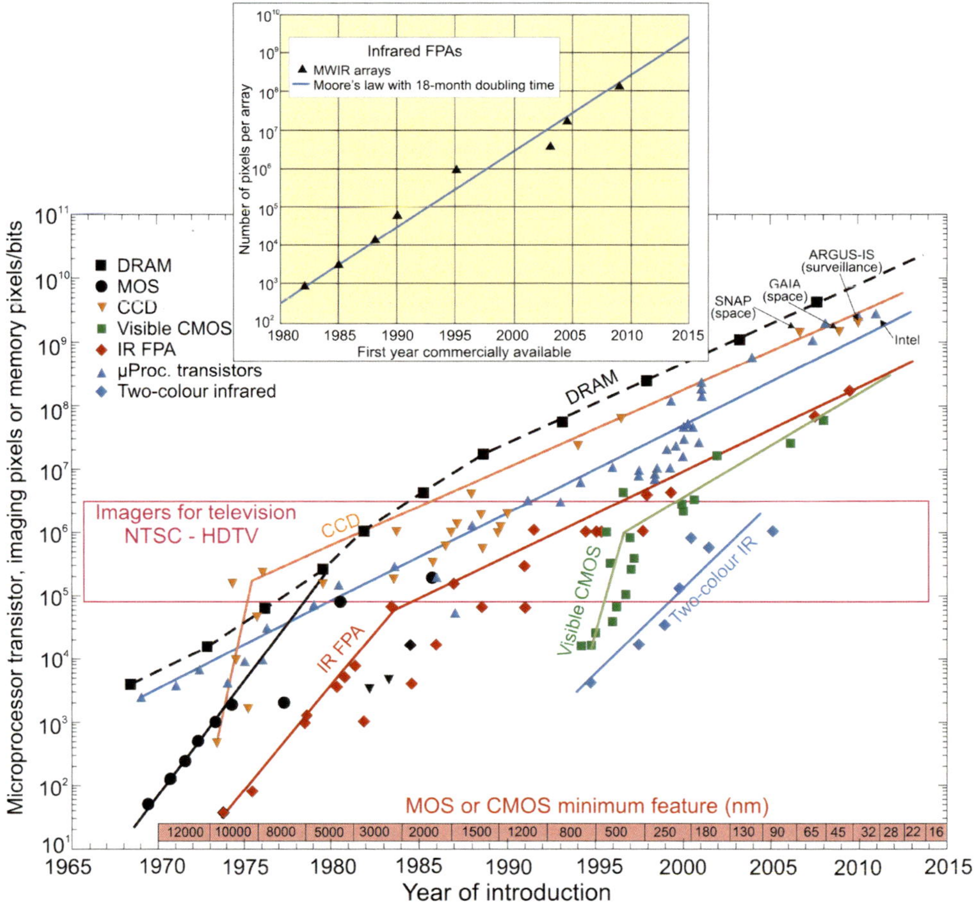

Figure 9.1 Imaging array formats compared with the complexity of silicon microprocessor technology and DRAM as indicated by transistor count and memory bit capacity (adapted from Ref. 1). The timeline design rule of MOS/CMOS features is shown at the bottom. Note the rapid rise of CMOS imagers, which are challenging CCDs in the visible spectrum. The number of pixels on an infrared array has been growing exponentially, in accordance with Moore's law, for 30 years with a doubling time of approximately 18 months. Infrared arrays with size above 100 megapixels are now available for astronomy applications. Imaging formats of many detector types have gone beyond that required for high-definition TV.

of pixels per array as a function of the first year of commercial availability of MWIR FPAs. CCDs above 3 gigapixels offer the largest formats.

IR array sizes will continue to increase but perhaps at a rate that falls below the Moore's law trend. An increase in array size is already technically feasible. However, the market demand for larger arrays is not as strong as it was before the megapixel milestone was achieved. In particular, astronomers were the driving force behind the effort for to make optoelectronic arrays match the size of the photographic film. Since large arrays dramatically multiply the data output of a telescope system, the development of large-format mosaic sensors of high sensitivity for ground-based astronomy is the goal of many astronomic observatories around the world. This is somewhat surprising given the comparative budgets of the defense market and the astronomical community.

A number of architectures are used in the development of IR FPAs. In general, they may be classified as hybrid and monolithic, but these distinctions are often not as important as proponents and critics state them to be. The central design questions involve performance advantages versus ultimate producibility. Each application may favor a different approach, depending on the technical requirements, projected costs, and schedule.

In the monolithic approach, both detection of light and signal readout (multiplexing) are done in the detector material rather than in an external readout circuit. The integration of detector and readout onto a single monolithic piece reduces the number of processing steps, increases yields, and reduces costs. Common examples of these FPAs in the visible and near-infrared regions are found in camcorders and digital cameras.

In the case of the hybrid approach, which dominates in infrared detector technology, we can optimize the detector material and multiplexer independently. Other advantages of hybrid-packaged FPAs are near-100% fill factors and increased signal-processing area on the multiplexer chip. Photodiodes with their very low power dissipation, inherently high impedance, negligible $1/f$ noise, and easy multiplexing via the ROICs, can be assembled in 2D arrays containing a very large number of pixels, limited only by existing technologies. Photodiodes can be reverse-biased for even higher impedance and can therefore better electrically match compact low-noise silicon readout preamplifier circuits. The photoresponse of photodiodes remains linear for significantly higher photon flux levels than that of photoconductors, primarily because of higher doping levels in the photodiode absorber layer and because the photogenerated carriers are collected rapidly by the junction.

Development of IR hybrid packaging technology began in the late 1970s (see Fig. 9.1) and took the next decade to reach volume production. In the early 1990s, fully 2D imaging arrays provided a means for staring sensor systems to enter the production stage. In the hybrid architecture, indium bump bonding with readout electronics allows the signals to be multiplexed from thousands or millions of pixels onto a few output lines, greatly simplifying the interface between the vacuum-enclosed cryogenic sensor and the system electronics.

Although FPA imagers are very common in our lives, they are quite complex to fabricate. Depending on the array architecture, the process can include over 150 individual fabrication steps. The hybridization process involves flip-chip indium bonding between the surfaces of the ROIC and the detector array. The indium bond must be uniform between each sensing pixel and its corresponding readout element in order to ensure high-quality imaging. After hybridization, a backside thinning process is usually performed to reduce the amount of substrate absorption. The edges off the gap between the ROIC and FPA can be sealed with low-viscosity epoxy before the substrate is mechanically thinned down to several microns. Some advanced FPA fabrication processes involve complete removal of the substrate material.

Innovations and progress in FPA fabrication are dependent on adjustments to the material growth parameters. Usually in-house growth has enabled manufacturers to maintain the highest material quality and to customize the layer structures for multiple applications. For example, since the HgCdTe material is critical to many principal product lines, and comparable material is not available externally, most global manufactures continue to supply their own wafers. Figure 9.2 shows the process flow for integrated infrared FPA manufacturing. As is shown, boule growth starts with the raw-material polycrystalline components. In the case of the HgCdTe FPA process, polycrystalline ultrapure CdTe and ZnTe binary compounds are loaded into a carbon-coated quartz crucible. The crucible is mounted into an evacuated quartz ampoule, which is placed in a cylindrical furnace. Large-crystal CdZnTe boules are produced by mixing and melting the

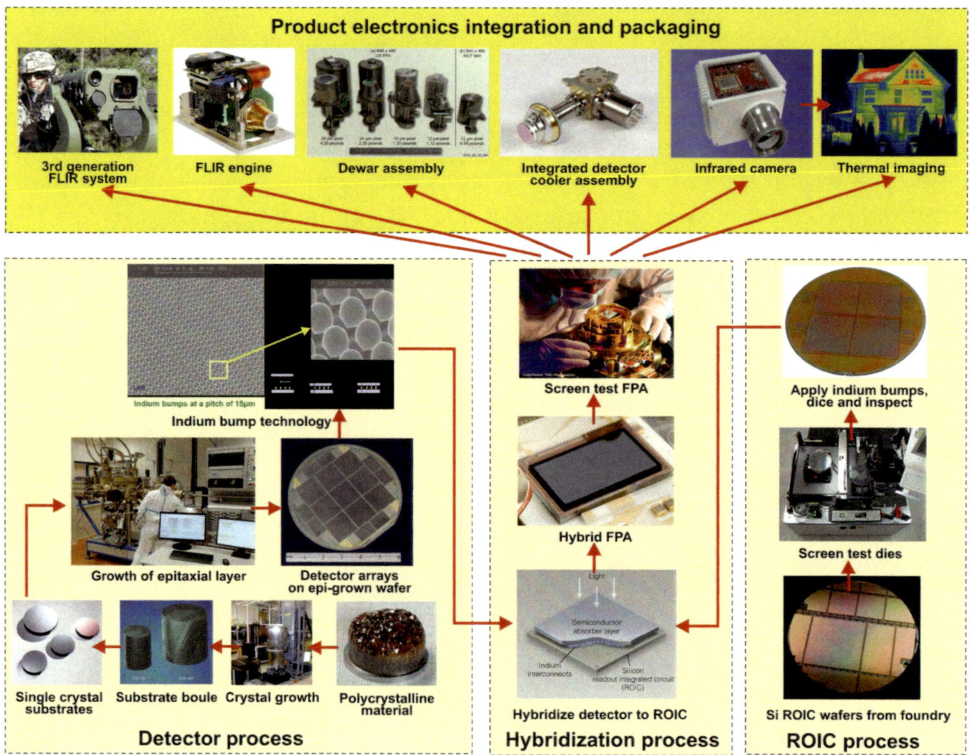

Figure 9.2 Process flow for integrated infrared FPA manufacturing.

ingredients, followed by recrystallizing with the vertical gradient freeze method. The standard diameters of the boules reach 125 mm. The boule substrate material is then cut into slices, diced into squares, and polished to prepare the surface for epitaxial growth. Typical substrate sizes up to 8 cm × 8 cm have been produced. The HgCdTe layers are usually grown on top of the substrate by MBE or MOCVD. In the case of MOCVD epitaxial technology, large-sized GaAs substrates are also used. The selection of substrate depends on the specific application. The entire growth procedure is automated, with each step having been programmed in advance.

MBE and MOCVD growth methods are well established for the III-V semiconductor materials. At present, MBE offers low-temperature growth under an ultrahigh-vacuum environment; *in situ* n-type and p-type doping; and control of composition, doping, and interfacial profiles. MBE is now the preferred method for growing complex, layered structures for multicolor detectors and for avalanche photodiodes.

After growing the detector epitaxial structures, the wafers are nondestructively evaluated against multiple quality specifications. They are then conveyed to the array processing line, where the sensing elements (pixels) are formed by photolithographic steps, including mesa etching, surface passivation, metal contact deposition, and indium bump formation. After wafer dicing, the FPAs are ready for mating to the ROICs. The ROIC branch of the process is shown in the lower right of Fig. 9.2. For each pixel on the detector array, there is a corresponding unit cell on the ROIC that collects the photocurrent and processes the signal. Each design is delivered to a silicon foundry for fabrication. Next, the ROIC wafers are diced and are ready for mating with the FPA. The most advanced flip-chip bonders, utilizing laser alignment and submicron-scale motion control, bring the two chips together (see the center of Fig. 9.2). At present FPAs with a pixel pitch size below 10 μm are aligned and hybridized with a high yield. Each FPA with attached ROIC is tested according to a defined protocol and is installed in a sensor module. Finally, associated packaging and electronics are designed and assembled to complete the integrated manufacturing process.

Detector FPAs have revolutionized many kinds of imaging from gamma rays to the infrared and even radio waves. More general information about the background, history, present stage of technology, and trends can be found in Refs. 2 and 3, for example. Information about the assemblies and applications can be found on different vendor websites.

9.2 Infrared FPA Considerations

It is well known that detector size d and $F/\#$ are the primary parameters of infrared systems.[4] These two parameters have a major impact on detection and identification ranges. Most current military systems have the classical design parameters represented in Fig. 9.3, where detector size ranges from

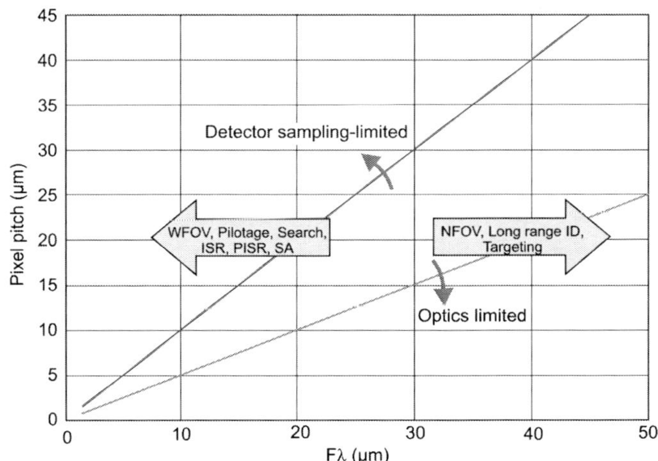

Figure 9.3 Classical infrared system design parameters (ISR – Intelligence, Surveillance, and Reconnaissance; PISR – persistant ISR; SA – situational awareness) (adapted from Ref. 6).

10 to 50 μm. For long-range identification systems, a high-$F/\#$ optics is used (for a given aperture) to reduce the detector angular subtense. On the other hand, wide-field-of-view (WFOV) systems are typically low-$F/\#$ systems with short focal lengths since the focal plane must be spread over wide angles. Recently published papers have shown that long-range identification does not need to be limited to high-$F/\#$ systems and that very small detectors enable high performance with a smaller package.[5,6]

The fundamental limit of pixel size is determined by diffraction. The size of the diffraction-limited optical spot or Airy disk is given by

$$d = 2.44\lambda F, \tag{9.1}$$

where d is the diameter of the spot, and λ is the wavelength. The spot sizes for f-numbers ranging from $F/1$ to $F/10$ are shown in Fig. 9.4. For typical $F/2.0$ optics at 4-μm wavelength, the spot size is 20 μm.

It is generally interesting to investigate pixel scaling beyond the diffraction limit using wavelength- and even subwavelength-scale optics that are enabled by modern nanofabrication (diffraction-limited pixel size is still relatively large compared with feature size, which can be achieved using state-of-the-art nanofabrication approaches).

FPAs of 1 cm^2 still dominate the IR market, while pixel pitch has decreased to 15 μm during the last few years, now reaching 12 μm,[7] 10 μm,[8,9] and even 5 μm in test devices.[10,11] This trend is expected to continue. Systems operating at shorter wavelengths are more likely to benefit from small pixel sizes because of the smaller diffraction-limited spot size. Diffraction-limited optics with low f-numbers (e.g., $F/1$) could benefit from pixels on the order of one wavelength across. Oversampling the diffractive spot may provide some additional

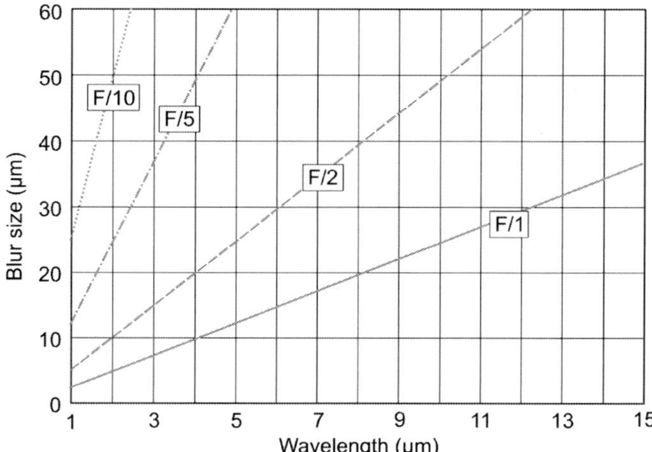

Figure 9.4   Optics diffraction limit. The spot size of a diffraction-limited optical system is the Airy disk diameter.

resolution for smaller pixels but saturates quickly as the pixel size is decreased. Pixel reduction is mandatory also for cost reduction of a system (reducing the optics diameter, dewar size and weight, together with the power, and increasing the reliability). In addition, smaller detectors provide better resolution.[12] Reduction of the focal plane proportionally to the detector size has not changed the detector field of view, so in the optics-limited region, smaller detectors have no effect on the system spatial resolution.

Figure 9.5 shows the influence of pixel shrinkage on the format enlargement of Sofradir's IR arrays. A catalog of detectors with pixel pitch of 15 µm [Epsilon (384 × 288), Scorpio (640 × 512), and Jupiter (1280 × 1014)] is compared with the Daphnis 10-µm product family.

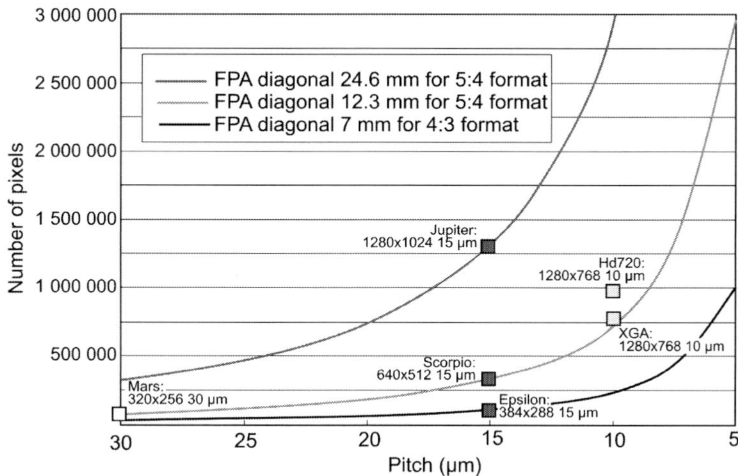

Figure 9.5   Number of pixels versus pitch size for Sofradir IR FPAs (reprinted from Ref. 8).

Recent progress in small-detector fabrication has raised interest in determining the minimum useful detector size.[4,5,12] The critical parameters that define the potential performance of an infrared FPA are modulation transfer function (*MTF*) and noise equivalent difference temperature (*NEDT*). *MTF* is a measure of the sharpness (or blurring) of images generated by any imaging system and is the ratio of the modulation of the image to the modulation of the object as a function of the space frequency (cycles per unit length) of the sine-wave pattern. The *NEDT* characterizes the thermal sensitivity of an infrared system, i.e., the amount of temperature difference required to produce a unity SNR. A smaller *NEDT* indicates better thermal sensitivity.

The system *MTF* is dominated by the *MTF*s of the optics, detector, and display, and can be cascaded by simply multiplying the *MTF*s of the components to obtain the *MTF* of the combination. Considering the optics and detector combination, $MTF_{DO} = MTF_{Optics} \times MTF_{Detector}$. Other blurs (aberrations, crosstalk, diffusion, etc.,) are considered negligible in this approach.

As shown by Holst,[13] FPA system performance can be described in the spatial domain, where the optical blur diameter is compared to the detector size, or in the frequency domain (*MTF* approach), where the optics cutoff is compared to the detector cutoff. Both comparisons provide a metric that is a function of $F\lambda/d$, where F is the focal ratio, λ is the wavelength, and d is the detector size. In other words, $F\lambda/d$ is the ratio of the detector cutoff frequency to the optics cutoff frequency and in space is a measure of the optical blur diameter relative to the detector size. The detector *MTF* exists for all frequencies from $-\infty$ to $+\infty$. The detector cutoff is defined as the first zero in $MTF_{Detector}$ and occurs when the spatial frequency equals F/d. The sampling frequency is determined by the detector pitch. Assuming that the pitch equals the detector size (100% fill factor array), the sampling frequency equals the detector cutoff and the Nyquist frequency, and equals $F/2D$, where D is the aperture diameter. The truth in this assumption is dependent on the quality of the detector reticulation achieved for the pixel pitch involved.

It can be shown that the range approximation can be estimated by[14,15]

$$Range \approx \frac{D\Delta x}{M\lambda}\left(\frac{F\lambda}{d}\right), \tag{9.2}$$

and the *NEDT* by[15]

$$NEDT \approx \frac{2}{C\lambda(\eta\Phi_B^{2\pi}\tau_{int})}\left(\frac{F\lambda}{d}\right), \tag{9.3}$$

where D is the aperture, M is the number of pixels required to identify a target Δx, C is the scene contrast, η is the detector collection efficiency, $\Phi_B^{2\pi}$ is the background flux into a 2π FOV, and τ_{int} is the integration time.

Equations (9.2) and (9.3) indicate that the parameter space defined by $F\lambda$ and d can be utilized in the optimum design of any IR system.

The detector-limited region occurs where $F\lambda/d \leq 0.41$, and the optics-limited region occurs where $F\lambda/d \geq 2$ (see Fig. 9.6). When $F\lambda/d = 0.41$, the Airy disk equals the detector size. A transition in the region $0.41 \leq F\lambda/d \leq 2.0$ is large and represents a change from detector-limited to optics-limited performance. The condition $F\lambda/d = 2$ is equivalent to placing 4.88 pixels within the Rayleigh blur circle. The lines presented a constant $F\lambda/d$ indicate a constant range and $NEDT$ [see Eqs. (9.2) and (9.3)]. For a given aperture D and operating wavelength λ, the detection range is given by the optimum resolution condition $F\lambda/d = 2$ and a minimum $NEDT$ for a given τ_{int} [see Eq. (9.2)]. From these considerations it can be concluded that the system $F/\#$ should be locked to the pixel size to predict the potential limiting performance of IR systems.

Figure 9.6 also includes experimental data points for various classes of thermal imaging systems that have been produced at DRS Technologies, including both uncooled thermal imagers and cooled photon imagers. The earliest uncooled imagers fabricated at the beginning of 1990s [barium strontium titanate (BST) dielectric bolometers and VO_x microbolometers] had large pixels of approximately 50-μm pitch and fast optics to achieve useful system sensitivities. With decreasing detector size, the relative apertures remained around $F/1$. Figure 9.6 shows that, as the pixel dimensions shrink

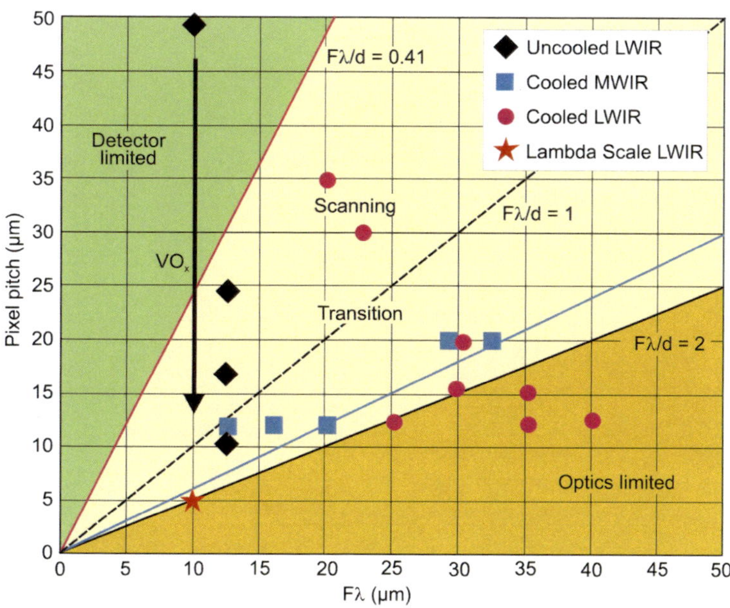

Figure 9.6 $F\lambda/d$ space for an infrared system design. Straight lines represent a constant $NEDT$. An infinite number of combinations provide the same range (adapted from Ref. 16).

over time, uncooled systems have steadily progressed from the detector-limited regime to the optics- limited regime. However, they are still far from the ultimate range capacity for $F/1$ optics.

The cooled thermal imagers include early LWIR scanning systems and modern staring systems operating in both MWIR and LWIR bands. LWIR imaging systems typically approach the $F\lambda/d = 2$ condition, whereas for MWIR systems, values of $F\lambda/d < 2$ are typically employed—lower available photon flux makes it difficult to maintain system sensitivity.

Figure 9.7 summarizes different behaviors of system MTF. The transition region can be further split by setting the optics cutoff frequency to equal the detector cutoff frequency, resulting in $F\lambda/d = 1.0$.[17] When $F\lambda/d = 1.0$, the spot size equals 2.44 times the size of the pixel. The optics-dominated region lies between the diffraction limit and this curve, while the detector-dominated region is located between this curve and the detector-limited curve. In the optics-dominated region, changes to the optics have a greater impact on the system MTF than the detector's impact. Likewise, for the detector-dominated region. Historically, most systems have been designed to have a resulting optics blur (to include aberrations) of less than 2.5 pixels ($\sim F\lambda/d < 1.0$). This is of course very dependent on the application and range requirements.

Table 9.1 provides the required $F/\#$ for $F\lambda/d = 2$ for various detector size. As is shown, with $F/1$ optics, the smallest useful detector size is 2 μm in the MWIR and 5 μm in the LWIR. With more realistic $F/1.2$ optics, the smallest useful detector size is 3 μm in the MWIR and 6 μm in the LWIR.

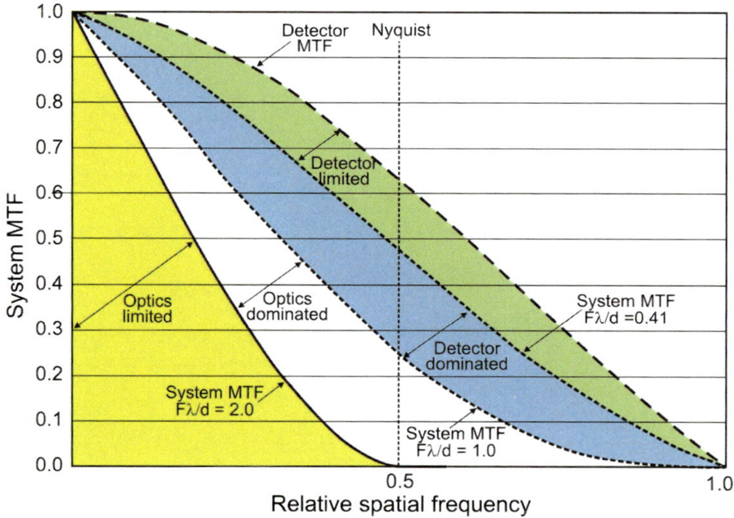

Figure 9.7 System *MTF* curves illustrating the different regions with the design space for various *Fλ/d* conditions. Spatial frequencies are normalized to the detector cutoff (adapted from Ref. 17).

Table 9.1 Required *F/#* for *Fλ/d* = 2.0. Real optics usually has *F/#* > 1 (data from Ref. 6).

d (μm)	MWIR (4 μm)	LWIR (10 μm)
2.0	1.0	–
2.5	1.25	–
3.0	1.33	–
5.0	2.5	1.0
6.0	3.0	1.3
12	6.0	2.4
15	7.5	3.0
17	8.5	3.4
20	10.0	4.5
25	12.5	5.0

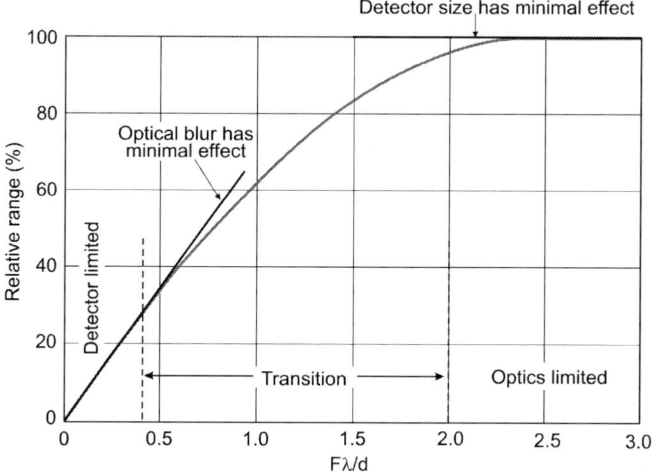

Figure 9.8   Relative range as a function of *Fλ/d*.

Holst and Driggers[6] have considered the influence of the optics infrared design on range approximation, which is illustrated in Fig. 9.8. When the IR system is detector limited, decreasing the detector size has a dramatic effect on range. On the other side, in the optics-limited region, decreasing the detector size has minimal effect on range performance. The acquisition range is reduced when atmospheric transmission and the *NEDT* are included.

As has been indicated by Kinch,[14,15] challenges that must be addressed in fabrication of small-pixel FPAs concern:

- pixel delineation,
- pixel hybridization,
- dark current, and
- unit cell capacity.

Above topics are considered in more detail in Ref. 18.

9.3 InSb Arrays

InSb photodiodes have been available since the late 1950s. They are used in the 1- to 5-μm spectral region and must be cooled to approximately 77 K. InSb photodiodes can also be operated in the temperature range above 77 K. Their applications include infrared homing guidance, threat warning, infrared astronomy, commercial thermal imaging cameras, and FLIR (forward-looking infrared) systems. One of the most significant recent advances in infrared technology has been the development of large 2D FPAs for use in staring arrays. Array formats are available with readouts suitable for both high-background $F/2$ operation and for low-background astronomy applications. Linear arrays are rarely used.

The earliest arrays, fabricated in the mid-1980s, were just 58×62 elements in size[19] compared to arrays that are up to 8192×8192 today, an increase of more than three orders of magnitude in pixel count (see Fig. 9.9). The array noise has improved over this time period from hundreds of electrons to as low as four electrons today.[20] Similarly, at the same time, detector dark current has decreased from about 10 electrons/second to as low as 0.004 electrons/second;[21] quantum efficiency has reached a level above 90%.

InSb material is far more mature than HgCdTe, and good-quality 6-inch-diameter bulk substrates are now commercially available. Below-10-μm-pitch gigapixel FPAs are set as a goal for realization within the next several years.

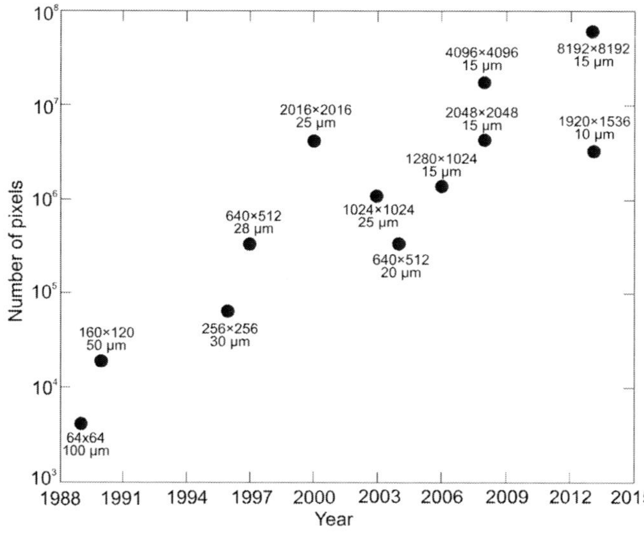

Figure 9.9 Progress in the development of InSb FPAs by L3 Cincinnati Electronics. The graph shows the total number of pixels during the year of first operation. Three million pixel arrays with a pixel dimension of 10 μm were first fabricated by SemiConductor Devices (Haifa) (adapted from Ref. 22 with additional data from Ref. 23).

Such large arrays are possible because the InSb detector material is thinned to less than 10 μm (after surface passivation and hybridization to a readout chip), which allows it to accommodate the InSb/silicon thermal mismatch. The InSb detector pixels are attached to a silicon substrate with a small 1-μm gap isolating the detector elements from each other and are essentially floating on the Si substrate. The gap between bumps is backfilled with epoxy. These improvements in yield and quality are due to the implementation of innovative passivation techniques, including specialized antireflection coatings and unique thinning processes. All of this processing is done at the wafer level. This is vitally important, since these devices are cooled from room temperature to 78 K several thousand times during their expected lifetime. The array operability (defined as the ratio of the number of nondefective pixels to total number of pixels in the array) is above 99.6% in over 15,000 cryogenic cycles.[22] At one cryogenic cycle per day, 15,000 cycles would take more than 40 years.

Large staring InSb FPA evolution has been driven by astronomy applications. Astronomers have funded large FPA development to dramatically improve telescope throughput. The first InSb array to exceed one million pixels was the ALADDIN array first produced in 1993 by Santa Barbara Research Center (SBRC) and demonstrated on a telescope by National Optical Astronomy Observations (NOAO), Tucson, Arizona, in 1994.[24] This array had 1024 × 1024 pixels spaced on 27-μm centers and was divided into four independent quadrants, each containing 8 output amplifiers. This solution was chosen due to the uncertain yield of large arrays at that time.

A chronological history of Raytheon Vision Systems' (RVS) astronomical FPAs is shown in Fig. 9.10.[25] The next step in the development of InSb FPAs for astronomy was the 2048 × 2048 ORION sensor chip assembly (SCA) (see Fig. 9.11). Four ORION SCAs were deployed as a 4096 × 4096 FPA in the NOAO near-IR camera,[26] currently in operation at the Mayall 4-m telescope on Kit Peak. This array has 64 outputs, allowing up to a 100-Hz frame rate. Many of the packaging concepts used on the ORION program are

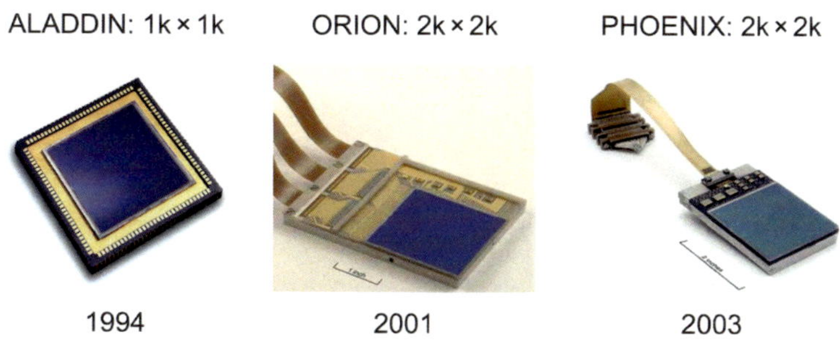

Figure 9.10 Timeline and history development of the RVS InSb astronomy arrays (adapted from Ref. 25).

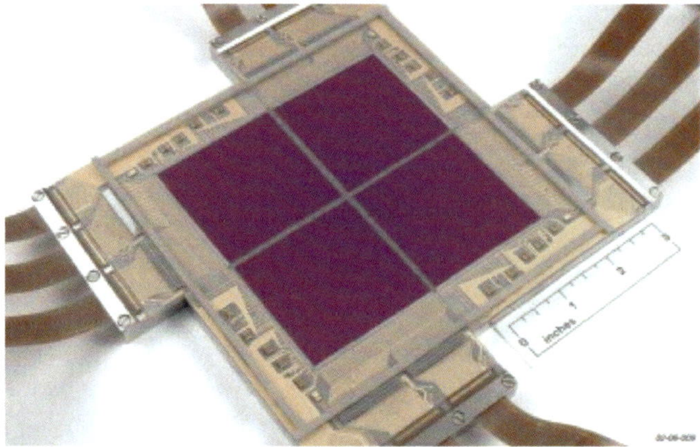

Figure 9.11 A demonstration of the 2-sided buttable ORION modules used to create a 4k × 4k FPA. One module contains an InSb SCA, while the others have bare readouts (reprinted from Ref. 26).

shared with the 3-side buttable 2k × 2k FPA InSb modules developed by RVS for the James Webb Space Telescope (JWST) mission.[27]

The PHOENIX SCA is another 2k × 2k InSb FPA that has been fabricated and tested. This detector array is identical to ORION (25-μm pixels); however, its readout is optimized for lower frames rates and lower power dissipation. With only 4 outputs, the full frame read time is typically 10 s. The smaller number of outputs allows a smaller module package that is 3-side buttable.[26] Three-side-buttable modules allow the possibility of realizing large detection areas.

Different formats of InSb FPAs have also found many high-background applications, including missile systems, interceptor systems, and commercial imaging camera systems. With an increasing need for higher resolution, several manufacturers have developed megapixel detectors. Table 9.2 compares the performance of commercially available megapixel InSb FPAs

Table 9.2 Performance of commercially available megapixel InSb FPAs.

Parameter	Configuration			
	2048 × 2048 (Raytheon ORION)	1024 × 1024 (L3 Cincinnati Electronics)	1024 × 1024 (Santa Barbara Focalplane)	1280 × 1024 (SCD)
Pixel pitch (μm)	25	25	19.5	15
Pixel capacity (electrons)	$>3 \times 10^5$	1.1×10^7	8.1×10^6	6×10^6
Power dissipation (mW)	<100	<100	<150	<120
NEDT (mK)	<24	<20	<20	20
Frame rate (Hz)	10	1 to 10	120	120
Operability (%)	>99.9	>99	>99.5	>99.5
Websites	www.raytheon.com	www.raytheon.com	www.sbfp.com	www.scd.co.il

fabricated by L3 Cincinnati Electronics, Santa Barbara Focalplane, and SCD. Santa Barbara Focalplane has developed a new large-format InSb detector with 1280×1024 elements and a pixel size of 12 μm.[28]

L3 Cincinnati Electronics is the manufacturer of large-format/wide-area surveillance sensors with 16.7 megapixels and an ultrawide FOV. This imaging engine is capable of detecting and identifying features that small-format sensors would miss. It is currently in use by U.S. assets in overseas combat zones. Table 9.3 presents typical performance specifications for this InSb sensor.

There are considerable efforts to decrease system size, weight, and power consumption (SWaP), and consequently reducing a system's cost, in order to increase the operating temperature in so-called HOTs detectors. Pixel reduction is also mandatory for increasing the detection and identification range of infrared imaging systems. The *MTF* functions of InSb FPAs with four different pitches (30, 20, 15, and 10 μm) are shown in Fig. 9.12.

SemiConductor Devices (Haifa) started in 1997 with the introduction of 320×256 format, 30-μm-pitch detectors, and continued with the larger-format arrays with pixel sizes of 25, 20, 15, and 10 μm. The shift to a smaller pixel dimension required the migration from a 0.5-μm CMOS process to a more advanced CMOS technology in order to allow a higher value of capacitance per unit area, a lower operating voltage for reduced power consumption, and a denser device layout for maintaining a high level of functionality. The trend to larger format and smaller pitch continued with Hercules, an InSb detector with 1280×1024 pixels of 15-μm pitch. The new

Table 9.3 Performance characteristics of sensors with large-format InSb arrays (IDCA – integrated detector cooler assembly).

	L3 Cincinnati Electronics (Large-Format/Wide-Area Surveillance Sensors)	**SCD (Blackbird IDCA)**
View of integrated detector		
Format	4096×4096	1920×1536
Pixel size	15×15 μm^2	10×10 μm^2
FPA power consumption	2/5 W	400 mW
Cooler power steady state	<55 W	20 W
Weight	~15 lb.	700 gr
NEDT	Dependent on integration time	<24 mK

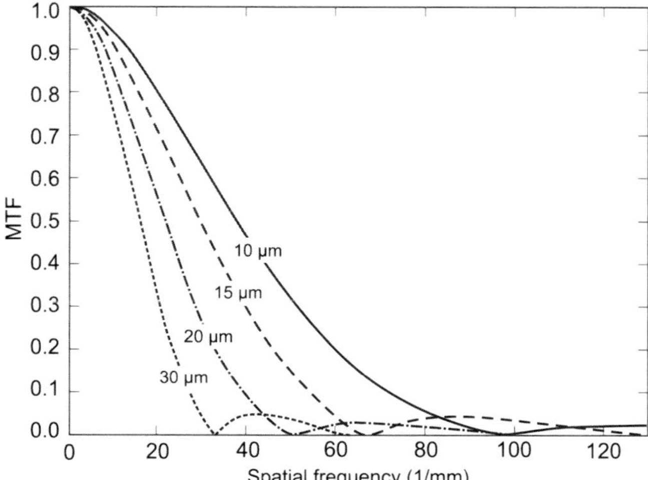

Figure 9.12 *MTF* curves of InSb FPAs with four different pitches: 30, 20, 15, and 10 μm, corresponding to SCDs Blue Fairy, Sebastian, Pelican, and Blackbird FPAs, respectively (adapted from Ref. 29).

Blackbird detector is a natural step in this roadmap with 3 million pixels in the FPA and a pixel dimension of 10 μm.

Table 9.3 presents typical performance characteristics for the Blackbird sensor module. Blackbird is packaged in a dewar that is integrated with a cryo-cooler and an electronic proximity board. This integrated detector cooler assembly (IDCA) makes a compact MWIR detector that generates 13-bit, 3M-pixel images at a frame rate of up to 120 Hz with a total power consumption of less than 30 W at 71 °C. Figure 9.13 shows an image made by this the new detector with high temperature and high spatial resolution.

9.4 InAsSb nBn Detector FPAs

MWIR nBn sensor arrays are manufactured by several companies. As mentioned in section 6.1, the nBn sensor design is self-passivating, decreasing leakage current and associated noise while improving reliability and manufacturability. Because of its simple design (see Fig. 9.14), the array technology is a major advance in the state-of-the-art for large-format infrared FPAs.

Because the nBn structure limits dark depletion current, the temperature at which it operates is increased. For example, Fig. 6.11 shows the temperature dependence of *NEDT* for the *F*/3.2 optics of the Kinglet digital detector. This sensor is based on SCD's Pelican-D ROIC and contains an nBn InAs$_{0.91}$Sb$_{0.09}$/B-AlAsSb 640 × 512 pixel architecture with a 15-μm pitch. The *NEDT* is 20 mK at 10-ms integration time, and the operability of nondefective pixels is greater than 99.5% after a standard 2-point nonuniformity correction.

Figure 9.13 Image from the Blackbird detector at *F*/3 taken 2 km away (reprinted from Ref. 29 with permission from SemiConductor Devices).

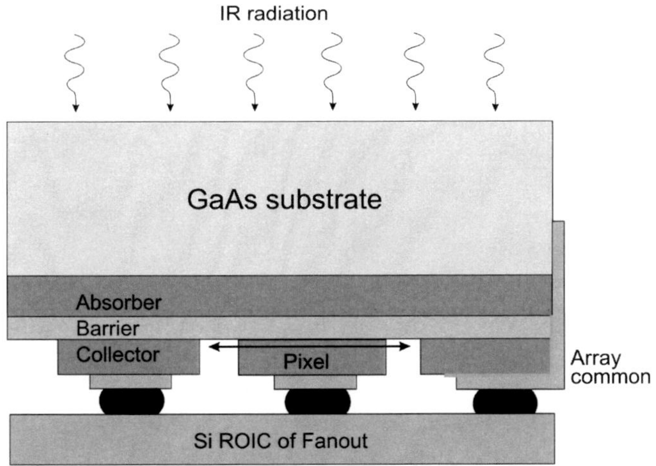

Figure 9.14 nBn detector array architecture.

The *NEDT* and operability begin to change above 170 K, which is consistent with the estimated BLIP temperature of 175 K.

The first commercially developed nBn InAsSb array sensor developed by Lockheed Martin Santa Barbara Focalplane operates at 145 to 175 K. In IRCameras' implementation (see Table 9.4) of Santa Barbara Focalplane's MWIR nBn sensor, a 1280×1024 format, 12-µm-pixel-pitch detector is packaged in a 1.4-inch-diameter dewar with an overall dewar housing length of about 3.8 inch, including the cooler. With a long-life, 25,000-h cryocooler consuming about 2.5 W and electronics adding another 2.5 W, the total

Table 9.4 nBn InAsSb FPA characteristics.

QuazIR™ HD+ Camera (IRCameras)   Hercules XBn IDCA (SCD)

Parameter	Performance	
Array format	1280×1024	1280×1024
Pixel pitch	12 µm	15 µm
Well capacity	2 Me$^-$	6 Me$^-$, 1 Me$^-$
Integration time	>500 ns to 16 ms	to 22 ms
Power consumption	5 W	5.5 W
Operating temperature	–40 °C / + 71 °C	–40 °C / +71 °C
Weight	~1 lb	~750 gr.
Size	2.35-inch W × 2.59-inch H × 2.75-inch L	Length (optical axis) of 149 mm

camera core power draw is only about 5 W total. A high-spatial-resolution image acquired with the nBn sensor running at 160 K is shown in Fig. 9.15.

Recently (in 2015), the first announcement of a high-definition FPA containing 5-µm pixels in a 2040×1156 array format was given by Caulfield et al.[31] Figure 9.16 shows an outdoor image from a laboratory camera with this array fabricated by Cyan Systems.

Figure 9.15 A MWIR nBn InAsSb FPA with 1280×1024 pixels images a scene from a baseball game when a player is attempting to steal second base (reprinted from Ref. 30 with permission from PennWell Corp.).

Figure 9.16   Image from a HOT MWIR nBn InAsSb array with 5-μm pixels in a 2040 × 1156 format (reprinted from Ref. 31).

9.5 Type-II Superlattice FPAs

The basic array technology of FPAs consists of photodiodes hybridized to a silicon focal plane processor (CMOS readout) by means of indium bumps, and then backside thinned with surface passivation and antireflection coating. The roadmap of T2SL photodetectors is shown in Fig. 9.17.

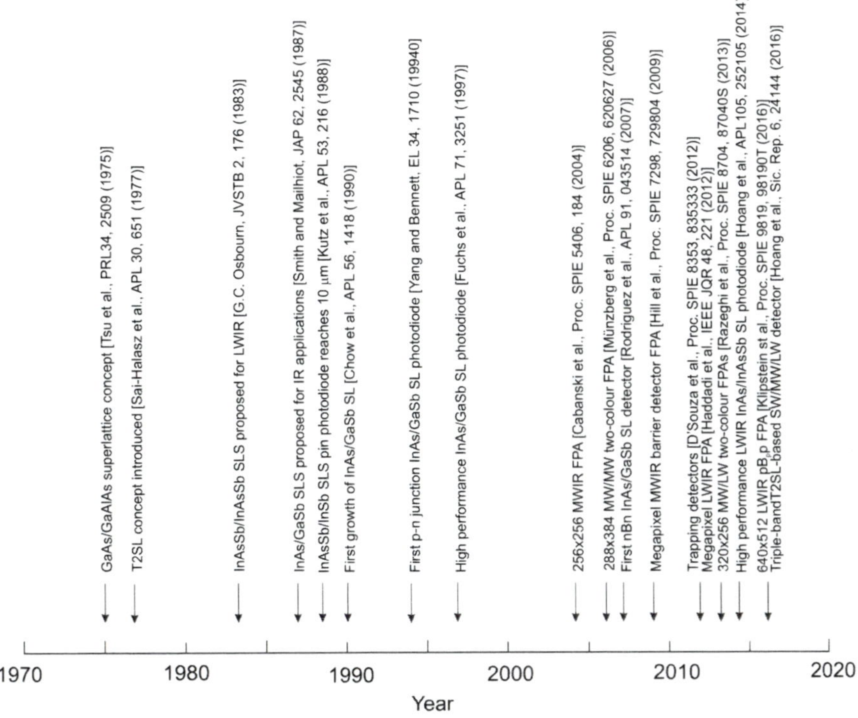

Figure 9.17   Roadmap development of type-II superlattice infrared photodetectors.

In 1987, Smith and Mailhiot proposed an InAs/GaSb T2SL for infrared detection applications.[10] Despite the positive theoretical predictions, high-quality growth of InAs/GaSb T2SL material had not demonstrated almost two decades later. Advances in the MBE superlattice material technology and device processing techniques have been gradually improved, making the fabrication of high-quality single elements and FPA more routine. During the last decade the first megapixel MWIR and LWIR type-II SL FPAs were demonstrated with excellent imaging.[32–40] At about 78 K, an *NEDT* value of below 20 mK was presented for MWIR arrays, and a value just above 20 mK was presented for LWIR arrays. Figure 9.18 shows images taken with with MWIR 640×512 nBn array and two (MWIR and LWIR) megapixel photovoltaic arrays.

The values of *NEDT* for MWIR type-II SL FPAs are shown in Fig. 9.19. An excellent *NEDT* value of approximately 10 mK measured with *F*/2 optics and integration time $\tau_{int} = 5$ ms has been presented for a 256×256 MWIR detector with a cutoff wavelength of 5.3 μm [see 9.19(a)].[32] Tests with reduced time down to 1 ms show that the *NEDT* scales inversely proportionally to the square root of the integration time. This means that even for short integration times the detectors are background-limited. Similar results have been demonstrated by Nortwestern University's group.[35] The minimum *NEDT* stays almost constant at 11 mK with an integration time of 10.02 ms up to 120 K, which suggests that the FPA is mainly dominated by temperature-insensitive noise, such as system noise or photon noise, since dark current noise should increase exponentially with temperature. From 130 K to 150 K, measurements are performed with an integration time of 4.02 ms to avoid saturation of the readout capacitors due to higher dark current levels. The increase in *NEDT* in this region might be related to the increase in the dark current. One very important feature of InAs/GaSb FPAs is their high uniformity. The responsivity spread shows a standard deviation of approximately 3%. It is estimated that the pixel outages are on the order of 1–2%

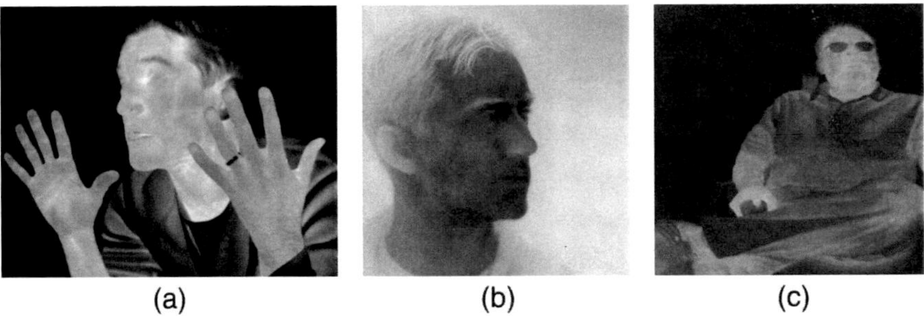

(a)   (b)   (c)

Figure 9.18 Images taken with (a) an nBn 640×512 MWIR FPA, as well as with megapixel (b) MWIR (p-i-n pixels) and (c) LWIR (CIRBD detector–see Table 6.1) Sb-based photovoltaic FPAs [parts (a) and (b) reprinted from Ref. 33; part (c) reprinted from Ref. 34 with permission from IEEE].

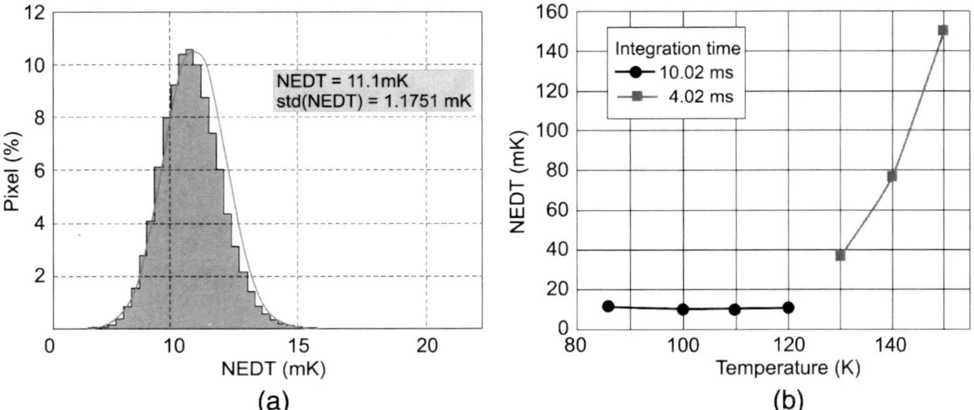

Figure 9.19 *NEDT* of MWIR type-II superlattice FPAs: (a) histogram of 256 × 256 FPA at *F*/2 optics, $\tau_{int} = 5$ ms, and 77 K (adapted from Ref. 32), and (b) temperature dependence for 320 × 256 FPA at *F*/2.3 optics; the integration time is reduced above 120 K to avoid saturation of the readout capacitor due to higher dark current levels (adapted from Ref. 35).

and the pixels are statistically distributed as single pixels without large clusters.[32]

Northwestern University's group has demonstrated a high-quality 1024 × 1024 LWIR FPA using the p-π-M-n pixel structure shown in Fig. 9.20(a). This device design combines both high optical and electrical performance. The M-structure and the double heterostructure design technique help to reduce both the bulk dark current and surface leakage current. A high quantum efficiency (>50%) was obtained thanks to a thick absorption region (6.5 µm). The device exhibited a dark current level below 5×10^{-5} A/cm^2 at –50 mV applied bias voltage at 77 K. Figure 9.20(b) shows a *NEDT* histogram after 2-point uniformity correction measured at 81 K with an integration time

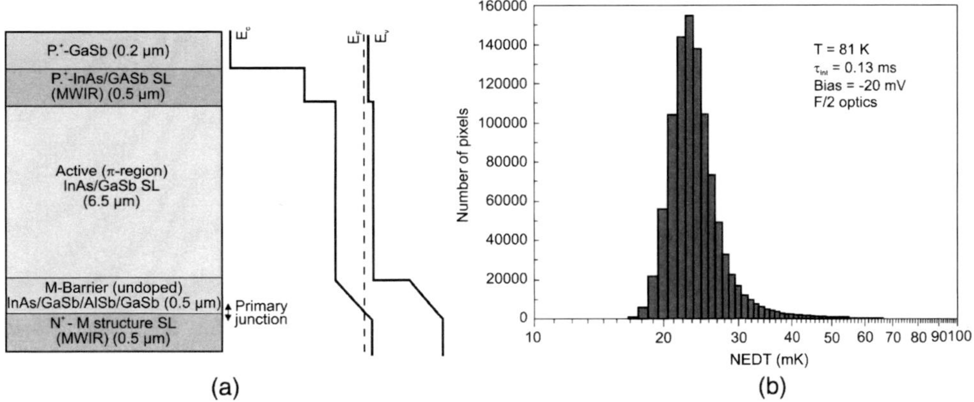

Figure 9.20 1024 × 1024 LWIR FPA: (a) p-π-M-n pixel structure and (b) *NEDT* histogram at 81 K (adapted from Ref. 36).

of 0.13 ms and *F*/2 optics. At a reverse bias of 20 mV, the median *NEDT* value was 27 mK.

Recently, SCD developed an advanced pB_pp T2SL barrier detector that enables diffusion-limited dark currents comparable to the HgCdTe "Rule 07" and has high quantum efficiency—above 50% (see Section 6.3).[39] Special attention was paid to eliminate the influence of the surface leakage current. A robust passivation process was developed that allows for glue underfills and substrate thinning after bonding the sensor array with indium bumps to the custom-designed silicon readout. Polishing the GaSb substrate down to a final thickness of about 10 μm is required in order to relieve stress during cooling. The 15-μm-pitch 640 × 512 FPAs are operated at 77 K with a 9.5-μm cutoff wavelength. The final IDCAs include a cold filter with a cutoff wavelength of 9.3 μm. Table 9.5 presents typical performance specifications.

Also recently, Fraunhofer IAF in cooperation with AIM Infrarot-Module GmbH, Heilbronn, Germany realized Europe's first InAs/GaSb T2SL imager for the LWIR with 640 × 512 pixels and 15-μm pitch. The demonstrator camera delivers good image quality (see Fig. 9.21) and achieves a thermal resolution at 55 K better than 30 mK with *F*/2 optics for a 300-K background scene.

Multicolor capabilities are highly desirable for advanced IR systems since gathering data in discrete IR spectral bands can be used to discriminate for both absolute temperature and unique signatures of objects in the scene. Recent trends in multispectral FPA development has leaned towards integrating multiband functionality into a single pixel, rather than combining multiple single-spectrum arrays, which requires spectral filters and spectrometers.

Table 9.5 Specifications of LW pB_pp T2SL array performance at 77 K (STD – standard deviation, DR – deviation ratio, MTTF – mean time to failure).[39]

Parameter	Value	Pelican-D LW IDCA
Format	640 × 512	
Pitch	15 μm	
Cutoff wavelength	9.3 μm (filter)	
Quantum efficiency	>50%	
Operability	>99%	
Residual non-uniformity (RNU)	<0.04% STD/DR @ 10–90% well fill capacity	
NEDT	15 mK @ 65% well fill capacity, 30 Hz (by averaging 8 frames)	
Response uniformity	<2.5% (STD/DR)	
Cooler	Ricor K548	
Weight	750 g	
Environment condition	−40 to +71 °C	
Total power at 23 °C	16 W	
Cool-down time	8 min	
MTTF (depends on mission profile)	15,000 hours	

Figure 9.21 Fraunhofer IAF image taken with a 640 × 512 LWIR heterojunction InAs/ GaSb T2SL camera with 15-μm pixel pitch (reprinted from Ref. 40).

Apart from HgCdTe photodiodes and QWIPs, the T2SL material system has emerged as a candidate suitable for multispectral detection due to its ease in bandgap tuning while retaining closely matched lattice conditions.[41,42] Three basic approaches to achieving multicolor detection have been proposed: multiple leads, voltage switching, and voltage tuning. These approaches are briefly described in Chapter 16 of Rogalski's monograph.[2]

The unit cell of integrated multicolor FPAs usually consists of several co-located detectors, each sensitive to a different spectral band (see Fig. 9.22). Radiation is incident on the shorter-band detector, with the longer-wave radiation passing through to the next detector. Each layer absorbs radiation up to its cutoff and hence is transparent to the longer wavelengths, which are then collected in subsequent layers. The device architecture is realized by placing a longer-wavelength detector optically behind a shorter-wavelength

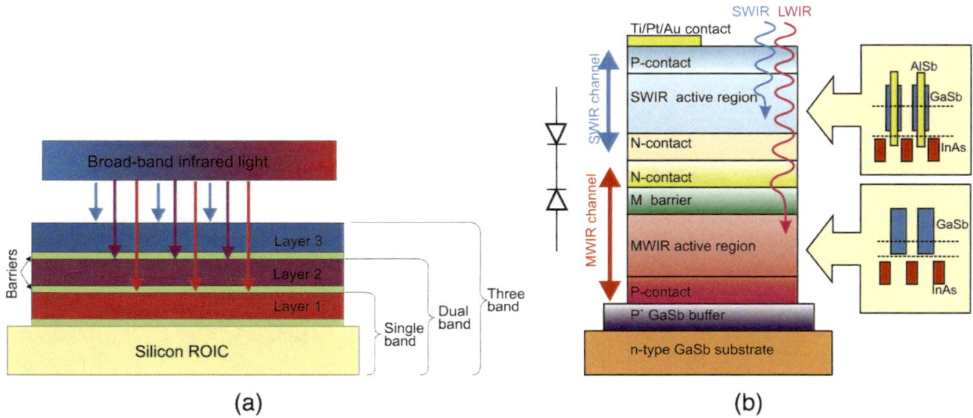

Figure 9.22 Multicolor detector pixels: (a) structure of a three-color detector pixel. IR flux from the first band is absorbed in layer 3, while longer-wavelength flux is transmitted to the next layers. Thin barriers separate the absorbing bands. (b) A dual-band SWIR/MWIR InAs/ GaSb/AlSb T2SL back-to-back p-i-n-n-i-p photodiode structure and schematic band alignment of superlattices in the two absorption layers (dotted lines represent the effective bandgaps of the superlattices) (reprinted from Ref. 43).

detector. The thickness of the entire vertical pixel structure is only about 5 μm, which significantly reduces the technological challenge in comparison to dual-band HgCdTe FPAs with a typical total layer thickness around 15 μm.

In 2005, AIM Infrarot-Module GmbH demonstrated the first bispectral InAs/GaSb superlattice IR camera. The manufacture of dual-color detectors via MBE and subsequent processing of FPAs were accomplished by Fraunhofer IAF. The pixel reduction to 30-μm pitch was achieved by restricting the number of contacts per pixel to two lands. A metallization grid deposited in the trenches and connected to the ROIC outside of the active array interconnects the common ground-contact vias. Fraunhofer's dual-color MWIR superlattice detector array technology with simultaneous, co-located detection capability is ideally suited for airborne missile threat warning systems.[44,45] Figure 9.23 illustrates a fully processed dual-color 288×384 FPA. With 0.2-ms integration time and 78-K detector temperature, the superlattice camera achieves an *NEDT* of 29.5 mK for the blue channel ($3.4 \ \mu m \leq \lambda \leq 4.1 \ \mu m$) and 14.3 mK for the red channel ($4.1 \ \mu m \leq \lambda \leq 5.1 \ \mu m$).

Figure 9.24 compares *NEDT* data for 384×288 dual-color InAs/GaSb SL detector arrays with a pitch size of 40 μm—each pixel with two back-to-back homojunction photodiodes detects a blue (3–4 μm) and a red (4–5 μm) wavelength band simultaneously.[46] The plots in the upper and lower rows represent the *NEDT* distribution of a typical dual-color FPA before (upper row) and after (lower row) implementation of a novel method for dielectric surface passivation. Pixels with high $1/f$ noise produce a tail in the root-mean-square (rms) noise distribution. While the blue channel histogram data is virtually unaffected by the modified process, the noisy pixels responsible for

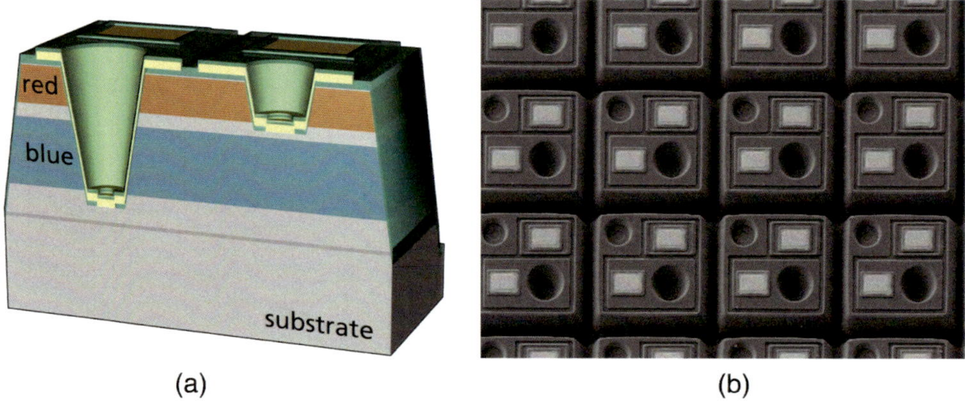

(a)   (b)

Figure 9.23 Dual-color InAs/GaSb SLS FPA: (a) schematic cross-section of detector pixel, and (b) photograph of dual-color pixel design that enable simultaneous co-located photon detection at 3–4 μm (blue channel) and 4–5 μm (red channel). At a pixel pitch of 30 μm, three contact lands per pixel permit simultaneous and spatially coincident detection of both colors [part (a) reprinted from Ref. 40; part (b) reprinted from Ref. 44].

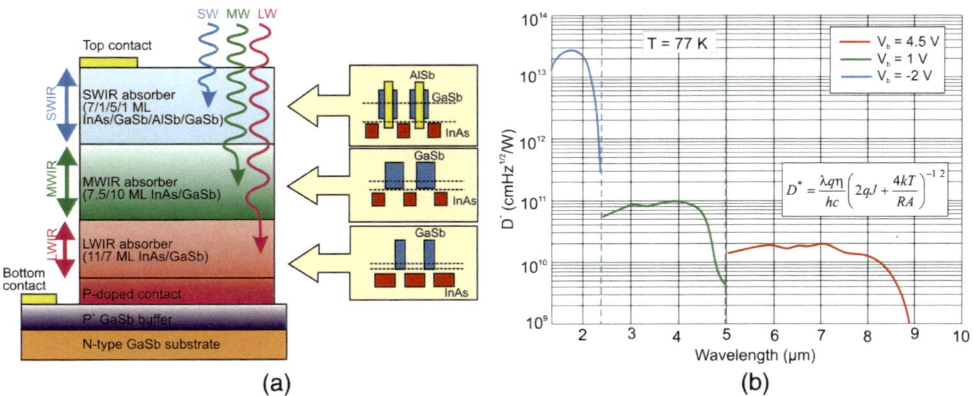

Figure 9.27 Triple-band SWIR/MWIR/LWIR T2SL photodiode: (a) structure with two terminal contacts and schematic band alignments; (b) calculated detectivity at 77 K using the equation in the inset. The SWIR detection operates at −2 V. MWIR detection operates at 1 V, and LWIR detection operates at 4.5 V positive bias voltages (reprinted from Ref. 49 under the terms of the Creative Commons CC-BY license).

Recently, a novel device design using a two-terminal, triple-band, T2SL-based SWIR/MWIR/LWIR photodetector similar to that presented in Fig. 9.27 has been demonstrated.[48] This device can perform as three individual single-color photodetectors operating sequentially according to the magnitude variation of applied bias (see Fig. 9.27). The triple-band SW-MW-LW photodiode design consists of a 1.5-μm-thick undoped active region in the SWIR, a 0.5-μm-thick *n*-doped SWIR layer ($n \sim 10^{18}$ cm^{-3}), a 2.0-μm-thick MWIR active region, and a 0.5-μm thick undoped layer followed by a 1.0-μm-thick p-doped LWIR active region ($p \sim 10^{16}$ cm^{-3}) and a 0.5-μm-thick bottom p-contact ($p \sim 10^{18}$ cm^{-3}). The total thickness of the device structure is 6 μm.

Figure 9.27(b) shows the calculated shot-noise-limited detectivity of the device in its three operation modes at 77 K based on the measured quantum efficiency, the dark current, and the resistance–area product. The device operates at biases of −2, 1, and 4.5 V and provides a D^* of 3.0×10^{13}, 1×10^{11}, and 2.0×10^{10} cm Hz$^{1/2}$/W at peak responsivity ($\lambda = 1.7$, 4.0, and 7.2 μm).

References

1. P. Norton, "Detector focal plane array technology," in *Encyclopedia of Optical Engineering*, edited by R. Driggers, Marcel Dekker Inc., New York, pp. 320–348 (2003).
2. A. Rogalski, *Infrared Detectors*, 2nd edition, CRC Press, Boca Raton, Florida (2010).
3. J. D. Vincent, S. E. Hodges, J. Vampola, M. Stegall, and G. Pierce, *Fundamentals of Infrared and Visible Detector Operation and Testing*, Wiley, Hoboken, New Jersey (2016).

4. G. C. Holst and T. S. Lomheim, *CMOS/CCD Sensors and Camera Systems*, JCD Publishing, Winter Park, Florida and SPIE Press, Bellingham, Washington (2007).

5. R. G. Driggers, R. Vollmerhausen, J. P. Reynolds, J. Fanning, and G. C. Holst, "Infrared detector size: How low should you go?" *Opt. Eng.* **51**(6), 063202 (2012) [doi: 10.1117/1.OE.51.6.063202].

6. G. C. Holst and R. G. Driggers, "Small detectors in infrared system design," *Opt. Eng.* **51**(9), 096401 (2012) [doi: 10.1117/1.OE.51.9.096401].

7. R. L. Strong, M. A. Kinch, and J. M. Armstrong, "Performance of 12-μm- to 15-μm-pitch MWIR and LWIR HgCdTe FPAs at elevated temperatures," *J. Electron. Mater.* **42**, 3103–3107 (2013).

8. Y. Reibel, N. Pere-Laperne, T. Augey, L. Rubaldo, G. Decaens, M. L. Bourqui, S. Bisotto, O. Gravrand, and G. Destefanis, "Getting small, new 10 μm pixel pitch cooled infrared products," *Proc. SPIE* **9070**, 907034 (2014) [doi: 10.1117/12.2051654].

9. Y. Reibel, N. Pere-Laperne, L. Rubaldo, T. Augey, G. Decaens, V. Badet, L. Baud, J. Roumegoux, A. Kessler, P. Maillart, N. Ricard, O. Pacaud, and G. Destefanis, "Update on 10 μm pixel pitch MCT-based focal plane array with enhanced functionalities," *Proc. SPIE* **9451**, 945182 (2015) [doi: 10.1117/12.2178954].

10. J. M. Armstrong, M. R. Skokan, M. A. Kinch, and J. D. Luttmer, "HDVIP five micron pitch HgCdTe focal plane arrays," *Proc. SPIE* **9070**, 907033 (2014) [doi: 10.1117/12.2053286].

11. W. E. Tennanat, D. J. Gulbransen, A. Roll, M. Carmody, D. Edwall, A. Julius, P. Dreiske, A. Chen, W. McLevige, S. Freeman, D. Lee, D. E. Cooper, and E. Piquette, "Small-pitch HgCdTe photodetectors," *J. Electron. Mater.* **43**, 3041–3046 (2014).

12. R. Bates and K. Kubala, "Direct optimization of LWIR systems for maximized detection range and minimized size and weight," *Proc. SPIE* **9100**, 91000M (2014) [doi: 10.1117/12.2053785].

13. G. C. Holst, "Imaging system performance based on $F\lambda/d$," *Opt. Eng.* **46**(10), 103204 (2007) [doi: 10.1117/1.2790066].

14. M. A. Kinch, "The rationale for ultra-small pitch IR systems," *Proc. SPIE* **9070**, 907032 (2014) [doi: 10.1117/12.2051335].

15. M. A. Kinch, *State-of-the-Art Infrared Detector Technology*, SPIE Press, Bellingham, Washington (2014) [doi: 10.1117/3.1002766].

16. J. Robinson, M. Kinch, M. Marquis, D. Littlejohn, and K. Jeppson, "Case for small pixels: system perspective and FPA challenge," *Proc. SPIE* **9100**, 91000I (2014) [doi: 10.1117/12.2054452].

17. D. Lohrmann, R. Littleton, C. Reese, D. Murphy, and J. Vizgaitis, "Uncooled long-wave infrared small pixel focal plane array and system challenges," *Opt. Eng.* **52**(6), 061305 (2013) [doi: 10.1117/1.OE.52.6.061305].

18. A. Rogalski, P. Martyniuk, and M. Kopytko, "Challenges of small-pixel infrared detectors: a review," *Rep. Prog. Phys.* **79**(4) 046501 (2016).

19. G. Orias, A. Hoffman, and M. Casselman, "58 × 62 indium antimonide focal plane array for infrared astronomy," *Proc. SPIE* **627**, 408–417 (1986) [doi: 10.1117/12.968118].

20. C. W. McMurtry, W. J. Forrest, J. L. Pipher, and A. C. Moore, "James Webb Space Telescope characterization of flight candidate NIR InSb array," *Proc. SPIE* **5167**, 144–158 (2003) [doi: 10.1117/12.506569].

21. A. W. Hoffman, E. Corrales, P. J. Love, J. Rosbeck, M. Merrill, A. Fowler, and C. McMurtry, "2K × 2K InSb for astronomy," *Proc. SPIE* **5499**, 59–67 (2004) [doi: 10.1117/12.461131].

22. M. Devis and M. Greiner, "Indium antimonide large-format detector arrays," *Opt. Eng.* **50**, 061016 (2011) [doi: 10.1117/1.390722].

23. G. Gershon, A. Albo, M. Eylon, O. Cohen, Z. Calahorra, M. Brumer, M. Nitzani, E. Avnon, Y. Aghion, I. Kogan, E. Ilan, and L. Shkedy, "3 Mega-pixel InSb detector with 10 μm pitch," *Proc. SPIE* **8704**, 870438 (2013) [doi: 10.1117/12.2015583].

24. A. M. Fowler, D. Bass, J. Heynssens, I. Gatley, F. J. Vrba, H. D. Ables, A. Hoffman, M. Smith, and J. Woolaway, "Next generation in InSb arrays: ALADDIN, the 1024 × 1024 InSb focal plane array readout evaluation results," *Proc. SPIE* **2268**, 340 (1994) [doi: 10.1117/12.185844].

25. E. Beuville, D. Acton, E. Corrales, J. Drab, A. Levy, M. Merrill, R. Peralta, and W. Ritchie, "High performance large infrared and visible astronomy arrays for low background applications: Instruments performance data and future developments at Raytheon," *Proc. SPIE* **6660**, 66600B (2007) [doi: 10.1117/12.734846].

26. A. W. Hoffman, E. Corrales, P. J. Love, J. Rosbeck, M. Merrill, A. Fowler, and C. McMurtry, "2K × 2K InSb for astronomy," *Proc. SPIE* **5499**, 59 (2004) [doi: 10.1117/12.461131].

27. A. M. Fowler, K. M. Merrill, W. Ball, A. Henden, F. Vrba, and C. McCreight, "Orion: A 1-5 micron focal plane for the 21st century," in *Scientific Detectors for Astronomy: The Beginning of a New Era*, edited by P. Amico, Kluwer, and Dordrecht, pp. 51–58 (2004).

28. http://www.sbfp.com/documents/FPA%20S019-0001-08.pdf

29. G. Gershon, A. Albo, M. Eylon, O. Cohen, Z. Calahorra, M. Brumer, M. Nitzani, E. Avnon, Y. Aghion, I. Kogan, E. Ilan, A. Tuito, M. Ben Ezra, and L. Shkedy, "Large format InSb infrared detector with 10 μm pixels," *OPTRO 2014 SYMPOSIUM– Optoelectronics In Defence and Security*, 28–30 January 2014.

30. A. Adams and E. Rittenberg, "HOT IR sensors improve IR camera size, weight, and power," *Laser Focus World*, January 2014, pp. 83–87.

31. J. Caulfield, J. Curzan, J. Lewis, and N. Dhar, "Small pixel oversampled IR focal plane arrays," *Proc. SPIE* **9451**, 94512F (2015) [doi: 10.1117/12.2180385].

32. W. Cabanski, K. Eberhardt, W. Rode, J. Wendler, J. Ziegler, J. Fleißner, F. Fuchs, R. Rehm, J. Schmitz, H. Schneider, and M. Walther, "Third generation focal plane array IR detection modules and applications," *Proc. SPIE* **5406**, 184 (2005) [doi: 10.1117/12.605818].

33. C. J. Hill, A. Soibel, S. A. Keo, J. M. Mumolo, D. Z. Ting, S. D. Gunapala, D. R. Rhiger, R. E. Kvaas, and S. F. Harris, "Demonstration of mid and long-wavelength infrared antimonide-based focal plane arrays," *Proc. SPIE* **7298**, 729804 (2009) [doi: 10.1117/12.818692].

34. S. D. Gunapala, D. Z. Ting, C. J. Hill, J. Nguyen, A. Soibel, S. B. Rafol, S. A. Keo, J. M. Mumolo, M. C. Lee, J. K. Liu, and B. Yang, "Demonstration of a 1024×1024 pixel InAs-GaSb superlattice focal plane array," *IEEE Phot. Tech. Lett.* **22**, 1856–1858 (2010).

35. M. Razeghi, H. Haddadi, A. M. Hoang, E. K. Huang, G. Chen, S. Bogdanov, S. R. Darvish, F. Callewaert, and R. McClintock, "Advances in antimonide-based Type-II superlattices for infrared detection and imaging at center for quantum devices," *Infrared Phys. & Technol.* **59**, 41–52 (2013).

36. M. Razeghi, H. Haddadi, A. M. Hoang, E. K. Huang, G. Chen, S. Bogdanov, S. R. Darvish, F. Callewaert, P. R. Bijjam, and R. McClintock, "Antomonide-based type-II superlattices: A superior candidate for the third generation of infrared imaging systems," *J. Electron. Mater.* **43**, 2802–2807 (2014).

37. P. Manurkar, S. Ramezani-Darvish, B.-M. Nguyen, M. Razeghi, and J. Hubbs, "High performance long wavelength infrared mega-pixel focal plane array based on type-II superlattices," *Appl. Phys. Lett.* **97**, 193505 (2010).

38. M. Razeghi and B.-M. Nguyen, "Advances in mid-infrared detection and imaging: a key issues review," *Rep. Prog. Phys.* **77**, 082401 (2014).

39. P. C. Klipstein, E. Avnon, D. Azulai, Y. Benny, R. Fraenkel, A. Glozman, E. Hojman, O. Klin, L. Krasovitsky, L. Langof, I. Lukomsky, M. Nitzani, I. Shtrichman, N. Rappaport, N. Snapi, E. Weiss, and A. Tuito, "Type II superlattice technology for LWIR detectors," *Proc. SPIE* **9819**, 981920 (2016) [doi: 10.1117/12.2222776].

40. R. Rehm, V. Daumer, T. Hugger, N. Kohn, W. Luppold, R. Müller, J. Niemasz, J. Schmidt, F. Rutz, T. Stadelmann, M. Wauro, and A. Wörl, "Type-II superlattice infrared detector technology at Fraunhofer IAF," *Proc. SPIE* **9819**, 98190X (2016) [doi: 10.1117/12.2223887].

41. A. Rogalski, J. Antoszewski, and L. Faraone, "Third-generation infrared photodetector arrays," **105**, 091101 (2009) [doi: 10.1117/12.2223887].

42. A. Rogalski, "New material systems for third generation infrared photodetectors," *Opto-Electron. Rev.* **16**, 458–482 (2008).

43. A. M. Hoang, G. Chen, A. Haddadi, and M. Razeghi, "Demonstration of high performance bias-selectable dual-band short-/mid-wavelength infrared photodetectors based on type-II InAs/GaSb/AlSb superlattices," *Appl. Phys. Lett.* **102**, 011108 (2013).
44. R. Rehm, M. Walther, J. Schmitz, F. Rutz, A. Wörl, R. Scheibner, and J. Ziegler, "Type-II superlattices: the Fraunhofer perspective," *Proc. SPIE* **7660**, 76601G (2010) [doi: 10.1117/12.850172].
45. F. Rutz, R. Rehm, J. Schmitz, J. Fleissner, and M. Walther, "InAs/GaSb superlattice focal plane array infrared detectors: manufacturing aspects," *Proc. SPIE* **7298**, 72981R (2009) [doi: 10.1117/12.819090].
46. R. Rehm, F. Lemke, M. Masur, J. Schmitz, T. Stadelman, M. Wauro, A. Wörl, and M. Walther, "InAs/GaSb superlattice infrared detectors," *Infrared Physics & Technol.* **70**, 87–92 (2015).
47. M. Razeghi, A. M. Hoang, A. Haddadi, G. Chen, S. Ramezani-Darvish, P. Bijjam, P. Wijewarnasuriya, and E. Decuir, "High-performance bias-selectable dual-band short-/Mid-wavelength infrared photodetectors and focal plane arrays based on InAs/GaSb/AlSb type-II superlattices," *Proc. SPIE* **8704**, 870454 (2013) [doi: 10.1117/12.2019145].
48. M. Razeghi, A. Haddadi, A. M. Hoang, G. Chen, S. Ramezani-Darvish, and P. Bijjam, "High-performance bias-selectable dual-band mid-/long-wavelength infrared photodetectors and focal plane arrays based on InAs/GaSb type-II superlattices," *Proc. SPIE* **8704**, 87040S (2013) [doi: 10.1117/12.2019147].
49. A. M. Hoang, A. Dehzangi, S. Adhikary, and M. Razeghi, "High performance bias-selectable three-color short-wave/mid-wave/ long-wave infrared photodetectors based on type-II InAs/GaSb/AlSb superlattices," *Sci. Rep.* **6**, 24144 (2016).

Chapter 10
Final Remarks

At present, III-V antimonide-based detector technology is under strong development as a possible alternative to HgCdTe detector material. The ability to tune the positions of the conduction and valence band edges independently in a broken-gap T2SL is especially helpful in the design of unipolar barriers. Unipolar barriers are used for implementing the barrier detector architecture for increasing the collection efficiency of photogenerated carriers and reducing dark current originating within the depletion region without inhibiting photocurrent flow. During the last decade, antimonide-based FPA technology has achieved a performance level close to that of HgCdTe. The apparent rapid success of the T2SL depends not only on the previous five decades of III-V materials, but mainly on innovative ideas recently emerging in the design of infrared photodetectors. However, a modern version of the technology is still in its infancy. The advent of bandgap engineering has given III-Vs a new lease on life.

For HOT MWIR operation, T2SL materials have demonstrated higher operating temperature (up to 150 K) compared to that of InSb ($\sim$80 K). In addition, T2SLs offer both performance and manufacturability, especially for large-format FPA applications. GaSb substrates with diameter up to 6 inches are commercially available.

From the considerations presented in this book, it can be concluded that despite the numerous advantages of III-V semiconductors (T2SLs and barrier detectors) over present-day detection technologies (including reduced tunnelling and surface leakage currents, normal-incidence absorption, and suppressed Auger recombination), the promise of superior performance of these detectors has not been yet realized. The dark current density is higher than that of bulk HgCdTe photodiodes, especially in MWIR range.

To attain their full potential, the following essential technological limitations such as short carrier lifetime, passivation, and heterostructure engineering, need to be overcome. Much improvement can be attributed to the identification and minimization of SRH traps. T2SL material has the potential to outperform HgCdTe if it can overcome current SRH defect limits. Introduction of barrier designs can considerably impede the flow of dark current without photocurrent impeding when a bias voltage is applied. It should be expected that future advances in III-V barrier detector technology would push the dark current down to "Rule 07" in a wider infrared spectral range.

From a performance perspective, III-V diffusion-current-limited FPAs can indeed operate at levels that approach that of HgCdTe, but always with a required lower operating temperature. It should however be reiterated that cost reduction of an IR system will ultimately be achieved only by room-temperature operation of depletion-current-limited FPAs with pixel densities that are fully consistent with background- and diffraction-limited performance due to system optics. In this context, the long SRH lifetime of HgCdTe mandates the use of these material systems for room-temperature operation.[1] Unfortunately, it will be rather difficult to improve the SRH lifetime to overcome the disadvantage of large InAs/GaSb T2SL depletion dark currents. The InSb SRH lifetime issue has been well known since it was first reported in the 1950s. Better conditions are observed in Ga-free InAs/InAsSb T2SLs due to the large values of carrier lifetimes, including SRH lifetime.

Kinch's recently published monograph[2] is largely devoted to the intense competition between HgCdTe and T2SLs for the future of infrared detector technology. It is clearly shown that "the ultimate cost reduction for an IR system will only be achieved by the room-temperature operation of depletion-current-limited arrays with pixel densities that are fully consistent with background- and diffraction-limited performance due to the system optics. This mandates the use of IR materials with a long S-R lifetime. Currently, the only material that meets this requirement is HgCdTe." Kinch predicted that large-area, ultrasmall-pixel, diffraction-limited and background-limited, photon-detecting, MW and LW HgCdTe FPAs operating at room temperature will be available within the next ten years.[3]

10.1 P-on-n HgCdTe Photodiodes

For lightly doped HgCdTe, the way to achieve very low carrier concentration in the detector absorber region is to apply sufficient reverse bias to fully deplete it, as shown in Fig. 10.1. In such conditions, free carrier and Auger recombination is eliminated. If GR currents are adequately low, the detector performance is limited by the background.

At the current stage of the technology, the above requirements are fulfilled in a p-on-n HgCdTe double-layer photodiode (DLPH)—see Fig. 10.2.[4] The absorber layer is surrounded by a wider-bandgap cap and buffer region in order to suppress dark current generation from these regions. The n-doping of the absorber is sufficiently low to allow full depletion at moderate bias. To suppress tunnelling current under reverse bias, a wide-bandgap cap is used. The planar nature of the structure is potentially self-passivating and is analogous to the pBn

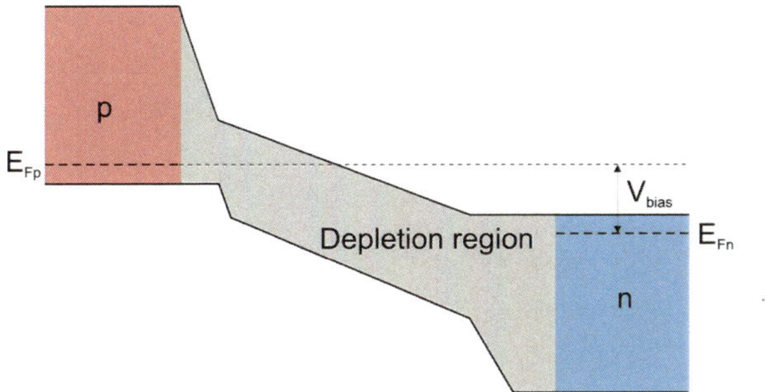

Figure 10.1 Band diagram for a reverse-bias p-i-n photodiode.

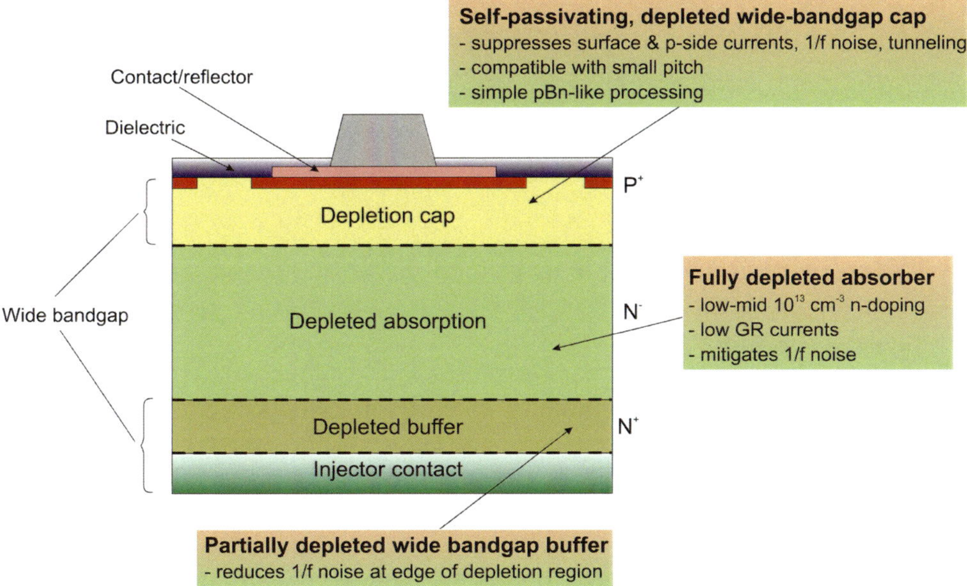

Figure 10.2 Heterojunction p-i-n photodiode architecture (reprinted from Ref. 4).

geometry of III-V barrier detectors. Moreover, as is discussed in Refs. 2 and 5, the fully depleted structure is compatible with small pixel pitch, fulfilling low crosstalk due to the built-in vertical electric field generated under detector reverse bias. As was mention in Section 5.4, both a fully depleted absorber and a wide-bandgap cap potentially reduce 1/*f* and random telegraph noise.

The GR current of a fully depleted detector can be estimated by following expression:

$$J_{GR} = q \frac{n_i}{\tau_{SRH}} w, \tag{10.1}$$

where w is the width of the depletion region, and τ_{SRH} is the SRH lifetime.

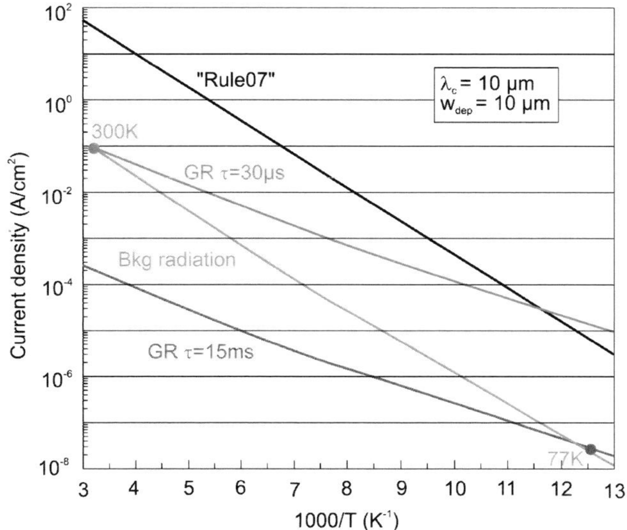

Figure 10.3 Arrhenius plot of dark current density for radiatively limited 10-µm-cutoff HgCdTe compared to the GR current and "Rule 07" (reprinted from Ref. 4).

Figure 10.3 compares the background-radiation-limited current density with that for "Rule 07" for 10-µm-wavelength HgCdTe and the GR current components calculated for two SRH lifetimes: 30 µs and 15 ms. The lifetimes are chosen to make GR and background radiation currents equal at 300 K and 77 K, respectively. From these estimations it can be concluded that SRH lifetimes greater than 30 µs and 15 ms are required to reach the background radiation limit at 300 K and 77 K, respectively.

Experimental data at 30 K indicate a very encouraging SRH lifetime for a 10.7-µm cutoff array pixel having 18-µm pixel pitch. A lower limit to the SRH lifetime is estimated to be 100 ms.[6] Assuming this lifetime, the current density for three different absorbing doping values is shown in Fig. 10.4 together with the current density from background blackbody radiation integrated over π steradians. As is shown, radiation current is observed to dominate for doping below approximately 10^{13} cm^{-3}.

10.2 Manufacturability of Focal Plane Arrays

Infrared system performance is highly scenario dependent and requires the designer to account for numerous different factors when specifying detector performance. This means that a good solution for one application may not be as suitable for a different application.[7] For example, while MWIR HgCdTe has a dark current several orders of magnitude lower compared to InSb, there are several applications in which InSb is preferred over HgCdTe due to the manufacturability arguments.

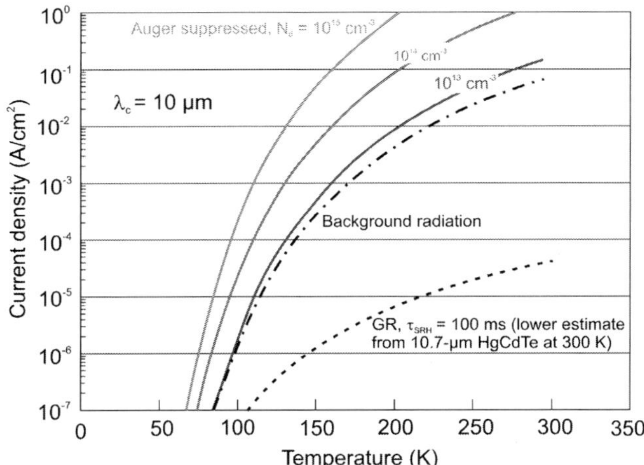

Figure 10.4 Comparison of GR current density extracted from 30-K lifetime measurements for 10-µm-cutoff HgCdTe photodiodes with Auger-suppressed and background radiation current densities (reprinted from Ref. 4).

HgCdTe is currently the most prevalent material system used in high-performance infrared detectors. Despite this standing, there are several drawbacks associated with HgCdTe devices. Being a II-VI semiconductor with weaker ionic Hg–Te bonds and high Hg vapor pressure, HgCdTe is soft and brittle, and requires extreme care in growth, fabrication, and storage. The common growth of HgCdTe epitaxial layers is more challenging than for typical III-V materials, resulting in lower yields and higher costs. HgCdTe material exhibits relatively high defect densities and surface leakage currents that adversely affect performance, particularly for LWIR devices. Also, composition uniformity is a challenge for HgCdTe devices, particularly for LWIR devices, leading to variability in cutoff wavelength. In addition, 1/*f* noise causes uniformity to vary over time, which is difficult to correct via image processing. As a result, LWIR HgCdTe detectors can only be fabricated in small FPAs.

In epitaxial growth of HgCdTe, the most commonly lattice-matched CdZnTe substrates are used. However, CdZnTe substrates are not lattice matched to Si readout integrated circuits—they are difficult to fabricate in large sizes with acceptable quality and are only available from limited sources.

Over the years, the market for HgCdTe has shrunk due to the success of the InGaAs, InSb, QWIPs, and uncooled microbolometers. HgCdTe is a II-VI material and has no other commercial leveraging. As a consequence, it has become difficult to maintain the entire industry base with limited demands on quantity. On the other hand, T2SL is a III-V material with an existing industry infrastructure for producing devices at a low cost. The existing facilities for III-V materials are supported by commercial products (e.g., cell phone chips and millimeter-wave integrated circuits), which present a less problem for the government to maintain the IR industry base.

An important advantage of T2SLs is the high-quality, high-uniformity, and stable nature of the material. In general, III-V semiconductors are more robust than their II-VI counterparts due to stronger, less ionic chemical bonding. As a result, III-V-based FPAs excel in operability, spatial uniformity, temporal stability, scalability, producibility, and affordability—the so-called "ibility" advantages.[8] The energy gap and electronic properties of T2SLs are determined by layer thicknesses rather than the molar fraction, as is the case for HgCdTe. The growth of T2SLs can be carried out with better control over the structure and with higher reproducibility. The spatial uniformity is also improved because the effects of compositional change due to flux and temperature nonuniformity are not as important as they are in ternary/quaternary bulk materials.

At present, the Vital Infrared Sensor Technology Acceleration (VISTA) U.S. government program is working on innovative approaches for infrared FPA technology to enhance infrared sensor capabilities. The VISTA program implemented a horizontally integrated model that significantly departed from the traditional vertical integration used by the HgCdTe industry. For example, HRL Laboratories served as an FPA foundry within the VISTA program. Using wafers grown by IQE and Intelligent Epitaxy Technology (IET) and based on designs from Jet Propulsion Laboratory (JPL), dual-band FPAs are fabricated and subsequently hybridized to a ROIC provided by Raytheon Vision Systems (RVS).[9]

Currently, very high-performance and very large-format LWIR FPAs do not exist with the state-of-the-art technology; there is an urgency to demonstrate and produce these arrays for systems where very large-format and high-performance LWIR FPAs are needed. VISTA's focus is on III-V superlattice epitaxial materials research for advanced infrared FPAs with large formats and reduced pixel size that are capable of MWIR/LWIR detection and dual-band sensing applications.

Recent progress in the VISTA program was presented during the *SPIE Defense + Security Symposium* in April 2017, in Anaheim, California (*Proc. SPIE* **10177**). This proceeding volume quotes the most impressive results presented at the symposium. Under the VISTA program, HRL has advanced the growth and fabrication of III-V bulk MWIR detector technology on GaAs substrates for HOT applications. It has been shown that small-pixel (5- to 10-μm pitch) technology is feasible as an attractive alternative to HgCdTe technology, mainly due to lower cost, ease of scalability to larger formats (e.g., $8k \times 8k/10$ μm), and better uniformity. Infrared FPAs with $2k \times 2k/10$-μm and $2k \times 1k/5$-μm formats have been demonstrated by developing high-aspect-ratio dry etching for mesa delineation (fill factor > 80%), proper device passivation by the dielectric layer, and high-aspect-ratio indium bump schemes (operability > 99.9%).

10.3 Conclusions

Summarizing the above discussions, we can state that:

- III-V materials have inherently short SRH lifetimes below 1 µs and require an nBn architecture to operate at reasonable temperatures; as such, they are diffusion current limited. This applies to both the simple alloy and the T2SL versions.
- HgCdTe alloys have long SRH lifetimes (>200 µs to 50 ms) depending on the cutoff wavelength.[3] They can thus operate with either architecture and may be diffusion or depletion current limited.
- III-Vs offer similar performance to HgCdTe at an equivalent cutoff wavelength but with a sizeable penalty in operating temperature due to the inherent difference in SRH lifetimes.

Presently, it is not clear whether the difference in SRH lifetimes between III-V alloy materials and HgCdTe is a fundamental property of the materials. The most popular III-V infrared material, InSb, has been resisting attempts to improve its SRH lifetime for fifty years—SRH lifetime is limited to a value of $\sim$400 ns.

In FPA fabrication, the major practical issue for T2SLs is a lack of stable passivation technique. Usually, surface donor contamination and insulator fixed positive charge are a common occurrence, which is not an issue for nBn architectures but for p-type embodiment of the T2SL presents a highly undesirable situation. The p-type T2SL barrier photodetector will also be sensitive to donor core dislocations; on the other hand, the nBn barrier detector is relatively immune to donor core dislocations.

Table 10.1 provides a snapshot of the current state of development of LWIR detectors fabricated from different material systems, including T2SLs. Note that TRL means technology readiness level. The highest level of TRL (ideal maturity) achieves a value of 10.[7] The highest level of maturity (TRL = 9) is credited to HgCdTe photodiodes and microbolometers. A little less maturity (TRL = 8) is credited to QWPs. The T2SL structure has great potential for LWIR spectral range applications with performance comparable to HgCdTe for the same cutoff wavelength, but requires a significant investment and fundamental material breakthrough to mature.

From an economical point of view and the perspective of future technology, an important aspect concerns industry organization. The HgCdTe array industry is vertically integrated; there are no commercial wafer suppliers because there are no diverse commercial applications of HgCdTe FPA that will be profitable. The wafers are grown within each FPA fabrication facility (or its exclusive partner). One important disadvantage of this integrity is the high cost. In the case of III-V semiconductors, horizontal

Table 10.1 Comparison of LWIR existing state-of-the-art device systems for LWIR detectors (TRL – technology readiness level).

	Bolometer	HgCdTe	QWIP	Type-II SLs
Maturity level	TRL 9	TRL 9	TRL 8	TRL 5–6
Status	Material of choice for applications requiring medium to low performance	Material of choice for applications requiring high performance	Commercial	Research and development
Operating temp.	Uncooled	Cooled	Cooled	Cooled
Manufacturability	Excellent	Poor	Excellent	Good
Cost	Low	High	Medium	Medium
Prospect for large format	Excellent	Very good	Excellent	Excellent
Availability of large substrate	Excellent	Poor	Excellent	Very good
Military system examples	Weapon sights, night-vision goggles, missile seekers, small UAV sensors, unattended ground sensors	Missile intercepts, tactical ground and air born imaging, hyperspectral, missile seekers, missile tracking, space-based sensing	Being evaluated for some military applications and astronomy sensing	Being developed in universities and evaluated in industry research environments
Limitations	Low sensitivity and long time constants	Performance susceptible to manufacturing variations. Difficult to extend to >14-μm cutoff	Narrow bandwith and low sensitivity	Requires a significant investment and fundamental material breakthrough to mature
Advantages	Low cost and requires no active cooling; leverages standard Si-manufacturing equipment	Near theoretical performance; will remain material of choice for minimum of the next 10–15 years	Low-cost applications; leverages commercial manufacturing processes; very uniform material	Theoretically better then HgCdTe; leverages commercial III-V fabrication techniques

integration is more profitable. This solution is especially effective for avoiding the heavy investment in capital equipment and subsequent upgrading and maintenance, as well as the cost of highly skilled engineers and technicians.

References

1. M. A. Kinch, *Fundamentals of Infrared Detector Materials*, SPIE Press, Bellingham, Washington (2007) [doi: 10.1117/3.741688].
2. M. A. Kinch, *State-of-the-Art Infrared Detector Technology*, SPIE Press, Bellingham, Washington (2014) [doi: 10.1117/3.1002766].
3. M. A. Kinch, "An infrared journey," *Proc. SPIE* **9451**, 94512B (2015) [doi: 10.1117/12.2183067].
4. D. Lee, M. Carmody, E. Piquette, P. Dreiske, A. Chen, A. Yulius, D. Edwall, S. Bhargava, M. Zandian, and W. E. Tennant, "High-operating temperature HgCdTe: A vision for the near future," *J. Electronic Mater.* **45**(9), 4587–4595 (2016).
5. A. Rogalski, P. Martyniuk, and M. Kopytko, "Challenges of small-pixel infrared detectors: a review," *Rep. Prog. Phys.* **79**(4) 046501 (2016).
6. C. McMurtry, D. Lee, J. Beletic, A. Chen, R. Demers, M. Dorn, D. Edwall, C. B. Fazar, W. Forrest, F. Liu, A. Mainzer, J. Pipher, and A. Yulius, "Development of sensitive long-wave infrared detector arrays for passively cooled space missions," *Opt. Eng.* **52**(9), 091804 (2014) [doi: 10.1117/1.OE.52.9.091804].
7. *Seeing Photons: Progress and Limits of Visible and Infared Sensor Arrays*, Committee on Developments in Detector Technologies; National Research Council, 2010, http://www.nap.edu/catalog/12896.html.
8. D. Z. Ting, A. Soibel, A. Khoshakhlagh, L. Höglund, S. A. Keo, B. Rafol, C. J. Hill, A. M. Fisher, E. M. Luong, J. Nguyen, J. K. Liu, J. M. Mumolo, B. J. Pepper, and S. D. Gunapala, "Antimonide type-II superlattice barrier infrared detectors," *Proc. SPIE* **10177**, 101770N (2017) [doi: 10.1117/12.2266263].
9. P.-Y. Delaunay, B. Z. Nosho, A. R. Gurga, S. Terterian, and R. D. Rajavel, "Advances in III-V based dual-band MWIR/LWIR FPAs at HRL," *Proc. SPIE* **10177**, 101770T (2017) [doi: 10.1117/12.2266278].

Index

Antoni Rogalski is a professor at the Institute of Applied Physics, Military University of Technology in Warsaw, Poland. He is a leading researcher in the field of infrared (IR) optoelectronics. During the course of his scientific career, he has made pioneering contributions in the areas of theory, design, and technology of different types of IR detectors. In 1997, he received an award from the Foundation for Polish Science (the most prestigious scientific award in Poland) for achievements in the study of ternary alloy systems for infrared detectors – mainly an alternative to HgCdTe new ternary alloy detectors such as lead salts, InAsSb, HgZnTe, and HgMnTe. His monograph *Infrared Detectors* (published by Taylor and Francis in 2011) was translated into Russian and Chinese. Another monograph entitled *High-Operating Temperature Infrared Photodetectors* (edited by SPIE Press, 2007), which he is the co-author, summarizes the globally unique Polish scientific and manufacturing achievements in the field of near room temperature long wavelength IR detectors. He was elected as a corresponding member (2004) and next as an ordinary member (2013) of the Polish Academy of Sciences. In June 2015, he was appointed as a dean of Division IV Polish Academy of Sciences: Engineering Sciences.

Małgorzata Kopytko received her M.Sc. degree in electronics and telecommunication from the Department of Electronics, Wrocław University of Technology, Wrocław, Poland, in 2005 and her Ph.D. degree in electronics from the Institute of Optoelectronics, Military University of Technology, Warsaw, Poland, in 2011. She is currently with the Institute of Applied Physics, Military University of Technology. Her research areas include device physics, design, technology, and modeling of different types of HgCdTe and InAsSb-based infrared detectors.

Piotr Martyniuk received his M.Sc. degree in applied physics from the Institute of Applied Physics, Military University of Technology in Warsaw, Poland, in 2001 and his Ph.D. degree in electronics from the Warsaw University of Technology, Warsaw, Poland, in 2008. He is currently working for Military University of Technology at the Institute of Applied Physics. His research areas include design, simulation, fabrication, and characterization of HgCdTe and T2SLs InAs/GaSb and InAs/InAsSb-based infrared detectors.